CONTROL OF HUMAN VOLUNTARY MOVEMENT

CONTROL OF HUMAN VOLUNTARY MOVEMENT

John C. Rothwell, PhD,

Department of Neurology,
Institute of Psychiatry,
University of London

AN ASPEN PUBLICATION
Aspen Publishers, Inc.
Rockville, Maryland
1987

Aspen Publishers, Inc.
1600 Research Boulevard, Rockville,
Maryland 20850

Library of Congress Cataloging-in-Publication Data

Rothwell, John C., 1954-
Control of Human Voluntary Movement.

+An Aspen publication+
Inclues index.
1. Human locomotion. 2. Human mechanics.
3. Muscles — Innervation. I. Title.
(DNLM: 1. Movement. WE 103 R848C)
QP301.R68 1987 612'.76 86-28722
ISBN 0-87189-311-8

CONTENTS

FOREWORD

The human motor system is unique. It talks, walks and can play the piano from a remarkably early age. But it is difficult to study. One cannot impale single neurones with electrodes or lesion discrete areas of the nervous system in man. However, data gleaned from such elegant experiments in lower species that walk on four feet may not reflect the organisation of human motor mechanisms.

John Rothwell is one of a small band of human-motor physiologists who have followed the dictum 'The proper study of mankind is man'. In this book, he brings together what is known about human motor physiology in an eminently readable and critical fashion. Of course, there is a stimulating symbiosis between animal and human experimental motor physiology, and this is effected by the integration of critical information that can only be obtained from work on animals with what is known about man. Many disciplines have interest in the mechanisms of human voluntary movement — physiologists, psychologists, physiotherapists and clinicians, be they neurologists or those working in orthopaedics, physical medicine or rehabilitation. All will find John Rothwell's book invaluable. To the beginner it provides an excellent introduction to the subject. To the expert it presents a coherent review of current knowledge and areas of uncertainty. What is abundantly clear is how much more remains to be discovered about how man controls movement. The stimulus provided by this volume will be invaluable to thought and experiment. Put it by the bedside with a red pencil and ideas will flow.

C.D. Marsden
Professor
Department of Neurology
Institute of Psychiatry
University of London

PREFACE

A careful study of authors' prefaces shows that they are invariably made up of two parts: excuses and thanks. First the excuses.

This book is intended as an introduction to the study of movement control in man. It is not a detailed source book of all current knowledge, but a guide to the subject which highlights some current areas of interest and controversy. Two large areas of movement control have been omitted: control of posture and locomotion, and control of eye movements. Both need a whole volume to themselves, and could not be done justice in a book of this size. The references at the end of each chapter are by no means a complete bibliography of the subjects covered. However, I have tried to include a number of very recent original papers in these lists. Together with the review articles, these should serve as a starting point for anyone wishing to pursue a topic in further detail.

Second, the thanks.

To the colleagues who have commented on parts of the manuscript, or on the topics covered in the text; Drs Berardelli, Burke, Day, Dick, Gandevia, Marsden, and Thompson; to the many authors and publishers who gave permission to reproduce their original diagrams; and to Pat Merton, whose idea it was that I should write the book in the first instance. Finally, my thanks to my wife, for her patience.

1 INTRODUCTION. PLANS, STRATEGIES, AND ACTIONS

The most remarkable thing about moving is how easy it is. It is only when we watch someone who cannot move, perhaps after a stroke, or someone whose movements continually go wrong, like the contortions of a child with athetoid cerebral palsy, that we are reminded of the problems of movement control with which our nervous system copes so uncomplainingly. This book will deal with voluntary movements of the limbs and their disorders, which are particularly well illustrated by movements of the human hand and arm.

It is sometimes fashionable to admire the exquisite anatomical perfection of our bodies, and to imagine that the complex arrangement of muscles, tendons and joints is responsible for the vast repertoire of movements which we can perform. Yet this is not entirely true. For example, the *anatomical* difference between the human hand and that of the Old World monkeys is small, but the abilities to write, type and play musical instruments remain characteristically human. The reason for this is that the nervous control of movement has been improved rather than there being any change in the design of the mechanical parts.

Unfortunately, little is known about the overall strategies which the brain uses to plan and execute a movement, but there are a number of observations which have been made which point towards some general principles of movement control.

For example, during the entire act of writing my signature I need only take one conscious thought, that is to take up my pen and write my name. The entire sequence of muscle contractions and movements which follow can occur entirely automatically, although I am, of course, free to intervene at will at any time. Even more remarkable is the fact that no matter which muscles I use to write, the writing turns out to be recognisably mine each time. The idea of the movement has been transformed into the appropriate action (Figure 1.1).

Thus, it is possible to distinguish the *idea* of the movement from the *plan* which is used to carry it out. The idea which I have of the form of my signature remains unchanged, while the motor plan is updated to suit the requirements of whatever muscles I use. But formation of the plan is only the first stage in movement. The nervous system has a good deal more work to do in timing the contraction of some muscles or the relaxation of others, and in checking the progress of the evolving movement before an action can be said to be successfully underway. The *execution* of the motor plan forms the last stage in the performance of a movement.

The essential stages in the production of any voluntary movement may be envisaged as following the sequence: (1) idea; (2) motor plan; (3) execution of programme commands; (4) move. The stages in this sequence form an

Figure 1.1: Two Specimens of Professor P.A. Merton's Handwriting. The sentence at the top was written large on a wall with a felt tip pen; the sentence at the bottom was written small on a piece of paper with a fine mapping pen. The writing on the wall is about ten times larger, and was performed mainly by moving the wrist, elbow and shoulder. The small writing mainly used the small muscles of the hand itself. Despite the differences in the muscles used, the character of the handwriting remains unmistakably the same (Merton, 1972; copyright Scientific American Inc., with permission).

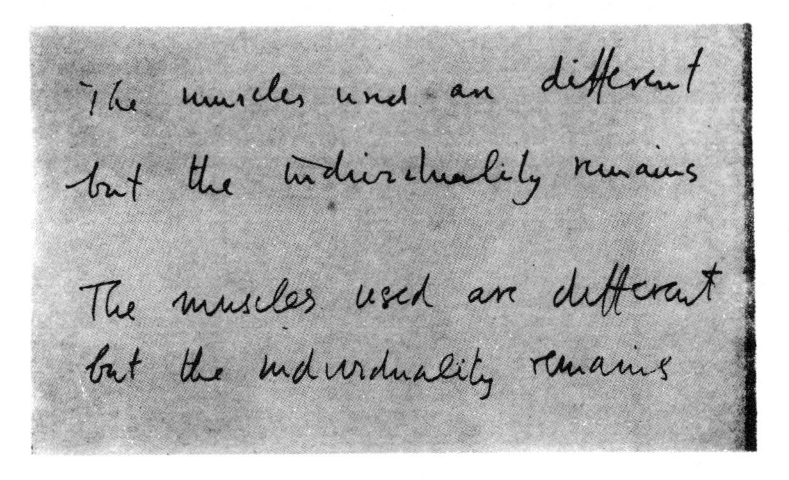

hierarchical chain of command, from the abstract concept of moving, to the practicalities of contracting the right muscles at the right time.

Categories of Muscles

We all know that movements are complicated: that there are many tens of muscles involved in almost any simple movement. We owe to two pioneers of muscle physiology, Duchenne and Beevor, the terms used to categorise muscle action and which help to describe a movement in detail.

The French neurologist, Duchenne, was perhaps the first to point out that a simple antagonistic relationship can only occur between two muscles at a hinge joint.

However, most muscles have more than one action, either because they span more than one joint or because they can produce rotation as well as flexion/extension at the same joint. Thus, the vastus medialis flexes the hip as well as extending the knee, the finger flexors flex both the wrist and the fingers, and biceps supinates the forearm as well as flexing the elbow, and hence is best displayed in the 'body-builder' pose. Although the combined action of the muscle may sometimes be useful, such as the role of biceps in picking up a piece of food and raising it to the mouth, it is often necessary to activate other muscles as *fixators* in order to limit the extent of the desired movement. A common example is the use of the wrist extensors to fixate the wrist while the finger flexors are used as prime movers during exploration and manipulation of an object.

Finally, it is important to remember that, as the song has it, the toe bone is connected to the foot bone, which is connected to the ankle bone, and so on up to the head bone. Movement is possible at almost all these joints, and it is a physical fact that forces exerted at any one joint are communicated through other intervening joints to the point of body contact with the supporting surface. Because of this, lifting a weight with the hand requires that the force be taken at the wrist, elbow, shoulder and back, down to the buttocks if seated, and down to the feet if standing. *Postural* muscle contractions occur which not only stabilise the individual joints, but also may produce subtle movements of the trunk to counterbalance movements of the arms and maintain the body's centre of gravity within the postural base.

In fact, it turns out that this categorisation of muscle types may have more profound implications than serving as a descriptive tool of movement types. Most muscles do not have one stereotyped function as agonist or fixator in all types of movement. They shift their function as the movement progresses. Thus, in reaching and grasping an object, the extensors of the wrist and fingers will be activated first as agonist to open the palm, and then as fixators to maintain wrist extension, as the fingers flex onto the object in a power grip.

The importance of this distinction between the use of the same muscle for different categories of movement is shown particularly well in some patients suffering from stroke. For example, a patient may be unable to use triceps to extend the elbow, but be perfectly capable of using the same muscle as an elbow fixator during supination of the wrist by biceps. It appears that there may be different anatomical pathways serving different categories of movement, or that different parts of the CNS may be used to calculate muscle activity under different conditions.

Problems of Moving

The major problem facing the motor control system, even in the simplest movements, is not only to contract the agonist, or prime moving muscle by the correct amount and at the appropriate time, but also to time and organise the pattern of antagonist, fixator and postural muscle contractions which are necessary to accompany its action.

As in the act of writing one's name, it is clear that there are some aspects of every action that never consciously enter into one's original *idea* of movement. But that is as far as introspection can take us. From there on it is a thorny problem of motor control as to what details are specified where and when in the chain of command producing a movement.

A working hypothesis, which will be used here and in the following chapters, is that the motor *plan*, like the initial *idea* of the movement, does not itself contain a complete description of the motor task. During execution, different parts of the brain and spinal cord are called upon to compute, for example, the exact postural/fixator activity needed to stabilise the primary movement. A computational analogy would be that the motor *plan* contains a general controlling programme and that various subroutines are handed over to other brain areas when necessary. These other areas may call up further subroutines *ad infinitum*.

From the point of view of our conscious voluntary control of movement, we tend to concentrate on the action of the prime moving muscle. Thus, when I stand to write on a blackboard, I concentrate at most on the motion of my fingers and hand. The delicate and equally important adjustments of my trunk, legs and shoulders all pass relatively unnoticed. Such details of movment are thought to be organised at low levels of the movement hierarchy, rather in the way that the Civil Service is expected to formulate the details of a minister's policy statement.

Nevertheless, convenient as this chain of command may be, it cannot be the only mechanism whereby voluntary control is exerted over the muscles. With practice, we can learn to use almost any muscle in our bodies as a prime mover. Indeed, we can even exert voluntary control over postural muscle activity. When we learn to lean outwards down the mountainside when skiing, against all our better instincts, we realise how important this more direct muscle control can be. In terms of the hierarchical model of motor control, it is necessary to allow for parallel pathways through which the conscious idea of movement may interact at all levels of movement control.

In the second half of the nineteenth century, John Hughlings Jackson devised a classification of types of movement that fits neatly into this description of movement control. He proposed that movements should be graded along a scale from the most to the least automatic. Those which could be least easily controlled by voluntary effort of will, such as the simple tendon jerk, were classified as the most automatic, whereas those most directly under conscious control of the individual were put into the least automatic category. The implication is that those movements which lie at low levels of the chain of command are the most difficult to activate via direct, parallel pathways.

A legacy

Most of these ideas of movement control were developed intuitively by the great clinical neurologists of the last century. They remain essentially the same today even when dressed up in the latest American computese. In fact, the idea of a chain of command is the only justification for treating the various areas of the CNS separately in discussion and investigation of movement control. It is assumed that there is some particular task or aspect of a movement that each area is best adapted to control, whether by virtue of its anatomical connections with other areas, or because of peculiarities of its internal circuitry. It is essentially a reductionist approach and follows very naturally from the hierarchical control of movement described earlier. However, useful as this approach may be, one most beware of pushing it to extremes. There is no simple answer to the question 'What does the cerebellum/basal ganglia **do**?' simply because neither structure ever does anything on its own.

The following chapters will describe each section of the CNS concerned with movement and what is known about the operation of each during different kinds of movements in man and animals. Unfortunately, it is easier to take the nervous system apart in this way than it is to reconstruct it later.

References and Further Reading

Merton, P.A. (1972) 'How We Control the Contraction of our Muscles', *Scientific American*, May, p. 32

2 MECHANICAL PROPERTIES OF MUSCLES

Review of Muscle Anatomy

Skeletal muscle is made up of long fibres, terminated at each end by tendinous material attached to the bone. These fibres are formed from a syncitium of cells whose walls fuse during development, and hence have many nuclei spread throughout their length. Groups of individual muscle fibres are gathered together into bundles called fascicles which are surrounded by a connective tissue sheath (Figure 2.1). The internal structure of the muscle fibre is quite complex. The main elements visible under the light microscope are the myofibrils. These run longitudinally throughout the fibre and constitute the contractile machinery of the muscle. Each myofibril is traversed by striations. Usually, the myofibrils are aligned so that the striations appear to be continuous right across the muscle fibre.

The myofibrils can be seen in more detail under the polarising light microscope. The basic structure consists of an alternating pattern of light and dark bands known as I (isotropic) and A (anisotropic) bands, respectively. In the centre of the I band is a dark Z line. The structures which repeat between adjacent Z lines are called sarcomeres. A paler line is sometimes seen to bisect the A band, and is known as the H zone (*Hell* is the German word for light) (see Figure 2.1).

The finest details of muscle structure only become clear under the electron microscope. Each myofibril consists of longitudinally orientated fine filaments called myofilaments. The bulk of these filaments is composed of two proteins, actin and myosin. Actin is present in the smaller diameter filaments, and myosin is found in the thicker filaments. Figure 2.1 shows the relationship between the filaments and the bands seen under the polarising light microscope. The actin filaments are attached at one end to the Z line, and are free to interdigitate at the other end with the myosin filaments. The A band represents the extent of the myosin filaments. The I and H bands represent the regions where there is no overlap between myosin and the actin filaments; the I band contains only thin filaments, and the H band only thick filaments. The M line is composed of thickenings which connect adjacent myosin filaments.

The Sliding Filament Hypothesis of Muscle Contraction

This hypothesis was formulated in the early 1950s after a single striking observation with the improved electron microscope (by H.E. Huxley and Hanson, 1954), and with the new polarising light miscroscope (by A.F. Huxley and Niedergerke, 1954). When a muscle contracts, the length of the thick and thin myofilaments remains constant, and they appear to slide over one another.

Figure 2.1: Structure of Skeletal Muscle. Panels A to M show gradually increasing details of structure from whole Muscle (A) to myofibrils (E); F, G, H, I show cross-sections of myofibrils at different points throughout their length; G shows a cross-section where only thick filaments are present; F is a section where only thin filaments are present.

The thin filaments (J) are made up of a core of actin molecules, together with troponin and tropomyosin. Thick filaments are made up of a myosin unit, as shown in L and M. Each myosin molecule consists of two parts, light meromyosin (LMM) and heavy meromyosin (HMM). The latter part includes the myosin heads, or S1 components (from Gordon, 1982; Figure 7.8; after Bloom and Fawcett (1970), *A Textbook of Histology*, 9th edn, W.B. Saunders, Philadelphia)

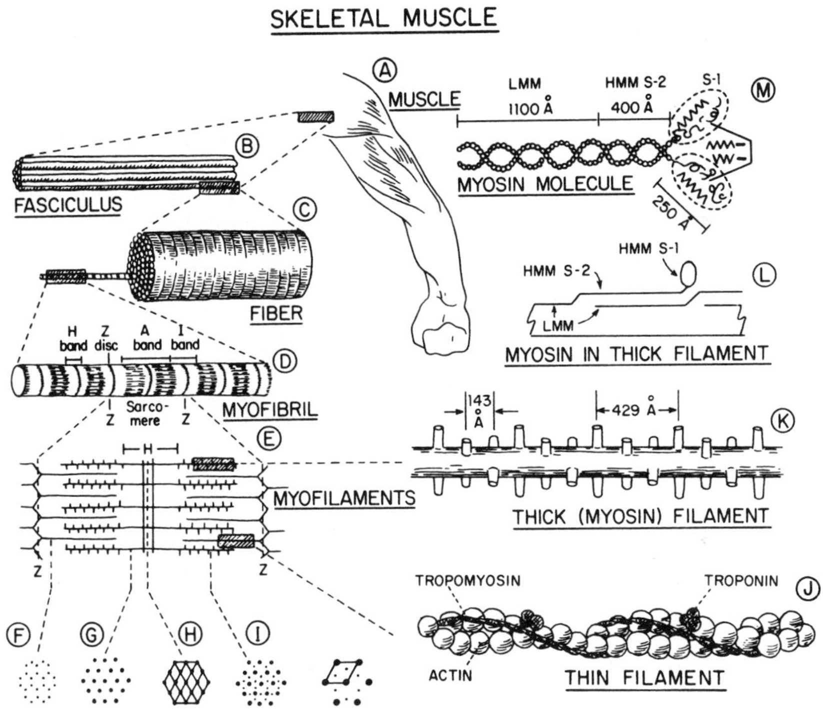

This observation solved the problem how muscle changes length. However, the problem of exactly how the filaments interact with each other and generate force is still a matter for debate.

The key force generating elements appear to be cross bridges between the actin and myosin filaments which can be seen in some preparations under the electron microscope. These consist of projections of the myosin filaments (called myosin heads) which extend towards adjacent actin filaments. In the presence of ATP, the myosin heads are thought to interact cyclically with special sites on the actin filaments. One possible mechanism is shown in Figure 2.2. In stage 1, ADP and Pi are bound to the myosin head. In this stage the heads are free to bind to sites on the actin filament, and form an acto–myosin–ADP–Pi complex (stage 2). Shortly after binding, this complex undergoes a conformational change. The myosin head rotates to a 45 degree angle, and relative force is exerted between the two filaments. This process leads to release of Pi and then ADP, and the formation of an acto–myosin complex (stage 3). The cycle is completed by binding of ATP to the complex. Myosin dissociates from the actin and at the same time hydrolyses ATP, although the reaction products (ADP and Pi) are not released but remain bound to the myosin head as a myosin–ADP–Pi complex (stages 1 and 4).

Figure 2.2: Postulated Mechanism of Cross-bridge Cycling by Actin and Myosin Filaments. The head of the myosin filament is stippled and forms the cross-bridge, or acto-myosin bond between the two filaments. Active sites on the actin filament are outlined by broken circles. (1) No bonds between filaments; (2) Initial attachment of myosin head to one of the active sites on the actin filament. The attachment takes place only in the presence of Ca^{2+} ions; (3) Formation of strong acto–myosin bond. This process causes a conformational change in the angle of the myosin head, which produces a relative movement of actin and myosin filaments across one another. ADP and phosphate ions are lost from the myosin; (4) If ATP is available, the acto–myosin bond can be broken. Subsequent hydrolysis of the ATP by the myosin ATPase returns the cycle to stage 1. If no ATP is available, the strong acto–myosin bond remains intact, as in muscular rigor (From Gordon, 1982; Figure 7, with permission)

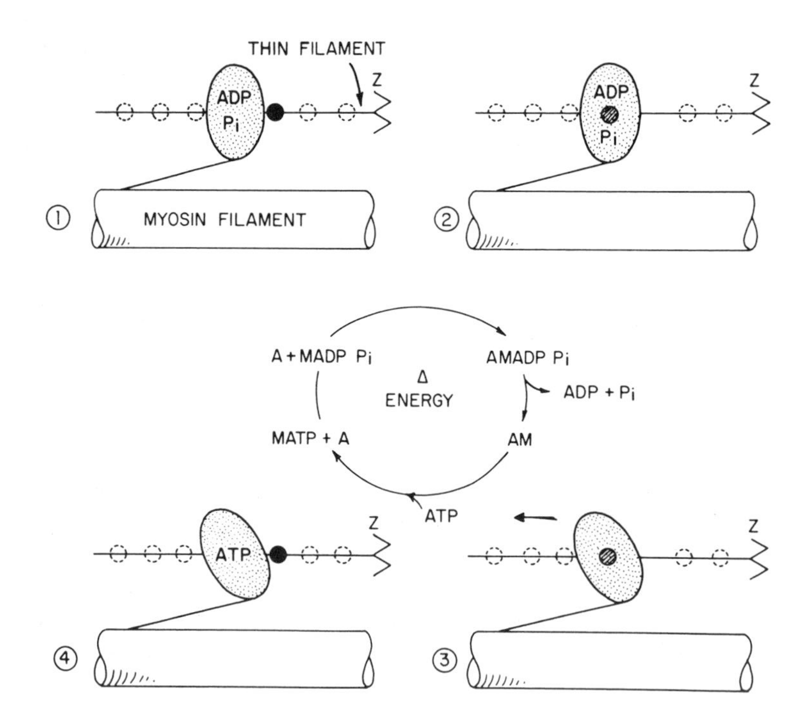

The critical stage of force generation is rotation of the myosin head. This causes the filaments to slide over one another if they are free to move (isotonic contraction). Otherwise an isometric force is generated.

In normal muscle, ATP is freely available, and the cycling process is controlled by the availability of actin binding sites. These are normally covered by a protein called tropomyosin, linked to a smaller protein known as troponin. The actin sites are revealed by changing the concentration of Ca^{2+} in the sarcoplasm. Troponin can bind Ca^{2+} when available, and in the process cause movement of the tropomyosin molecule to reveal active sites on the actin. At rest, the concentration of Ca^{2+} in the sarcoplasm is very low so that the majority of acto–myosin bonds remain unformed (state 1 in Figure 2.2). The Ca^{2+} ions are stored in special membranous structures called the sarcoplasmic reticulum, separate from the external fluid which surrounds the muscle.

Control of Ca^{2+} concentration occurs in the following way. Throughout its length, the cell membrane of the muscle fibre (sarcolemma) has numerous infoldings which form a system of membranes within the muscle fibre called the transverse tubular system. This forms regions of close opposition with the sarcoplasmic reticulum known as triads. Release of neurotransmitter at the neuromuscular junction initiates an action potential along the sarcolemma. This depolarises the membrane of the transverse tubular system, which acts as a device to transmit the depolarisation to the interior of the fibres. Depolarisation of the transverse tubular system causes release of Ca^{2+} from the sarcoplasmic reticulum into the sarcoplasm and allows formation of acto–myosin bonds. Ca^{2+} is then actively pumped back into the sarcoplasmic reticulum and contraction ceases.

After death, Ca^{2+} is no longer pumped back into the sarcoplasmic reticulum and ATP sysnthesis ceases. The acto–myosin molecules in muscle become bonded together in rigor mortis (see Pollack, 1983; Eisenberg and Greene, 1980; Huxley, 1974; Huxley and Hanson, 1954; Huxley and Niedergerke, 1954, for further details).

Mechanical Properties of Muscle

The reason that the detailed mechanical properties of muscle needs to be known is to be able to predict how much force a muscle will produce under given conditions. The same command, in terms of action potentials along the muscle nerve, can generate varying levels of force which depend on the muscle length, the velocity of shortening and the state of activation of the muscle before receiving the command. Thus the CNS cannot assume that the same command always will produce the same movement. All these factors must be taken into account.

Studies on Tetanically Activated Isolated Muscle

In these, now classical, studies, single muscles were isolated from an animal.

No reflex arcs remained intact, and to remove the complicating factor of activation history, the experiments were performed either with the muscle completely inactive, or during maximum activation of the muscle with tetanic stimulation.

A useful model used to describe the mechanical behaviour of such a preparation is shown in Figure 2.3. It consists of three elements: the series and parallel elastic elements are purely passive components acting like mechanical springs, whereas the contractile element actively generates muscle force. The contractile element has its own mechanical properties, namely viscosity and stiffness, but these properties change with the level of muscle contraction, unlike those of the two passive components.

Figure 2.3: Mechanical Model of Muscle. It consists of a contractile element and two (series and parallel) elastic elements. The properties of the elastic elements are constant, whereas the mechanical properties (that is, the stiffness and viscosity) of the contractile component vary with the state of muscle contraction (from Roberts, 1978; Figure 2.4, with permission)

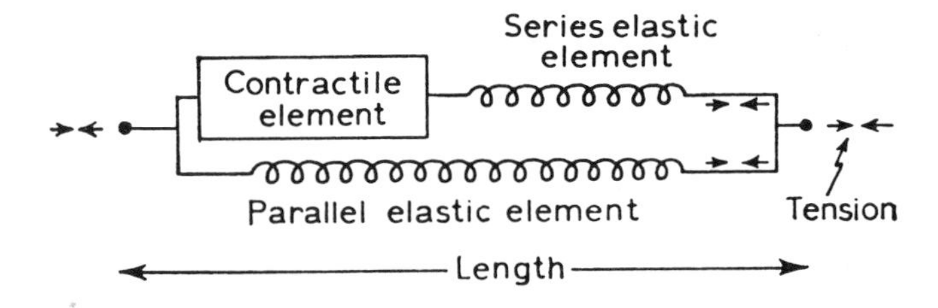

It is not so simple to divide the actual structure of muscle into three components as required in this simplistic model. In fact, as seen below, the model only helps to describe the mechanical properties of muscle, and is not intended to imply specific relationships between hypothetical springs and muscle structure. This is because the model was formulated well before the sliding filament theory of muscle contraction, and before detailed electron microscope pictures of muscle were available. Nevertheless, there are some relationships that can be drawn which help in understanding how the muscle works. Thus, the contractile mechanism can be said to reside in the interaction between the actin and myosin filaments; the series elastic element in the tendinous insertions of muscle, and in the acto–myosin cross bridges themselves; and the parallel elastic element in the sarcolemma of the muscle cells and the surrounding connective tissue.

The Length–Tension Relationship

To determine the effect of length on the force exerted by a muscle, the factors of velocity and time are held constant. The muscle is held at different lengths, either shorter or longer than its physiological 'resting' length, with one tendon attached to a stiff force transducer. It is then stimulated tetanically via the muscle nerve; and the force exerted on the transducer is recorded. The upper dotted curve in Figure 2.4 describes this behaviour for the soleus muscle of the cat. As the muscle length is increased, the force exerted rises and often reaches a peak before falling slightly at the end of its normal length range.

Figure 2.4: The Effect of Muscle Length on the Isometric Tension Exerted by Isolated Cat Soleus Muscle. The upper dotted line shows the total tension measured after 1.5s tetanic stimulation of the muscle nerve at 50Hz. The lowest dotted line shows the passive tension recorded in inactive muscle. Subtraction of these values from the upper dotted curve gives the true action tension generated by tetanic stimulation of the muscle (continuous line). The middle curves show similar data for peak twitch tensions produced by single supramaximal nerve stimuli. Changes in muscle length are shown in centimetres with the corresponding angle of the ankle joint in the intact animal below. Mean sarcomere lengths equivalent to the ankle positions have been added above the figure (from Rack and Westbury, 1969; Figure 3, with permission)

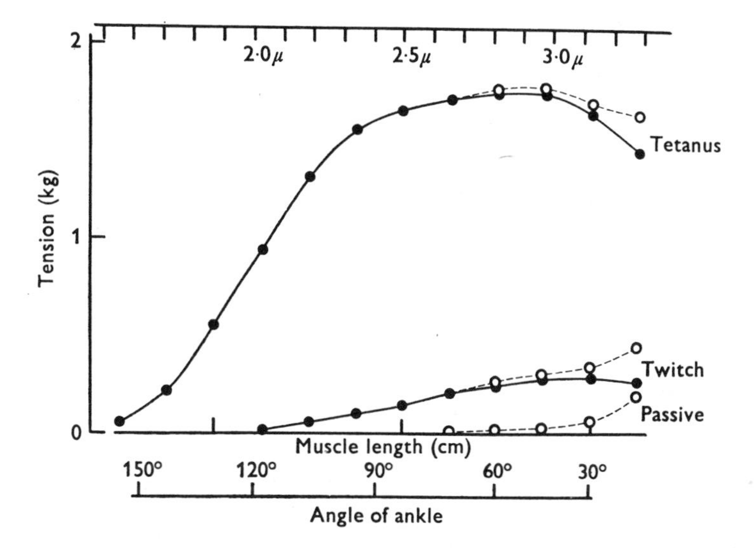

Figure 2.3 shows that the tension recorded in these experiments is due not only to the activity of the contractile element, but also to passive stretch of the parallel elastic element. An estimate of the force produced by the parallel elastic element can be obtained by applying stretch to a relaxed muscle (the lowest dotted line in Figure 2.4). Under these conditions, the contractile element exerts no force, and all the tension is exerted across the parallel elastic element. Subtraction of this passive tension from the total tension recorded in active muscle gives the true force produced by the contractile element (upper continuous curve in Figure 2.4).

The reason for the influence of length on the force produced by the contractile mechanism lies in the way actin and myosin filaments interact in single sarcomeres. The length–tension relationship for a single muscle fibre is shown in Figure 2.5, with the overlap between the thick and thin myofilaments in a single sarcomere drawn above. As the overlap between the filaments increases, more force is produced because more cross bridges can form. However, as the sarcomere spacing decreases further, the actin filaments begin to 'overlap'. It is believed that this interferes with cross-bridge formation and causes the force to decline. In whole muscle, the transition points between each part of the curve occur at slightly different lengths for each sarcomere and each fibre, so that the length–tension diagram is more rounded than for a single fibre (see Gordon, Huxley and Julian, 1966; Huxley, 1974, for further details).

Figure 2.5: Relationship Between Striation Spacing and the Tension Generated in a Single Frog Muscle Fibre. The upper diagrams (1–6) show the degrees of overlap of thin and thick filaments within each sarcomere, and the spacing between the Z lines. The bottom curve shows the length–tension relationship for the fibre. Maximum tension is generated with maximum overlap between the filaments (2 and 3). Tension drops if the overlap is less or if the thin filaments collide with each other (from Gordon, Huxley and Julian, 1966, with permission)

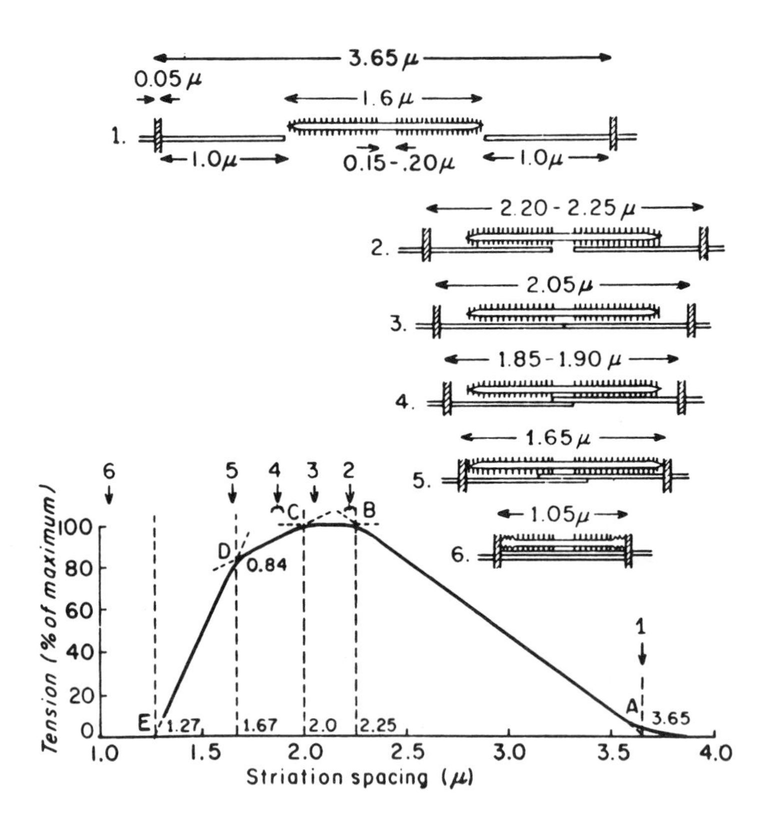

Twitch Contractions at Different Muscle Lengths

The classical length–tension diagram of Figure 2.4 shows how the maximum isometric force of muscle varies at different muscle lengths. The peak tension attained in a single maximal twitch follows a similar curve (see Figure 2.4, middle curves). However, measurements of peak tension do not give the complete picture of how muscle twitches are affected by muscle length. Figure 2.6 gives a typical example of raw data from the isolated cat soleus muscle. With increasing muscle length, the time course of the twitch changes: both time taken to reach peak tension and duration of the relaxation phase are prolonged. The reason for this is thought to be that at short muscle lengths, there is a less efficient activation of the contractile machinery. A single nerve impulse may give a smaller and perhaps shorter period of activation (for example, by a decrease in the amount of Ca^{2+} released from the sarcoplasmic reticulum). The implication is that the minimal stimulation rates needed to achieve a fused contraction will vary with muscle length.

Figure 2.6: Records of Isometric Twitches from Isolated Cat Soleus Muscle Held at Different Lengths. The longest duration twitch was at a muscle length equivalent to an angle of 30 degrees at the angle. The others were recorded after shortening the muscle by successive 5mm steps. Note how the initial tension in the relaxed muscle changes with length (baseline of the twitch). The twitches reach peak tension earlier and relax faster when the muscle is short (from Rack and Westbury, 1969; Figure 4, with permission)

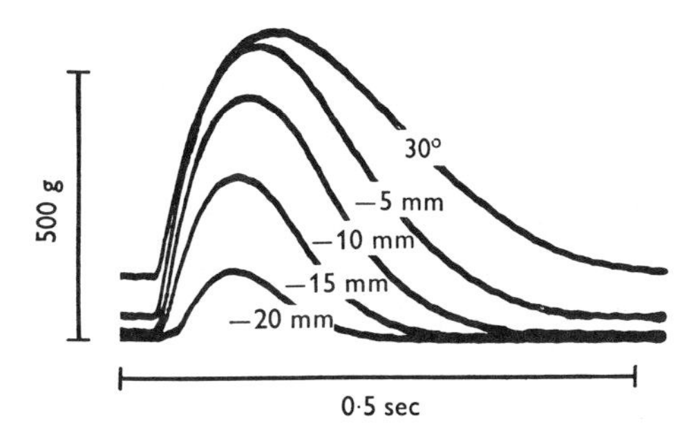

Force–Velocity Relationship

The rate at which a muscle can shorten depends on the force it is exerting. The isometric force exerted by a muscle is always greater than the force exerted during shortening. The method used to measure force–velocity relationships is shown in Figure 2.7. The muscle is stimulated tetanically, and when tension has built up, a catch is released and the muscle shortens against a preset load. Some records of the shortening profile are shown in Figure 2.7. After releasing the catch, the muscle shortens very rapidly and then settles down to a slower, steady rate of contraction. The initial rapid phase of shortening can be explained by reference to the mechanical analogue of muscle in Figure 2.3. In the isometric state, the contractile element exerts its maximum tension through the series elastic element. When the catch is released, the contractile element can only shorten at a finite speed, whereas the series elastic element shortens immediately to take up a length appropriate to the new load. It is only later that the slower contraction of the contractile element becomes evident in the position records as the ramp phase of shortening. The smaller the load, the faster the muscle shortens.

Figure 2.7: Determination of Force–Velocity Curve for a Contracting Muscle. The figure below shows diagrammatically how the curve is built up. The muscle is held fixed at a constant length and stimulated tetanically via its nerve. When maximum isometric tension is achieved, the muscle suddenly is allowed to shorten against a load. The smaller the load (that is, trace 4), the faster the final velocity of shortening. The figure opposite shows the force–velocity relationship for a single muscle fibre (upper graph) and for a whole muscle (lower graph). Force is plotted as a fraction of the maximum isometric tension produced by the muscle at its initial length. Speed of shortening is in muscle lengths per second. In the upper graph, the different symbols refer to experiments carried out with the muscle fibres held at different starting lengths of (open circles: sarcomere length of 2.03 μm, closed circles: 2.23 μm and triangles: 2.43 μm). The starting length defines Po, the maximum isometric tension (c.f. tension–length relationship). The lower graph, from a whole muscle, shows what happens when a load greater than Po is applied to the muscle. Rather than shortening, the muscle lengthens. However, the velocity at which it is stretched is lower than expected. The dotted line is extrapolated from the relationship of force and velocity at loads less than Po. The solid line shows the true (experimental) relationship. It lies above the dotted line. The findings of a lower lengthening velocity than predicted indicates that the muscle is stiffer when lengthening than it is when shortening. The muscle gives way at forces above about 1.5 Po (from Gordon, 1982; Figures 7.35 and 7.36, with permission)

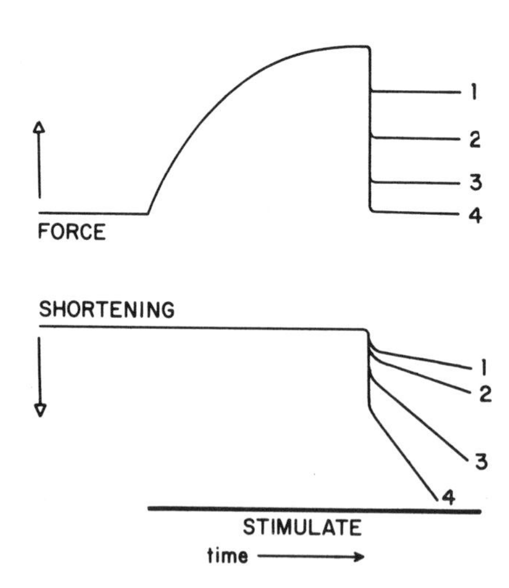

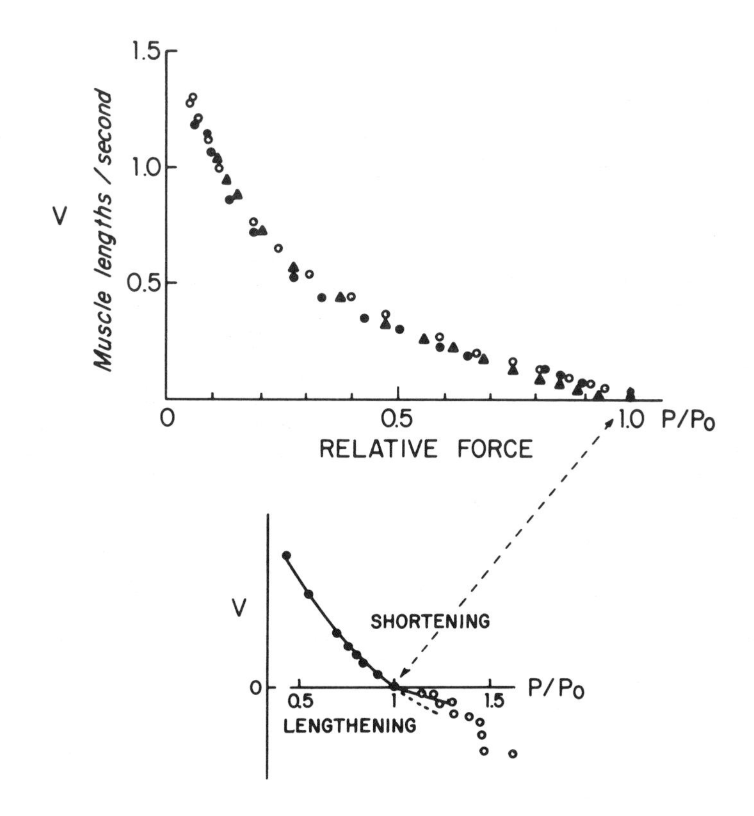

The force–velocity curve also can be extended to lengthening contractions, that is, attempted shortening against a load greater than isometric. In this case, the muscle is extended by the load with a speed related to the excess tension. These points are plotted in Figure 2.7 as negative shortening velocities.

The shortening region of the force–velocity curve has been fitted by many equations. The best-known is that of Hill:

$$(T+a)\ (V+b) = (T_0+a)b = \text{a constant}$$

where a and b are constants, and T is the tension exerted by the muscle. T_0 is the maximum isometric tension of the muscle and V the rate of shortening. The curve is of interest when one considers the amount of work that a muscle can perform on a load. Work is defined as 'force × distance' moved by the load perpendicular to the direction of the force. Thus, in an isometric state, although force is a maximum, no work is done since the load does not move. Similarly, at the maximum velocity of shortening, the muscle cannot produce any force against an external load, and again the work done is zero. Maximum work is performed at intermediate velocities of shortening.

Hill's equation (nor any of the other postulated functions) does not describe the lengthening part of the force–velocity curve. This is usually taken to mean that the process of force generation is different during lengthening than it is during shortening. During lengthening, a tetanically stimulated muscle can exert a force greater than isometric. As described later, this extra force is often used during normal movement.

There have been several attempts to explain the force–velocity curve on the basis of the cross bridge theory of muscle contraction. The best-known is that developed by Huxley in 1957. There were two important assumptions in this model: (1) that the cross bridges themselves were elastic (that is, they represent part of the series elastic element of muscle); and (2) that myosin heads could interact with a given actin site over a relatively wide range of distances (about ± 10 nm in the model).

If myosin heads interacted with a relatively distant actin site and underwent a conformational change as shown in Figure 2.3, the elastic element would be stretched, and force developed between the filaments. If a myosin head interacted with a relatively close actin site, then the conformational change might produce no stretch of the elastic element, and no force would be developed. Finally, if a myosin head remained attached to an actin site beyond the point of zero force, then by compressing the elastic element in the cross bridge, it could exert a negative force between the filaments. Thus the force developed depends on both the number of cross bridges and the distance over which each one interacts.

To fit the force–velocity curves of tetanically activated muscle, Huxley assumed that in the isometric state, the rate of cross bridge detachment was lower than the rate of attachment. Thus, in the steady state, all the cross bridges were formed and the force maximum. However, it was suggested that during shortening, there was an increase in the rate at which the cross bridges detached, and also an increase in the number of attached cross bridges which exerted negative force. The result would be a decrease in the total force exerted by the muscle. At maximum shortening velocities, very few bonds remained attached, and those that did remain were equally distributed between states exerting positive and negative force. No external force could be developed.

The sliding filament theory of muscle contraction also accounts for the difference between muscle behaviour during lengthening and shortening contractions. During lengthening contractions, the rate of detachment of acto-myosin bonds is slower than during shortening at the same velocity. The effect is for cross bridges to remain attached and to be stretched forcibly until detachment finally occurs. Such stretching is beyond the normal range of myosin attachment, and lengthens the cross bridge elasticity to increase the force exerted by the muscle.

The Effect of the Rate of Muscle Stimulation

The response of skeletal muscle to a single nervous impulse is a twitch contraction. The features of this contraction differ from muscle to muscle, and will be dealt with in Chapter 3. If more than one impulse is given at an interval which is less than the contraction time of the muscle, the forces produced by each impulse summate. A fused tetanus is produced when the force fluctuations to each individual impulse can no longer be distinguished.

The force developed depends also on the pattern of stimulation. An interesting example of this is the so-called 'catch' enhancement of muscle tension which is produced by the insertion of a single short interval burst into a low frequency stimulus train (Figure 2.8). As expected, the extra stimulus results in an immediate increase in muscle tension. Unexpectedly, however, the effect outlasts the stimulus for some time: the force produced by a low frequency train is elevated above normal for many seconds. The mechanism of this effect is unknown, although it may affect Ca^{2+} sequestration following activation.

Figure 2.8: Catch Property in a Slow Motor Unit From a Cat Medial Gastrocnemius Muscle. Upper traces are isometric force records, lower traces (dots) represent timing of muscle nerve stimulation. Arrows show points at which extra impulses have been inserted into the trains. In a and b, an extra impulse was inserted into the train at the arrow point, some 10 ms after the first. This produces the rapid rise in initial force, far greater than that seen in trace c. The extra force level is then 'caught' throughout the rest of the train. In c, an extra impulse was inserted a short time after the seventh impulse of the train, resulting in extra force production, that was again 'caught' at the new level. In b, an impulse interval was purposely lengthened, with the opposite effect (from Burke, Rudomin and Zajac, 1970; copyright American Association for the Advancement of Science, 1970; Figure 3, with permission)

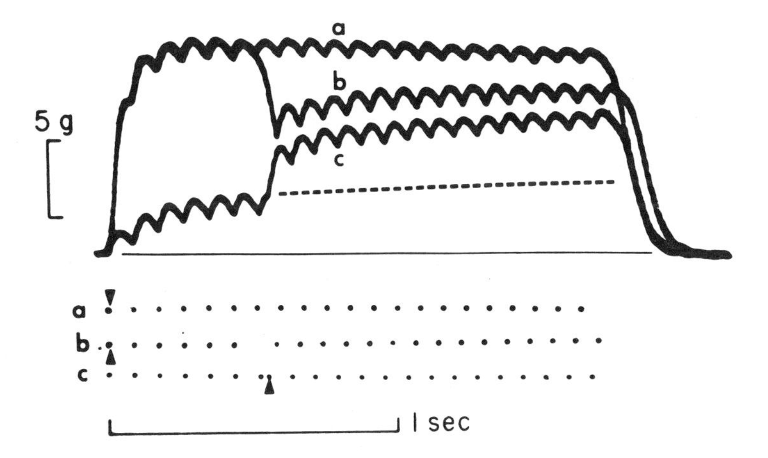

Behaviour of Isolated Muscle Stimulated at Subtetanic Rates

The previous section described the classical experiments on mechanical properties of fully activated isolated muscle. However, under normal conditions, muscle is never activated in a pattern resembling that of synchronous tetanic stimulation. First, the individual motor units tend to be activated asynchronously and, second, the stimulation frequency is seldom as high as that needed to produce a full tetanic contraction. When muscle is stimulated in a more physiological fashion, its mechanical properties are rather different than those seen during synchronous tetanic contractions.

Asynchronous Activation of Muscle

Rather than stimulating the whole muscle nerve tetanically, it is possible to produce a more 'physiological' contraction by subdividing the nerve into several filaments, so that each filament can be stimulated separately and asynchronously. Figure 2.9 shows the difference in tension produced by asynchronous stimulation of several nerve filaments, and synchronous stimulation of the whole muscle nerve. Synchronous stimulation at 5Hz produces an unfused contraction, in which the variation in tension between each stimulus amounts to about 50 per cent of the mean. Asynchronous stimulation of separate filaments at the same rate produces a contraction which has less ripple on the tension record (in Figure 2.9 it is only of the order of 5 per cent mean tension). More surprisingly, however, asynchronous stimulation produces a more forceful contraction in which the mean level of tension is some 50 per cent greater than with synchronous stimulation. This difference between the force produced by the two types of stimulation is seen only at low rates of stimulation. When the rate of synchronous stimulation is rapid enough to produce a fused contraction, then there is no difference in the amount of ripple or the mean level of the tension record (see Figure 2.10).

Figure 2.9: Isometric Tension Produced by Isolated Cat Soleus Muscle Stimulated at 5Hz. The grossly unfused contraction was recorded during synchronous stimulation of the muscle nerve. The smooth contraction was recorded when five filaments of the motor nerve were stimulated asynchronously at the same frequency (from Rack and Westbury, 1969; Figure 5, with permission)

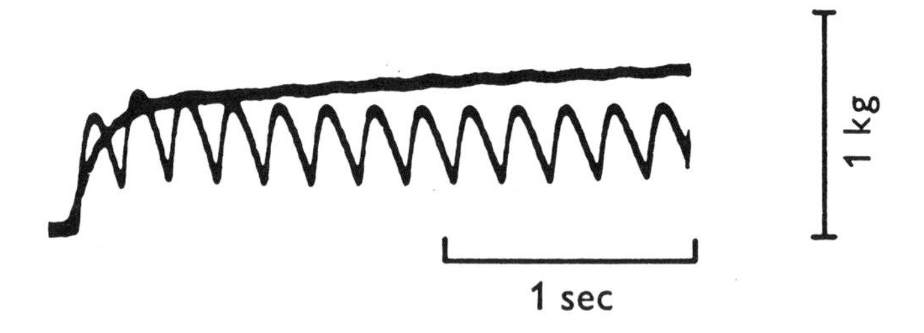

The reason for this difference is as follows. During an isometric contraction, the muscle fibres are not stationary; any increase in tension stretches the tendon and allows the fibres to shorten. The tension fluctuations in an unfused tetanus are accompanied by visible movement of the muscle fibres. This is much smaller during asynchronous stimulation of several nerve filaments. It is thought that at low stimulus rates, muscle is best able to generate tension when the fibres are stationary. If this is true then, at low stimulus rates, whenever a muscle contraction is made smoother by distributing the stimulation pulses, then more force is developed.

Figure 2.10: The Effect of Muscle Length on the Active Isometric Tension Produced by Isolated Cat Soleus Muscle Stimulated at Different Rates. Asynchronous stimulation of five motor nerve filaments was used to plot points joined by the continuous line. The vertical lines show the limits of tension fluctuations during synchronous stimulation of all five nerve roots. Changes in muscle length are shown in centimetres; the corresponding angle of the ankle in the intact animal is drawn below (from Rack and Westbury, 1969; Figure 8, with permission)

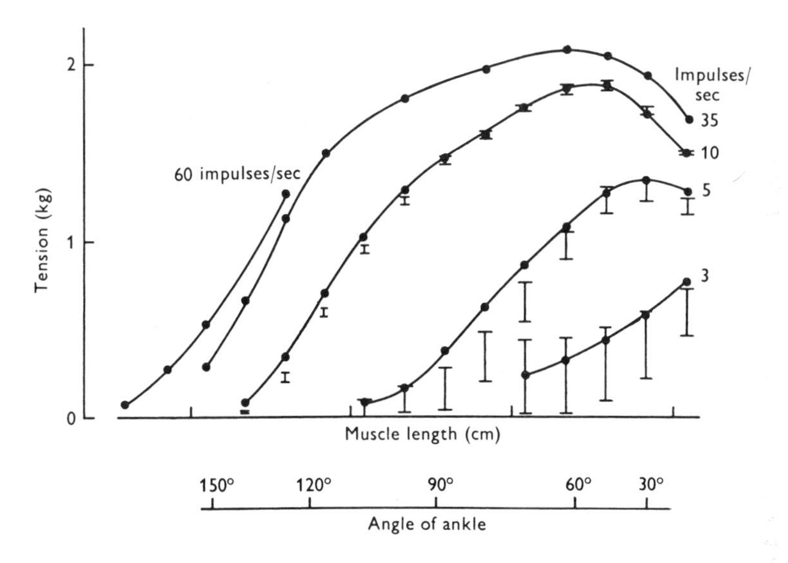

Length–Tension Relationship at Different Levels of Muscle Activation

The length–tension curve of partially activated muscle is very similar in form to those obtained during maximal tetanic stimulation. In Figure 2.10, the continuous lines show the length–tension relationship measured during distributed asynchronous stimulation of five separate filaments of a muscle nerve. Vertical bars show the limit of tension fluctuation during synchronous stimulation of the whole nerve. The graphs illustrate two important points:

1. The previous section described how the differences in mean tension produced by the two methods was most pronounced at low stimulus rates (for example, curve at 5Hz). In addition, Figure 2.10 shows that this difference also depends on muscle length. At short muscle lengths, twitch contraction time is short (see above), so that synchronous stimulation of the whole muscle nerve at low rates produces unfused contractions. Under these conditions there is much more movement within the muscle than with asynchronous stimulation, and the mean tension is correspondingly low. If the muscle is lengthened, the twitch

contraction time increases, and the contractions produced by synchronous nerve stimulation become fused. There is less internal movement in the muscle, so that the tensions generated by synchronous and asynchronous stimulation become similar.

2. Although the shapes of the length–tension curves are similar at all levels of activity, they are displaced from each other along the x axis. That is, the steep part of the curve lies at different muscle lengths, depending on the level of activation. The probable explanation for this is that at short muscle lengths, the process of activation may function less effectively (see also section above on twitch contractions at different muscle lengths). Thus, a high stimulation rate is needed to produce maximum tension at short muscle lengths, while at longer lengths, relatively low stimulation rates produce high tensions.

Forced Lengthening of Partially Activated Muscle

Force–velocity curves for submaximally activated shortening muscle can be constructed during asynchronous stimulation of several nerve filaments. In many respects, they are similar to those seen in maximally activated muscle. However, the behaviour during lengthening contractions can be strikingly different. When maximally activated musles are forcibly stretched, they exert a tension greater than isometric (see Figure 2.7). When submaximally active muscles are stretched, tension may fall below isometric. Figure 2.11 shows tension changes recorded in isolated cat soleus muscle during forced lengthening at constant velocity. On the left of the picture, the muscle was stimulated maximally at 50Hz; on the right, several muscle nerve filaments were stimulated asynchronously at 7Hz to produce a smooth submaximal contraction. In each case, tension rose steeply during the first 0.5mm of movement, and in each case it rose gradually during the later part of the movement. However, the initial rise in tension at 7Hz stimulation was followed by a fall which took the tension during lengthening below the isometric tension corresponding to the new muscle length. Thus, under these conditions, the muscle was being extended by a force smaller than it could have withstood during a sustained isometric contraction.

The explanation for this behaviour is thought to be as follows. In the initial isometric state, the rate at which myofibrillar cross bridges break down is relatively slow. At the onset of forced lengthening, they remain attached over a short distance before they begin to dissociate at a rate faster than occurs during isometric contractions. The rapid rise of tension at the start of forced lengthening is due to stretch of attached cross bridges. While the cross bridges remain attached, the stiffness of the muscle is very high. Over this range (up to about 1mm in a muscle 20cm long) the muscle is said to show *short range stiffness*. The behaviour after the point at which cross bridges begin to dissociate depends on the level of muscle activity. When the muscle is maximally activated, the rate at which cross bridges form may be so high that the increased rate of destruction during lengthening might not seriously diminish the number of cross bridges remaining intact. When the level of muscle activation is lower, then the cross

Figure 2.11: Tension Recorded in Isolated Cat Soleus Muscle During Forcible Lengthening from a Steady Isometric Contraction. Length records are below and tension records above. In a, five separate motor nerve filaments were stimulated asynchronously at 50Hz; in b, the filaments were stimulated asynchronously at 7Hz. The dashed lines indicate the isometric tension obtained during sustained contractions at some of the lengths that the muscle shortened through. Note the fall in tension during lengthening in b, to a level below isometric (from Joyce, Rack and Westbury, 1969; Figure 2, with permission)

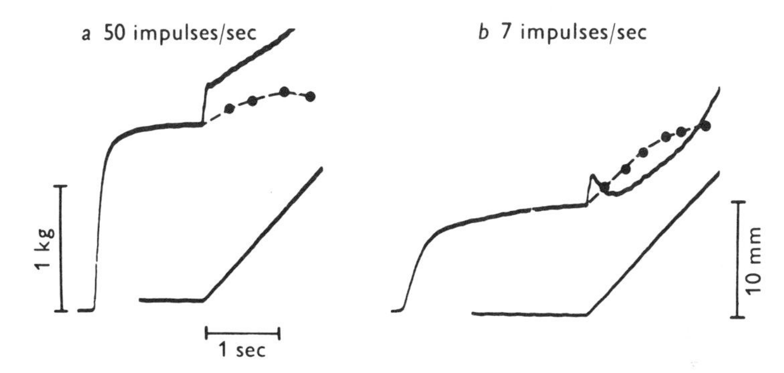

bridges may form more slowly so that during lengthening, the number of attached cross bridges might diminish and the force falls below isometric.

The implication is that submaximally activated muscle may resist lengthening with a very high initial short range stiffness. However, if the muscle is stretched beyond the point at which cross bridges become dissociated, then the stiffness falls dramatically. Such effects are important not only in describing the resistance of extrafusal muscle to external disturbances, but also affect the response of muscle spindles to applied stretch (see Chapter 4) (see articles by Rack and colleagues for details of the work covered in this section).

Muscle Mechanics in Intact Man

It is not so easy to study the mechanical properties of single muscles in man as it is in animal experiments. There are two main problems. First, it is usually very difficult to stimulate a single muscle tetanically without causing severe discomfort to the subject and without the stimulating current spreading to activate other muscles in the vicinity. Second, it is not possible to measure directly the force exerted by a muscle. Because muscles act across one or more joints, it is only possible to measure the torque exerted about the joint or the force exerted by the limb against a force transducer. If the characteristics of a single muscle are to be studied, then corrections must be applied to account for the changing mechanical advantage of the muscle at different joint angles.

Despite this, some useful studies have been made in man which confirm the basic animal data described above. For example, it has proved possible to measure the length–tension characteristics of the biceps brachii muscle. Tetanic

Figure 2.12: Effect of Joint Position on Torque Developed During Twitch (Tw) and Tetanic Contractions of Tibialis Anterior, and During Maximum Voluntary Contraction of the Ankle Dorsiflexors (MVC). Stimulation applied to the motor point of tibialis anterior at frequencies indicated by the numbers at the end of each curve. Mean values are plotted with standard error limits (from Marsh, Sale, McComas and Quinlan, 1981; Figure 3, with permission)

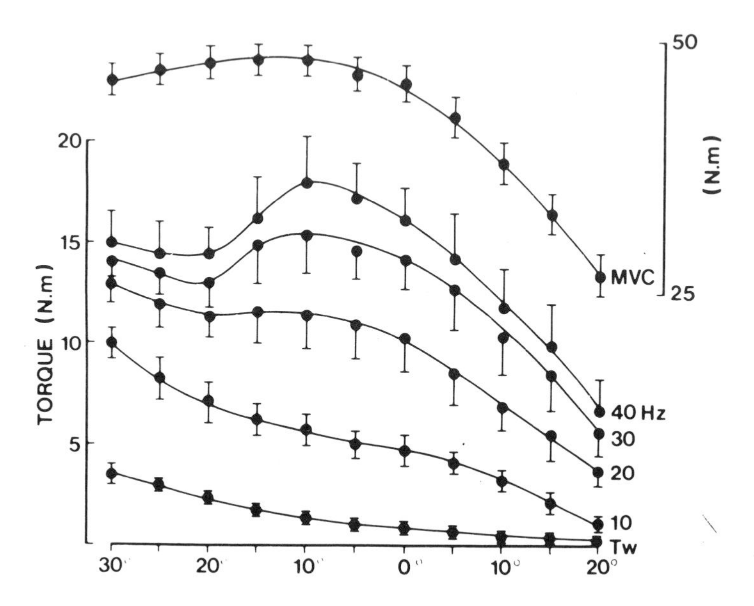

stimulation can be applied to the motor point of biceps without danger of excessive spread of activity to other muscles, and being superficial it is relatively easy to measure the anatomical parameters of the muscle in intact man. After corrections are made for the angle of insertion of the tendon at different elbow angles, it is found that the biceps changes length by about 6cm over the full range of elbow angles, and has a peaked length–tension curve with maximum tension being exerted at a length corresponding to an elbow angle of 100–120 degrees.

However, it can be argued that it is not necessary to extract the length–tension characteristics by trigonometric corrections, since it is the force produced by the limb, or the torque exerted by the muscle around the joint that is the important mechanical characteristic for the CNS. Thus, other studies do not correct for the angle of insertion of the muscle or calculate the actual muscle length. They simply measure the angle–torque relationship around a joint. A typical curve from the tibialis anterior muscle is shown in Figure 2.12. In these experiments, the tibialis anterior was stimulated at its motor point at several different frequencies, and the ankle torque measured at several different angles. At higher stimulation frequencies, the angle–torque relationship is peaked like that of the length–tension curves described above. It is similar to the angle–torque relationship measured during maximal voluntary dorsiflexion of the foot.

Figure 2.13: Torque Developed at the Ankle During Stimulation of the Motor Point of Tibialis Anterior Muscle at Different Frequencies. Curves plotted with the ankle held isometrically at mid and extremes of joint position. A shows the relative torques achieved at each position; B shows the actual torques achieved. P, plantarflexion; D, dorsiflexion (from Marsh, Sale, McComas and Quinlan, 1981; Figure 5, with permission)

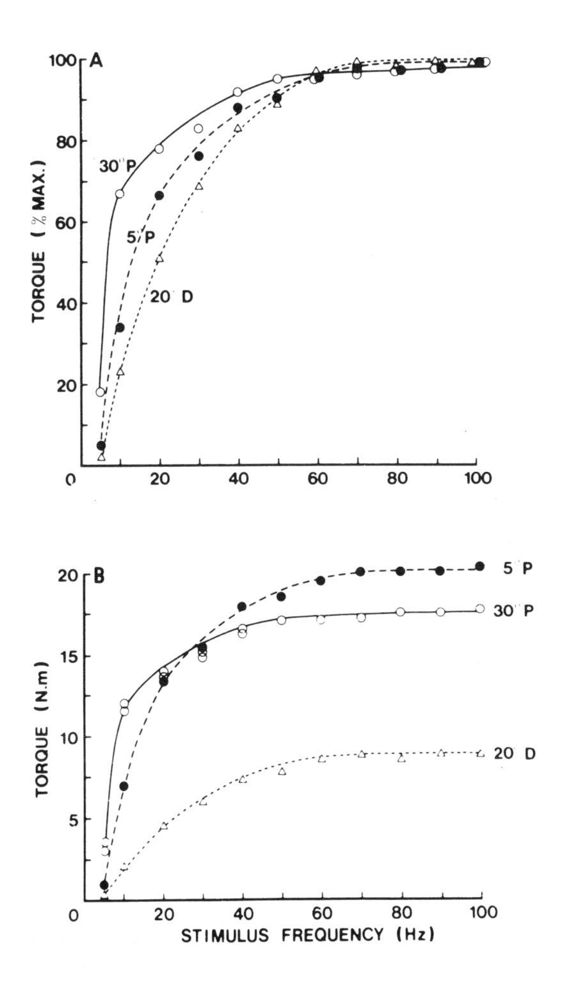

Peak voluntary torque is much higher than that seen during electrical activation because, during voluntary activation, force is produced by the long extensors of the toes in addition to tibialis anterior.

These experiments on tibialis anterior also illustrate the effect of changing muscle length on the twitch contraction time. When the ankle is dorsiflexed and the muscle short, the duration of a single twitch contraction is short; when the ankle is plantarflexed and the muscle extended, the twitch contraction time is longer. The result is that the twitches summate more readily when the muscle is long. Thus, for a given frequency of activation, the muscle exerts a larger proportion of its maximum torque when the ankle is plantarflexed (see Figure 2.13).

Force–velocity curves for a single human muscle have not been obtained. However, curves for all the flexor muscles acting at the elbow have been calculated by releasing the elbow during maximal voluntary activation. The force–velocity curves obtained in this way can be fitted, like those from isolated animal preparations, by Hill's equation (see articles by Marsh et al, 1981; Ismail and Ranatunga, 1978; Wilkie, 1950 for further details of the work covered in this section).

Effects of Muscle Properties on Control of Movement

Effects Arising from the Length–Tension Relationship

1. Force Production at Different Muscle Lengths. A striking example of the effect of muscle length on the force developed is the position which the hand and fingers adopt during a power grip. The wrist is held extended by contraction of the extensor muscles in the forearm, while the fingers flex powerfully onto an object in the palm of the hand. A more flexed wrist allows the flexors to shorten further, but at this length they are unable to produce as much force. As every hero in a gangster movie knows, the way to make the villain release his grip on the gun, is to flex the wrist forcibly to reduce the power of the finger flexors. The gun will thereupon fall dramatically to the floor, to be kicked neatly away by the hero's foot. The importance of this example is that it provides evidence that the nervous system must somehow take into account the length–tension relationship of muscle during normal movement, in this instance by extending the wrist when a maximal flexor force is required.

2. Stiffness of Muscle. Relaxed muscles present little or no resistance to changes in muscle length. When active, however, stiffness can be surprisingly high, particularly the short-range stiffness seen during small changes in muscle length, or at the start of larger muscle stretches. Stiffness increases with the force of muscle contraction. This stiffness plays the initial part in the response of a limb to a disturbing force. Thus if muscles cocontract at a joint, the mechanical stiffness of the limb is increased, helping to resist forces that might be expected to be encountered, for example, in carrying a full glass across a crowded party room.

Effects Arising from the Force–Velocity Relationship

1. Efficiency of Muscle Contraction. As shown above, maximum work will be performed by a muscle shortening at intermediate muscle lengths. Power, which is the rate at which work is done, also is maximum at these lengths. Thus, in order to achieve maximum efficiency of muscle, it is necessary to arrange for the contraction to take place at about these velocities and lengths. A good

example of the importance of these effects in normal movement is given by the gearing on a bicycle. Load and speed are matched to the properties of the muscle so that power output can be maintained regardless of the steepness of the incline.
2. *Lengthening Contractions.* Lengthening contractions also are a common feature of normal movement. They are sometimes known as eccentric contractions as opposed to the more usual shortening or concentric contractions. In fact, lengthening contractions occur regularly in activities such as walking, jumping or hopping. Hopping is a particularly instructive example since the nervous system seems to make purposeful use of a lengthening contraction to provide extra force output from the muscle.

Hopping can be initiated voluntarily or reflexly. If the force under the foot is measured during a voluntary hop, the first thing to happen is that there is a momentary decrease in the force, followed by a rapid increase which produces the upthrust of the hop (see Figure 2.14). This preceding relaxation allows the body to fall a little, stretching the extensor muscles at a time when they are being reactivated for take-off. A voluntary hop always is initiated in this way, and subjects find it almost impossible to hop without first bending the knee.

Figure 2.14: Time Course of Force Changes on Platform During Reflex and Voluntary Hopping. Reflex hopping was induced by sudden pulls to the shoulder of the standing subject which made him off-balance. Voluntary hops were self-initiated and were always preceded in the force record by a momentary decrease in force (from Roberts, 1978; Figures 10.22 and 10.23, with permission)

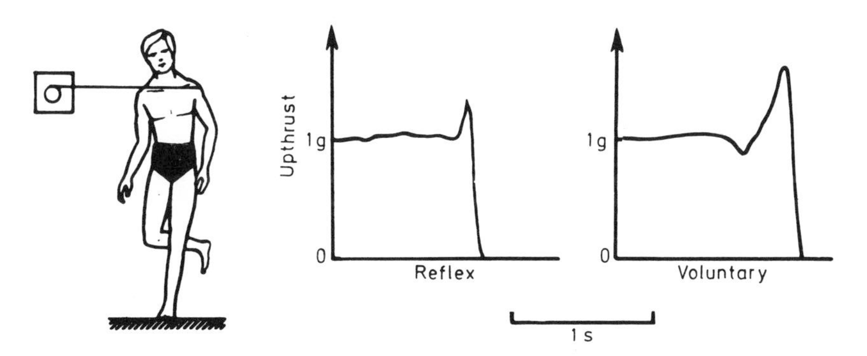

Reflex hopping is produced in a subject standing on one leg by pulling him briskly off-balance. In this instance, the hop never shows the initial dip in the force record; there is only an uncomplicated contraction of the extensor muscles. Reflex hopping, unlike voluntary hopping, does not make use of the extra force available from the muscle in a lengthening contraction. Perhaps the nervous system, in attempting to restore balance as fast as possible, 'trades off' the loss of contractile force against the reduction in delay in starting the hop.

Intrinsic Feedback Control of Muscle Contraction

Both the length–tension and force–velocity relationships can contribute to compensation for unexpected disturbances of movement. For example, a sudden increase in load can produce increased muscle tension not only via reflex pathways (see Chapter 6), but also by the nature of the length–tension characteristics of the muscle.

Figure 2.15: Response of a Normal Subject and a Deafferented Man to an Unpredictable Five-fold Increase in the Viscosity of the Load Opposing a Thumb Flexion Movement. Control trials shown by thick lines, trials with unexpected increase in viscosity by thin lines. The traces are (top) thumb position, (middle) rectified electromyogram from the long flexor of the thumb and (bottom) torque generated by motor against which subject is pressing, which approximates the torque exerted by the thumb on the lever. Note the large increase in torque during the addition of increased viscosity (from Rothwell, Traub, Day *et al.*, 1982; Figure 14, with permission)

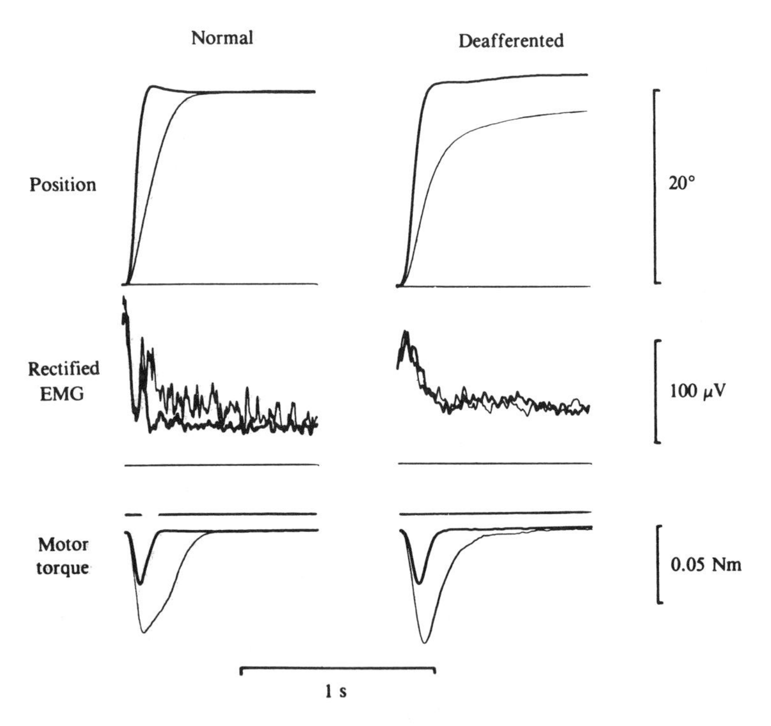

The force–velocity relationship of active muscle also can produce surprisingly large effects on the force exerted by a muscle. An example is shown in Figure 2.15. On the left are the responses from a normal subject, and on the right are responses from a patient who was deafferented by a severe peripheral neuropathy of unknown origin. The responses from this individual are uncontaminated by any reflex contractions of the muscle. Both subjects were trained to perform fairly rapid thumb flexion movements to a given end position against a small force without visual feedback. On random trials, the viscosity of the load was suddenly increased at the start of the movement. To a normal person this feels as if the load has to be pushed through a jar of treacle, and slows the movement considerably. In the normal subject, the sudden introduction of this unexpected viscosity produced extra reflex and voluntary electromyogram activity in the thumb flexor muscle, and the thumb finally reached the intended end position. The deafferented man had no clues that the load had changed and his electromyogram was the same as in the control trials. However, the force output from his thumb still changed considerably, although not as much as normal. Slowing the movement increased the amount of force that the muscle could exert, and the area under the curve describing the force output of the muscle increased by a factor of five. The same effects are seen if the inertia of the load is changed.

A Theory of Movement Control Which Makes Use of the Mechanical Properties of Muscle

Emilio Bizzi has argued that the length–tension relationships of agonist and antagonist muscles acting around a single joint could be used by the nervous system to specify particular joint angles. If, for example, both muscles are activated equally, then the tension in the agonist will increase with movement of the joint, which increases the length of the muscle. By the same reasoning, such a movement will shorten the antagonist and reduce its tension. The point of equilibrium will be given by the intersection of the respective length–tension curves, where the forces generated in each muscle are equal and opposite (see Figure 2.16). Activation of a muscle can be regarded as changing the length–tension relationship of the muscle, since it will increase the stiffness. Thus, if the agonist muscle is activated, its length–tension curve will be shifted upwards, and the intersection point of equilibrium with the antagonist curve will be shifted likewise. The joint will move automatically to take up the new position of equilibrium. Movement to any joint angle can be viewed in the same way, that is by changing the balance between the length–tension curves of each muscle.

The attraction of looking at muscle activation in terms of shifting length–tension relationships is that this provides a mechanism whereby the nervous system can specify the end (equilibrium) position of a movement without calculating the separate steps to get there. In other words, this is a way of specifying a movement to a particular place without necessarily calculating a specific trajectory. As long as the length–tension relationships show no prominent peaks, there can be only one point of joint equilibrium. (If there were peaks in the curves, then more than one position would be stable, and the initial starting position would determine the end point assumed by the joint.)

Bizzi and his colleagues (1982) have carried out some elegant experiments on monkeys which indicate that this may be the way in which some movements are organised. The experiment involved training monkeys to flex their elbows and point to a target light when illuminated in order to achieve a food reward

Figure 2.16: Diagram of How Length–Tension Relationships of Extensor and Flexor Muscles About a Single Joint Might Interact to Specify a Particular Joint Angle. Length–tension curves for flexor (T_f) and extensor (T_e) muscle are plotted against the angle of the joint (θ), rather than the length of the muscle. The point of intersection of the length–tension curves is the equilibrium position of the joint (vertical lines). The slope of the length–tension curves can be shifted by voluntary activity. For example, increasing flexor activity increases flexor muscle stiffness and shifts the flexor curve upwards. In normal circumstances this might be accompanied by decreased activity and hence decreased stiffness of extensor muscles. Thus, the two new length–tension curves then intersect at a different joint angle, that is, the equilibrium angle has been shifted (from Bizzi, Accornero, Chappele *et al.*, 1982; Figure 1, with permission)

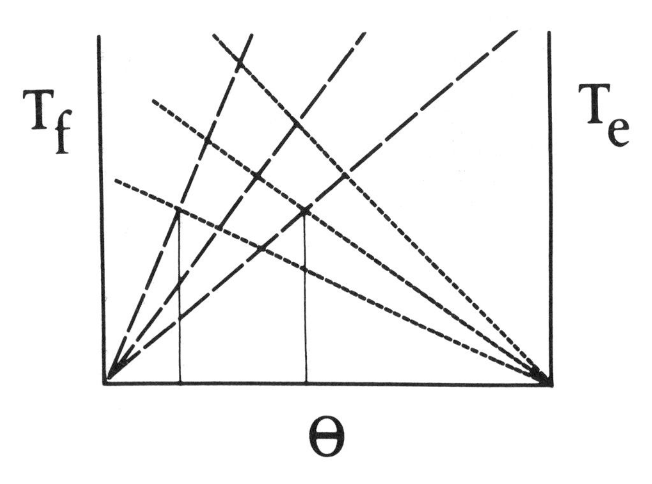

(Figure 2.17). In random trials, the arm movement was perturbed by the experimenter either loading or assisting the movement for a short period. The monkeys could not see their arm, but they could feel the perturbation. They achieved the final end position with reasonable accuracy despite the disturbance. Of course, in these monkeys, both reflex and voluntary muscle activity might have helped them to compensate for the disturbance.

To determine to what extent these corrections were necessary, the animals then were deafferented by surgical section of the dorsal spinal roots in order to remove any sensory feedback from the arm, and allowed to relearn the task. After retraining, the monkeys were deprived of vision of the arm, and the accuracy of pointing in the absence of any sensation was assessed. Again in a number of trials, the movement of the forearm was interrupted briefly by passive manipulation by the experimenter. Since the monkey was deafferented, it could not feel the interruption of movement and no reflex events intervened to correct for them. Despite this, the monkeys reached the final end position with little loss of accuracy (Figure 2.17), suggesting that the final end position of the movement was planned rather than the trajectory necessary to achieve it.

These elegant experiments are not the final answer to movement control. It is obvious that in some instances, the trajectory of movement, as well as its final

Figure 2.17: Movement of the Elbow to a Target Position Indicated By an Illuminated Lamp. The monkey cannot see its own arm (from Polit and Bizzi, 1979; Figure 1, with permission). In the bottom part of the figure are records of arm position before and after surgical deafferentation of the arm. Records in A are control trials in which no disturbance was given to the movement. In B, the movement was assisted briefly (arrows) and in C, it was reversed (arrows). Despite these disturbances, the monkey reaches the final target without substantial loss of accuracy even after deafferentation (from Polit and Bizzi, 1978; copyright American Association for the Advancement of Science, 1978; Figure 1, with permission)

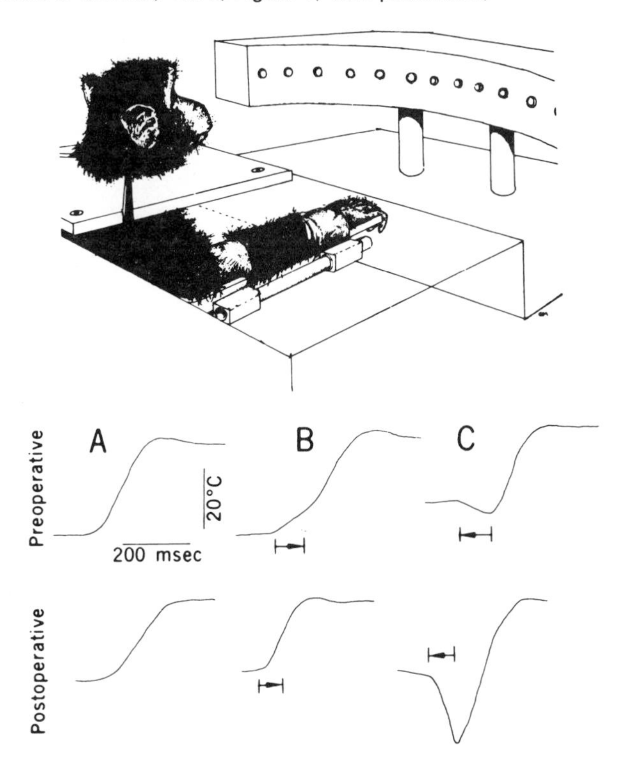

position, is of importance to the movement. We need to move slowly on some occasions and rapidly on others, and in such circumstances, muscle activity will have to modulate during the movement according to the velocity at which we wish to move. Another objection is the assumption that the point of equilibrium between the length–tension relationships of different muscles will be specified with enough accuracy to allow accurate movement to take place. For example, in Figure 2.15, it is clear that the deafferented man did not reach the final intended end position when the viscosity of the load was increased. In fact, if the limb had moved according to the spring hypothesis, introduction of extra viscosity would have had no effect on the final end position of the movement. The fact that it had implies that the general applicability of the theory is still under question (see articles by Bizzi *et al.*, 1982; Day and Marsden, 1982, for further details of the 'spring' hypothesis).

References and Further Reading

Review Articles

Eisenberg, E. and Greene, L.E. (1980) 'The Relation of Muscle Biochemistry to Muscle Physiology', *Ann. Rev. Physiol.*, *42*, pp. 293–309

Gordon, A.M. (1982) 'Muscle' in T. Ruch and H. Patton (eds.) *Physiology and Biophysics*, vol. IV, Saunders, Philadelphia, pp. 170–260

Hill, A.V. (1970) *First and Last Experiments in Muscle Mechanisms*, Cambridge University Press, London

Huxley, A.F. (1974) 'Review Lecture. Muscular Contraction', *J. Physiol.*, *243*, p. 1–43

Partridge, L.D. and Benton, L.A. (1981) 'Muscle, the Motor' in V.B. Brooks (ed.), *Handbook of Physiology*, sect. 1, vol. 2, part 1, Williams and Wilkins, Baltimore, pp. 43–106

Pollack, G.H. (1983) 'The Cross-bridge Theory', *Physiol. Rev.*, *63*, pp. 1049–113

Roberts, T.D.M. (1978) *Neurophysiology of Postural Mechanisms*, Butterworths, London, Chapters 2 and 10

Original Papers

Bizzi, E., Accornero, N., Chappele, W. *et al.* (1982) 'Arm Trajectory Formation in Monkeys', *Exp. Brain Res.*, *46*, pp. 139–43

Burke, R.E., Rudomin, P. and Zajac F.E. (1970) 'Catch Property in Single Mammalian Motor Units', *Science*, *168*, pp. 122–4

Day, B.L. and Marsden, C.D. (1982) 'Accurate Repositioning of the Human Thumb Against Unpredictable Dynamic Loads is Dependent Upon Peripheral Feedback', *J. Physiol.*, *327*, pp. 393–407

Gordon, A.M., Huxley, A.F. and Julian, F.J. (1966) 'The Variation in Isometric Tension with Sarcomere Length in Vertebrate Muscle Fibres', *J. Physiol.*, *184*, pp. 170–92

Hill, A.V. (1938) 'The Heat of Shortening and the Dynamic Constants of Muscle', *Proc. Roy. Soc. B.*, *126*, pp. 136–95

Huxley, A.F. (1957) 'Muscle Structure and Theories of Contraction', *Prog. Biophys. Chem.*, *7*, pp. 255–318

Huxley, A.F. and Niedergerke, R. (1954) 'Structural Changes in Muscle During Contraction,' *Nature*, *173*, pp. 971–7

Huxley, H.E. and Hanson, J. (1954) 'Changes in the Cross-Striations of Muscle During Contraction and Stretch and Their Structural Interpretation', *Nature*, *173*, pp. 978–87

Ismail, H.M. and Ranatunga, K.W. (1978) 'Isometric Tension Development in a Human Skeletal Muscle in Relation to Its Working Range of Movement: The Length–Tension Relationship of Biceps Brachii Muscle', *Exp. Neurol.*, *62*, pp. 595–604

Joyce, G.C., Rack, P.M.H. and Westbury D.R. (1969) 'The Mechanical Properties of Cat Soleus Muscle During Controlled Lengthening and Shortening Movements', *J. Physiol.*, *204*, pp. 461–74

Marsh, E., Sale, D., McComas, A.J. *et al.* (1981) 'Influence of Joint Position on Ankle Dorsiflexion in Humans', *J. Appl. Physiol.*, *51*, pp. 160–7

Polit, A. and Bizzi, E. (1978) 'Processes Controlling Arm Movements in Monkeys', *Science*, *201*, pp. 1235–7

——— (1979) 'Characteristics of Motor Programs Underlying Arm Movements in Monkey', *J. Neurophysiol.*, *42*, pp. 187–94

Rack, P.H.M. and Westbury, D.R. (1969) 'The Effect of Length and Stimulus Rate on the Tension in the Isometric Cat Soleus Muscle', *J. Physiol.*, *204*, pp. 443–60

Rothwell, J.C., Traub, M.M., Day, B.L. *et al.* (1982) 'Manual Motor Performance in a Deafferented Man', *Brain*, *105*, pp. 515–42

Wilkie, D.R. (1950) 'The Relation Between Muscle Force and Velocity in Human Muscle', *J. Physiol.*, *110*, pp. 249–80

3 THE MOTOR UNIT

The previous chapter discussed some of the properties of whole muscles and the importance of these properties in the overall control of movement. The present chapter is concerned with the internal organisation of single muscles. A muscle does not consist of an homogenous population of muscle fibres: there are several different types of fibre within muscle, each of which have different mechanical properties. Thus the range of operation of the whole muscle is extended beyond that expected from any one fibre type alone.

The Concept of the Motor Unit

In mammalian skeletal muscle, the primary organisation of the muscle is imposed by the CNS. During development, or during re-innervation following nerve injury, each nerve axon forms synaptic contacts with many muscle fibres. At this stage each fibre also receives terminals from many different axons, but as time passes, the terminals from one axon predominate and the others degenerate (but see section on tonic muscle fibres below). In the final state, each muscle fibre is contacted by terminals from a single parent axon. The combination of motoneurone axon, terminal branches and the muscle fibres which they innervate is known as the *motor unit*, a term coined by Leyton and Sherrington in 1925 (Figure 3.1). This is a basic 'quantal' unit of muscular contraction and represents the smallest number of fibres that can be activated by the CNS at any one time.

Twitch and Tonic Muscle Fibres

Striated muscle contains two types of muscle fibre. The great majority of human striated muscle is made up of *twitch* fibres, although *tonic* fibres have been demonstrated in the extraocular, laryngeal, and middle ear muscles. It is also thought that some intrafusal fibres of the muscle spindle may be of the tonic type. The two types are similar in the basic molecular structure, but differ in their nerve innervation and electrical properties. In twitch fibres, synapses with the motor nerve are found only at one end-plate region, where there is a complex infolding of the underlying fibre membrane. This region is organised to produce large end-plate potentials, which can trigger off all-or-none propagated action potentials. These travel to either end of the fibre and the depolarisation causes a near synchronous contraction, or twitch, of the whole fibre.

In tonic fibres, there is more than one end-plate region per fibre, but each

Figure 3.1: Reconstruction of the Territory of a Single Motor Unit in the Medial Gastrocnemius Muscle of the Cat. A single motor axon, innervating a type FF unit, was stimulated repetitively so as to fatigue the muscle fibres which it innervated. Fatigue depletes fibres of their glycogen reserve (see text). The muscle was then quickly excised and frozen and stained for fibres showing glycogen depletion using the periodic acid-Schiff method. The position of identified fibres is shown in the sections of muscle taken on the right. This was a large unit, with some 750 fibres distributed over a large area of the left part of the muscle (stippled area on plan view) (from Burke and Tsairis, 1973; Figure 1, with permission)

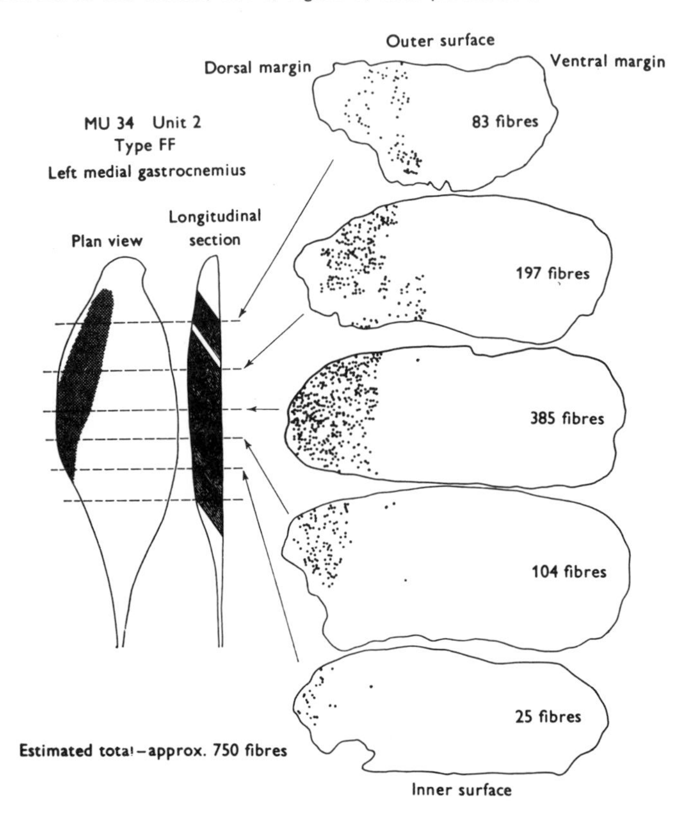

is simpler than that of the twitch fibre, not having the complex infolded membrane or such a highly branched motor axon. In general, tonic fibres do not conduct action potentials; depolarisation spreads by local circuits from each end-plate. Thus, tonic fibre contraction can be graded by the amount of membrane depolarisation, unlike that of the twitch fibres in which the action potential always depolarises the fibre by the same amount. Indeed, tonic fibres rarely produce any tension in response to a single stimulus. Tonic fibres are sometimes known as *slow* fibres, but they should not be confused with the *slow twitch* fibres described below. The remainder of this chapter will be concerned only with twitch fibres (see Lewis, 1981, for further information of the work in this section).

Physiological Investigation of the Motor Unit

Territory and Size of Motor Units

In animal experiments individual motor units can be stimulated electrically via their motor axon in teased-out strands from ventral root filaments, or by microelectrodes inserted into the motoneurone cell body in the ventral horn of the spinal cord. The distribution of muscle fibres which make up the motor unit can be visualised directly using the glycogen depletion method, which was devised by Edstrom and Kugelberg in the late 1960s.

Single units are stimulated repetitively at high frequency until the fibres within the unit become fatigued and depleted of glycogen. They are then stained selectively by the PAS (periodic acid-Schiff) method. Rather than showing the fibres of a motor unit to be in close proximity, this reveals that they are spread throughout a fairly large territory of muscle (10 to 30 per cent) and show considerable overlap with fibres from other units (Figure 3.1). Even within a single muscle fascicle, the fibres belong to many different motor units, there being only two to five fibres from each unit. Most fibres do not have direct contact with other fibres of the same unit.

The same method allows counts to be made of the number of fibres per unit. However, a more common way of estimating the *average* number of fibres per unit in different muscles is to count the total number of fibres in the muscle and divide by the number of motor axons in the muscle nerve. In animal experiments, counts of motor axons usually are made one to two weeks after deafferentation by section of the dorsal roots. Obviously, in human material, such pre-treatment is not possible: estimates of the number of motor fibres are made by counting the number of large axons (greater than 7–8 μm diameter) in intact muscle nerves, and then assuming 40 to 50 per cent of these are sensory axons. In general, motor units are largest in those muscles which act on the largest body masses. There are over 1,000 fibres per unit in the medial gastrocnemius, but less than 100 fibres per unit in the extraocular muscles (see Edstrom and Kugelberg, 1968; Burke and Tsairis, 1973; Buchthal and Schmalbruch, 1980; Burke, 1981).

Differences in the Contraction of Motor Units

Ever since the time of Ranvier, it has been known that whole muscles may have different contractile properties. In 1873, he showed that red muscles, such as the soleus, were in general slower to contract following an electrical stimulus than pale muscles like the gastrocnemius. This difference between muscles extends to the level of the motor unit. Indeed, an excess of one type of unit gives a muscle its characteristic overall properties.

Speed of Contraction and Fatiguability (Figure 3.2). Two main categories of motor unit can be distinguished on the basis of their contraction speed following a single stimulus to the nerve axon. This is known as the *twitch test*. Fast units

Figure 3.2: Physiological Identification of Motor Unit Type. Each row illustrates the tension records produced by activity in a particular type of motor unit. The columns show the different responses of these units to three tests, tetanus, twitch and fatigue. Note the calibration marks for tension are different for each unit. Tetanic tension is smallest for type S units: type FF and FR units show 'sag' (arrow). Twitch size is smallest and twitch duration longest in type S units. Repeated tetani do not produce fatigue in type S and type FR units, but do in type FF units (three superimposed records of the first, 60th and 120th tetanus) (from Jami, Murthy, Petit *et al.*, 1982; Figure 1, with permission)

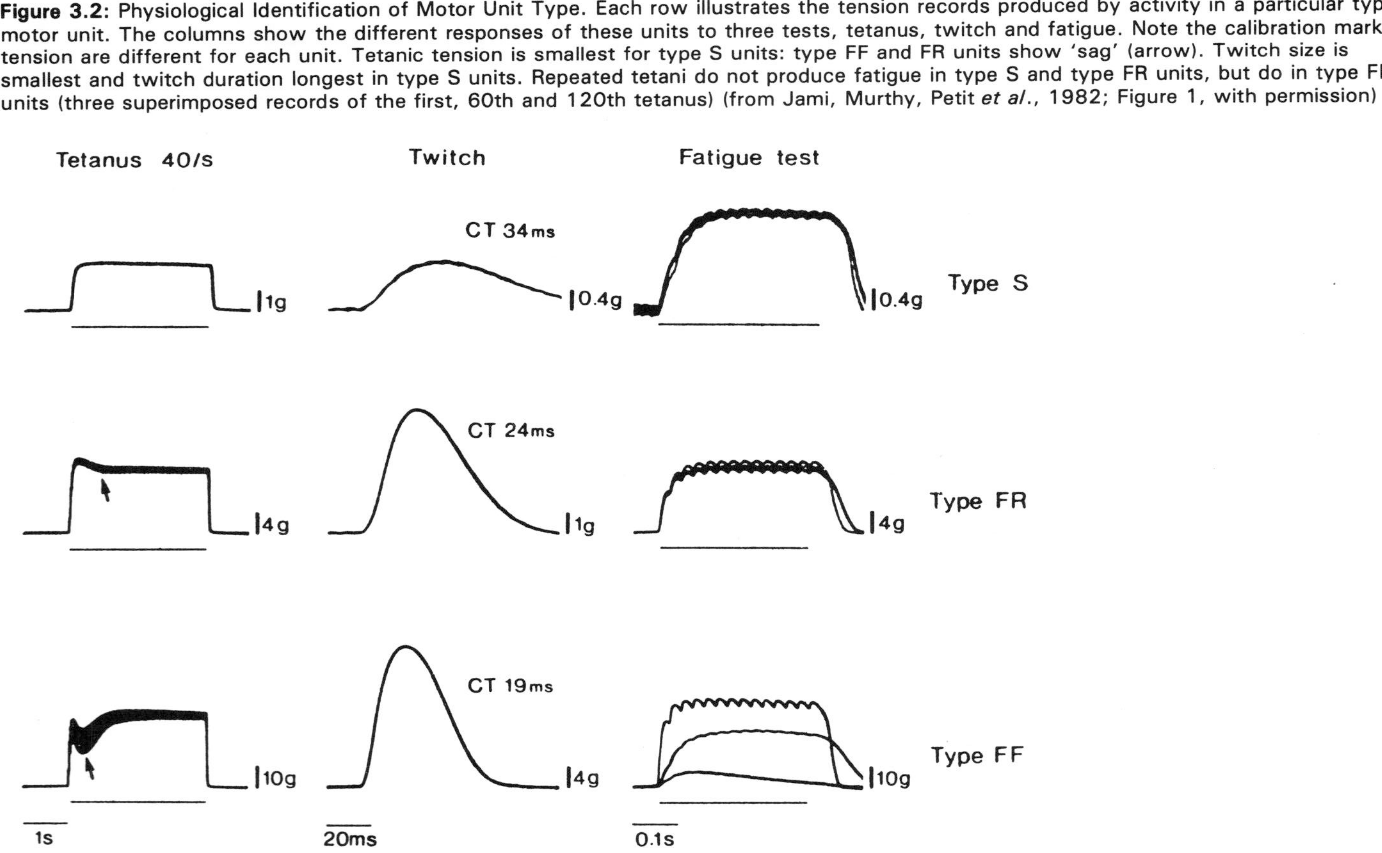

(F) reach their peak tension and relax more rapidly from it than slow (S) units. These two categories of unit also behave differently in the *sag test*. Brief unfused tetani are given by stimulating the axon at intervals of 1.25 times the twitch contraction time. Fast units show a slow decline in the peak tension over the duration of the tetanus ('sag'), whereas slow units do not.

The group of fast units can be further subdivided on the basis of resistance to fatigue. Repeated tetanic stimuli are given briefly every second and the tension output monitored for each contraction. Slow units produce almost the same output for each contraction. Fast units do not. There is a subgroup with fast twitch contraction times (FF, fast fatiguable) which show a rapid decline in tension output over the first few minutes of the test. Another subgroup (FR, fast fatigue-resistant) has a high resistance to fatigue similar to the S units. Sometimes, another category of fast units is added: F(int), with properties lying between the FF and the FR subgroups. It should be noted that muscle fibres may be fatigued without any failure of electrical transmission at the neuromuscular junction. That is, the fibre action potential can be normal, while twitch force is reduced, indicating that there is fatigue of the force generating mechanism (see section on human fatigue below).

Twitch Tension and Specific Twitch Tension. The major way of classifying motor units relies on the speed of contraction and fatiguability tests above. However, there are many other differences between units, although they are not so useful in distinguishing motor unit *type*. The peak twitch or tetanic tension is larger in fast units than in slow units and, in general, the peak tension declines in the order FF > FR > S (see the three-dimensional relationship in Figure 3.3).

The reason for this is a combination of several factors. First, estimates of the average number of fibres per unit have shown that type S units have a smaller number of muscle fibres than type FF or FR. This is done by counting the total number of muscle fibres with each characteristic histochemical profile (see below) and multiplying by the relative frequencies of each type of unit within the same muscle. In addition, it can be seen microscopically that type S units have fibres with smaller diameter than type FF or FR units.

However, even when these two factors of innervation ratio and fibre size are taken into account, they are still insufficient to explain completely the difference between the force production of fast and slow units. That is, when expressed as the twitch tension per unit cross-sectional area of muscle (which takes into account the differences above), the S units still generate less force than the F units. The type S units are said to have a low *specific twitch tension*. This is thought to be due to differences in the mechanism of force production within the fibrils of the type S units.

Post-tetanic Potentiation. Following a tetanus, the peak twitch tension of fast units is increased for a short period. The contraction of slow units in different muscles is either less potentiated or may even be diminished. The reason for this

Figure 3.3: Three-dimensional Relationship Between Twitch Tension, Twitch Contraction Time and the Fatigue Index (the Ratio Between Contraction Strength in the First and 120th Tetanus of a Series Given Regularly at 1s Intervals) in the Medial Gastrocnemius Muscle of the Cat. Units showing 'sag' (that is, FR and FF) denoted by open circles, whereas units without 'sag' denoted by stippled circles (that is, S). Note the clear division of the units plotted by open circles into two groups: the type FF with a very low fatigue index, and the type FR with high fatigue index. Two units fall between these groups and may be of the F_{int} category (from Burke, Levine, Tsairis *et al.*, 1973; Figure 5, with permission)

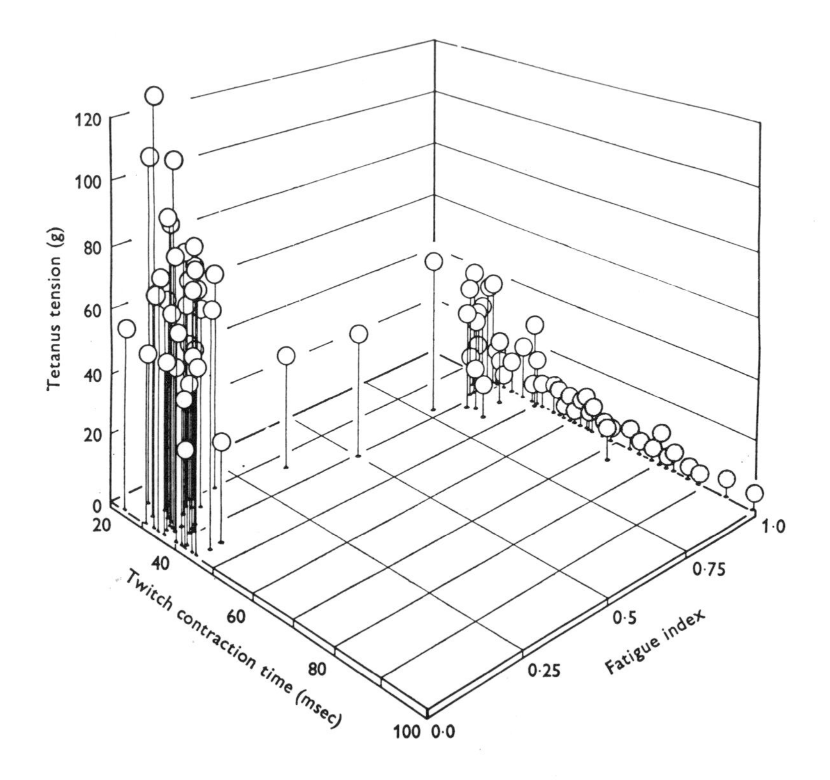

effect and its difference between unit types is unknown. It may be related to the kinetics of Ca^{2+} binding and sequestration following muscle activation.

Size and Conduction Velocity of Motor Axons. Although the motor axons supplying extrafusal muscle are all of the alpha category, they have a range of sizes (in the cat, from 10–20 μm) and conduction velocity (cat, 40–100 ms^{-1}). It has been known for many years that the motor nerves supplying slow muscles have a lower conduction velocity than those to fast muscles. The same applies at the level of the motor unit. The average conduction velocity of axons supplying the different types of motor unit follows the order FF > FR > S. In addition, it is also known that large diameter axons have large cell bodies within the ventral horn of the spinal cord. Thus, fast units have motoneurones with large cell bodies, while slow units are innervated by smaller motoneurones.

Summary. Muscles are made up of various proportions of these different types of motor unit. Units with a fast contraction time are composed of fibres with a relatively large diameter and are supplied by a fast conducting axon from a large motoneurone. They produce high twitch tensions and are relatively easy to fatigue. Slow units are slow to contract and are made up of rather fewer fibres of small diameter. The motor axons conduct slowly and the units produce only small amounts of force, although they are fatigue-resistant.

The advantage of having the subgroups is that the total range of operation of the muscle is extended beyond that of any single unit type. The relative number of each unit can give distinctive properties to whole muscles which makes them suitable for different kinds of movement. In the cat, prototypical examples of this are the soleus and gastrocnemius muscles. The soleus has a large percentage of type S units. It has a slow contraction time and is not readily fatigued. Gastrocnemius has more fast units and can exert a greater tension than soleus, but only over short periods. Consistent with this difference in contractile properties is the role played by each muscle in normal movement. Soleus is used almost continuously in walking and standing, whereas gastrocnemius is particularly useful in jumping and sprinting.

Even blood flow through the muscle is related to the properties of the motor units. In the cat, the flow to gastrocnemius has a capacity 2.5 times greater than that to soleus. Because the oxygen consumption per twitch of slow units is less than that of fast units, this means that the blood supply to soleus is sufficient to deliver enough oxygen even during a long tetanus at 20 per second. In contrast, oxygen delivery fails in gastrocnemius at low rates of activation above four per second. At the level of the single unit, capillary density is greater next to the fibres of slow than fast units (see Burke *et al.*, 1973; Buchthal and Schmalbruch, 1980; Burke, 1981).

Histochemical and Biochemical Classification of Muscle Fibres

Most mammalian muscles contain a mixture of muscle fibre types which can be distinguished by biochemical methods. It is probable that much of the physiological differences between motor unit types can be explained on the basis of differences in the biochemical profile of their constitutent muscle fibres (see below). For many years it had been known that there were different numbers of mitochondria within muscle fibres, suggesting differences in their oxidative capacities. More recently, histochemical methods have been used, which allow visualisation of enzyme content, metabolic substrates or structural proteins within individual fibres. Differences in the content of metabolic enzymes suggests preferential use of particular metabolic pathways. Examples of some of the presently used stains are given in the following list:

1. ATPase activity of myofibrillar material. Myosin ATPase has various forms or isoenzymes which break down ATP at different rates. Testing in an alkaline pH of 9.4 separates two main types of muscle fibre. Those which stain darkly are predominant in pale muscle, and those which stain lightly are more common in red muscle. Sometimes acid preincubation at pH4.6 is used to separate out three different categories.

2. SDH (succinic dehydrogenase) and NADH-D (NADH-dehydrogenase). Both these enzymes are involved in major pathways for oxidative metabolism in the mitochondria.

3. Glycogen and phosphorylase stains. Both stains relate to the capacity for anaerobic metabolism, used when the oxygen supply of muscle is limited. Phosphorylase is an enzyme involved in the breakdown of glycogen.

By applying selective stains for several enzymes to serial sections of muscle, a particular 'histochemical profile' can be obtained for each fibre. Unfortunately, the nomenclature for this classification has been changed repeatedly and a classification evolved for the muscles in one animal may not be appropriate to other species. In this chapter discussion is limited to fibre types in guinea pig, cat and man; rat muscle fibres have a different histochemical profile.

Two main histochemical nomenclatures have been developed: types I, IIA and IIB; and SO (slow, oxidative), FOG (fast, oxidative glycolytic) and FG (fast, glycolytic). The latter nomenclature employs terms relating to the physiological properties of the fibres, and is discussed in the next section. Table 3.1 shows the typical histochemical profile of each of the fibre types. ATPase levels and the capacity for anaerobic respiration separate the fibres clearly into two groups. Type I fibres have a low capacity for anaerobic respiration and a low level of ATPase staining, whereas type II fibres have a high anaerobic capacity and high levels of ATPase staining. Levels of oxidative enzymes can be used to separate two subtypes of group II fibres: IIA have high levels of both oxidative and glycolytic enzymes; IIB have high levels of glycolytic enzymes but are poor in oxidative capacity. (See articles by Lewis, 1981; Kugelberg and Edstrom, 1968; Burke *et al.*, 1973; Burke, 1981; for further details of the work covered in this section.)

Correlation between Histochemical and Physiological Classifications of Motor Units

Ever since Ranvier's observation that the colour and histological feature of whole muscle were in many cases correlated with the speed of contraction, it has become generally accepted that the increasingly well-studied histochemical characteristics of individual muscle fibres must be related in some way to the physiological properties of the same fibres. However, this is technically rather difficult to prove, since it involves being able to analyse the histochemistry of individual muscle fibres which comprise a physiologically identified motor unit.

An important advance was made with the introduction of the glycogen

Table 3.1: Relationship between Fibre Type, Motor Unit Type and Histochemical Profiles of Muscle Fibres

Fibre Type	I SO (Slow, oxidative)	IIA FOG (fast, oxidative glycolytic)	IIB FG (fast, glycolytic)
Motor Unit Type	S	FR	FF
Histochemical Profiles			
Myofibrillar ATPase (pH9.4)	Low	High	High
NADH dehydrogenase	High	Medium-High	Low
SDH	High	Medium-High	Low
Glycogen	Low	High	High
Phosphorylase	Low	High	High
Capillary Supply	Rich	Rich	Sparse
Fibre diameter	Small	Medium-Small	Large

depletion technique. In an elegant series of experiments on the gastrocnemius muscle, Burke *et al.* (1973) studied individual motor units for the physiological properties. The unit was then fatigued by repetitive stimulation of its axon, and serial sections were cut from the muscle. Serial sections were then stained for glycogen depletion, or tested for the levels of oxidative (SDH), glycolytic (phosphorylase) or ATPase enzymes. Thus the fibres of an individual unit could be identified in one section by their lack of glycogen and characterised in the other sections for their histochemical properties. It should be noted, however, that the results from slow twitch units were a little more difficult to interpret because of the problems in fatiguing all the fibres of each unit so that they could be identified by the glycogen depletion method.

There were two important findings: the first was that all the fibres from a single motor unit were of the same histochemical type; the second was that the histochemical characteristics were intimately related to the physiological contractile properties of the unit. Table 3.1 summarises the important relationships between histochemistry and physiology.

The *speed* of contraction was related to the ATPase and phosphorylase activity of the fibres. Fast twitch fibres had high levels of phosphorylase, and stained darkly for ATPase after alkaline incubation. They were type II fibres. The *fatiguability* of contraction was related to the SDH activity. FF units were made up of fibres which had low levels of SDH (type IIB, or FO), whereas S units had fibres with high levels of SDH (type I, or SO). FR units were intermediate (type IIA, or FOG) (see Table 3.1). As a general rule, the relative susceptibility to fatigue was related to the relative amounts of oxidative and glycolytic enzymes. The intensity of stain for ATPase reflected the twitch contraction time of the unit.

The precise nature of the connection between speed of contraction and ATPase staining is not clear. Although contraction speed is proportional to the intensity of staining in a wide range of muscles, there are instances in which two fibres in a muscle may stain with the same intensity, yet have contraction times which differ by a factor of two. Part of the problem arises because the stain is not specific for myofibrillar ATPase but also stains mitochondrial and sarcotubular ATPases. Thus, differences between fibre types may not necessarily indicate differences in myofibrillar ATPase. Also, it is not clear, as yet, whether contraction speed is a function of myofibrillar ATPase activity. Increased speed of contraction may mean that the sarcomeres slide over each other more quickly, but the ATPase activity may not be the rate-limiting step in the process. It is possible that the rate of Ca^{2+} release from the sarcoplasmic reticulum may limit contraction velocity. In this connection it is interesting to note that the rapid relaxation time of fast units in a single twitch probably is due to an increased rate of reabsorption of Ca^{2+} into the sarcoplasmic reticulum (see Burke, 1981, for a full review of this topic).

Some Electrophysiological Properties of Motoneurones

Motoneurones are the largest cells in the ventral horn of the spinal cord and may have dendritic trees which extend for distances of over 1 mm. In order to understand better the ways in which the nervous system regulates the firing frequency and pattern of these cells, it is important to have some knowledge of their intrinsic electrophysiological properties. The features have been analysed in detail.

Synaptic Inputs to Motoneurones

The main characteristics of excitatory postsynaptic potential (EPSP) and inhibitory postsynaptic potential (IPSP) generation in motoneurones are well known, since these cells were the subject of the first microelectrode studies in the spinal cord. Synapses are formed on the whole of the soma and dendritic membrane, but action potentials can be generated only on the somatic membrane and at the axon hillock.

Of these two sites, the axon hillock has the lowest threshold for spike initiation. The dendrites do not appear to have any of the voltage-sensitive ion channels necessary to produce regenerative action potentials.

When a cell is impaled by a microelectrode, the penetration is invariably made into the soma, since the dendrites and axon are far too small to allow insertion of a needle. Recordings made here of the size and time course of EPSP give an indication of the dendritic location of the active synapses. Electrical models of motoneurones first suggested, and later it was confirmed, that EPSPs produced by terminals on distal portions of the dendritic tree tend to be smaller and to have slower rise and decay times (Figure 3.4) than synapses nearer the soma. The reason for this is the increased resistance and membrane capacitance between distal dendrites and soma, which attenuates and slows the EPSPs. However, absolute size of an EPSP is not a useful parameter to provide unequivocal evidence for synaptic location, since the absolute size of an EPSP

Figure 3.4: Upper Part, Schematic Diagram of Motoneurone Cell Body and Dendritic Tree. Circles below indicate the mathematical transformation of this motoneurone into a series of linked compartments described by circuitry above. Lower part, comparison of EPSP shapes recorded at soma. Dotted line shows assumed time course of excitatory conductance change. A is the EPSP expected from equal input into all compartments of the model. For curve B, input was into compartment 1 alone, for C into compartment 4, and for D into compartment 8 alone. Curves are scaled so that their peak amplitudes are equal to 1.0 on the y-axis. The absolute sizes would, of course, be quite different, being, for example, larger in B than in D. The time course is plotted in terms of the membrane time constant τ (from Rall, 1967; Figures 1 and 2, with permission)

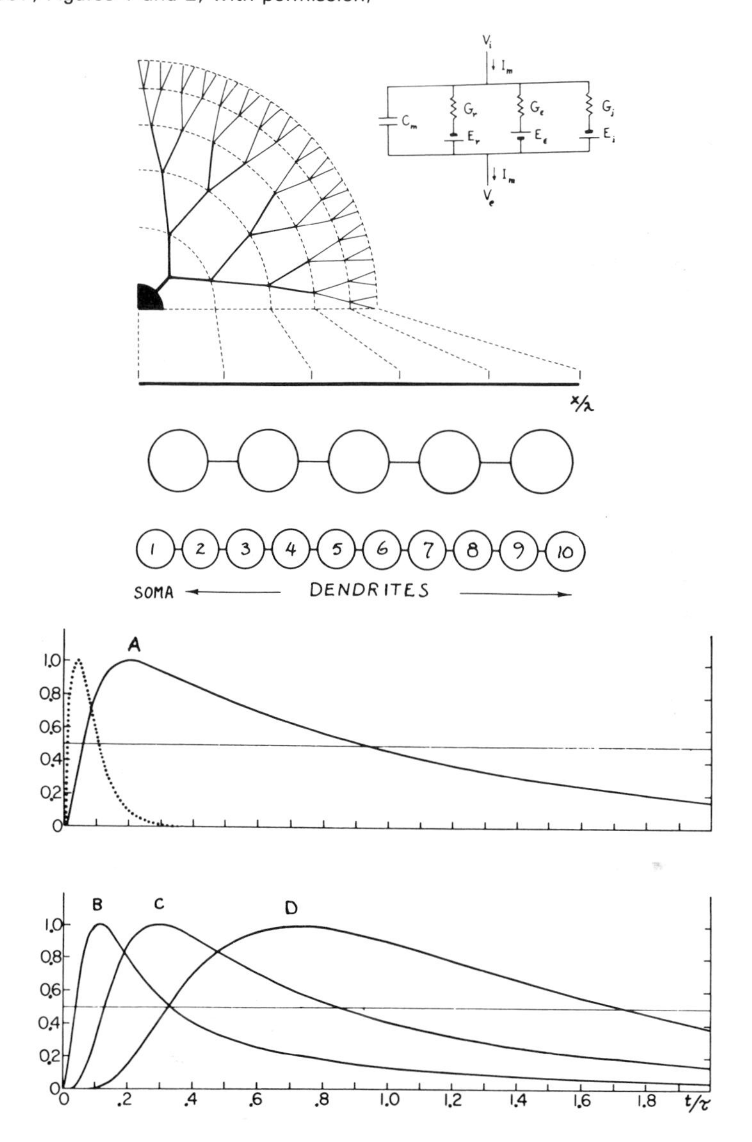

depends on the number and efficiency of active synaptic contacts as well as distance from the soma membrane.

Usually, the EPSPs recorded at the soma are analysed only in terms of their rise times to peak amplitude. Longer rise times indicate more distal synaptic inputs.

Input Resistance of Motoneurones

The input resistance of a whole cell gives a measure of the voltage change produced across the cell membrane by a known current applied at the soma. The usual procedure is to record the cell membrane potential with a microelectrode, at the same time as a current is injected into the soma. Large cells, with large diameter axons (and high conduction velocities), have a lower input resistance than small ones. This means that for a given current, the small motoneurones will be depolarised more than large cells. If the voltage threshold for firing is the same in all cells, then the small cells will reach threshold first. The reason for the different input resistances is caused mainly by the different surface areas of the cells. The specific resistivity (the resistance per unit area) of the cell membrane probably does not vary greatly between motoneurones so that the effect of increasing the cell surface area can be imagined to be analogous to adding more and more resistors *in parallel* between the interior and exterior of the cell. The nett effect is, of course, to *decrease* the total resistance of the pathway. An alternative and more common way of looking at this is to use the *conductivity* of the cell membrane, which is the reciprocal of resistivity. The specific conductivity is expressed in mho per unit area. Conductivities add linearly when the membrane area is increased by a unitary amount.

Firing Patterns of Motoneurones

In order to understand how fast and how many impulses a motoneurone will fire in response to particular synaptic inputs, cells can be depolarised by a known current applied to the soma through an intracellular microelectrode. The firing pattern can then be analysed in response to carefully controlled input currents. Examples of the response to step changes of current to different levels are shown in Figure 3.5. All the recordings are taken from an intracellular microelectrode. The three traces in row 1 show the response to antidromic invasion of the cell following stimulation of the motor axon. The action potential rapidly repolarises and is followed by a small delayed depolarisation which sits as a small wavelet on the potential before recovering to resting potential.

This contrasts with the spikes recorded following current injection into the soma through the microelectrode. These spikes, seen in row 2, do not have any delayed depolarisation. They behave differently to an antidromic spike, and are followed by an after-hyperpolarisation. If two potentials follow one another, the after-hyperpolarisations can summate. After-hyperpolarisation is also seen following action potentials initiated by normal synaptic inputs, and is the usual mode of cell firing. Lack of an after-hyperpolarisation with antidromic cell

Figure 3.5: Action Potentials Recorded from a Microelectrode Inserted into the Soma of a Motoneurone. In row 1 are three records of potentials produced by antidromic invasion of the motoneurone. The first two records are on a much shorter time-base than the third (note the 50 Hz signal) and show that the spike is followed by a delayed depolarisation before returning back to resting potential. The third trace shows a repetitive series of antidromic potentials. Row 2 shows the spikes recorded following depolarising current injection into the soma. These spikes are followed by an after-hyperpolarisation, which can be seen to summate (in the first record) with that following a second spike. The third record of the row illustrates repetitive firing produced by prolonged, steady, current injection into cell (shown as downwards step in the upper trace). At onset of current injection, the cell fires two impulses in very rapid succession and then settles down to a regular firing rate. Summation of the after-hyperpolarisation from the first two impulses can be seen. Rows 3 and 4 should be read vertically: 3 shows steady firing of motoneurone caused by current injection into soma, in which the after-hyperpolarisations after each spike are quite clear; 4 shows in detail what happens at the onset of current injection into the cell (downwards step in upper trace). The first two spikes occur rapidly and then firing settles down to a steady rate. After-hyperpolarisations summate as in row 2 (from Granit, Kernell and Smith, 1963; Figure 9, parts 4–7, with permission)

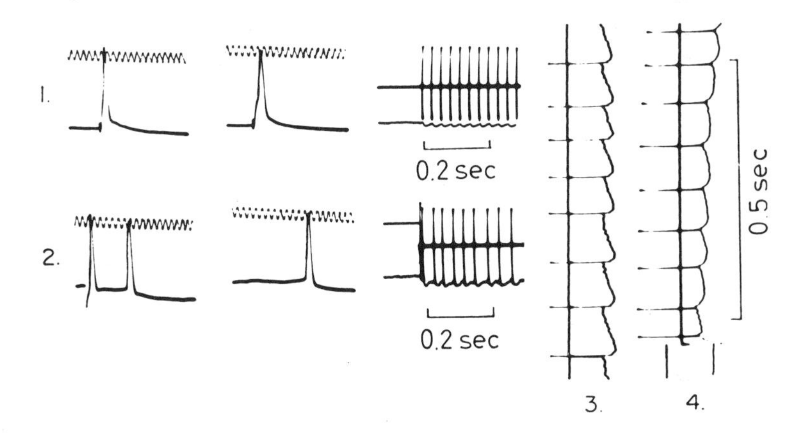

firing represents an unusual mode of activation. The delayed depolarisation is thought to be caused by a wave of depolarisation spreading out into the dendrites.

The third record in row 2 illustrates what happens to the firing rate of the cell immediately following a step depolarisation of the soma membrane. The cell fires off two impulses in rapid succession and then settles down to a slower steady rate of firing. The same can be seen on a longer time scale in part 4 of Figure 3.5. The relationship between steady firing frequency and applied current (F-I curve) is shown in Figure 3.6 and was first described by Kernell (1965). The curve consists of two approximate straight line segments known as the primary and secondary range of firing. It shall be seen below that the steady discharge of most motoneurones invariably is restricted to the primary range (between 10–50 Hz in different motoneurones). The secondary range, in which large depolarising currents must be applied, is rarely achieved in normal

Figure 3.6: Relationship Between Current Injected into Soma and Firing Frequency of a Cat Motoneurone. On the left is the relation for the steady-state firing frequency of the cell, on the right are the relations for the first, second and subsequent intervals after onset of stimulation in a different cell. Because neurones fire rapidly at onset of current injection, the initial firing rates are much higher than those recorded later in the train of action potentials. Note the two straight line segments of each curve: these are the primary and secondary ranges of motoneurone firing described in the test (from Kernell, 1965; Figures 4 and 5, with permission)

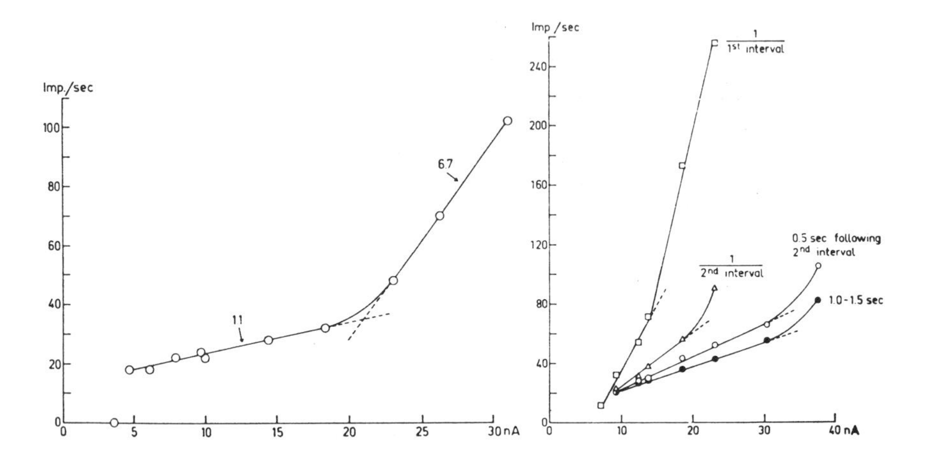

activation. Entry into the secondary range probably is caused by partial inactivation of the spike-generating mechanism due to maintained depolarisation of the cell. This reduces the height of the action potentials and also produces a change in the after-hyperpolarisation (see below).

The initial rapid adaptation of firing frequency to that seen in the steady state is caused mainly by development of after-hyperpolarisation which follows each spike. The after-hyperpolarisation builds up over the first two to three impulses and serves to stabilise the motoneurone against excessive depolarisation. One consequence of this is that initially it is very easy to push the firing frequency of the motoneurone into the secondary range. However, as the after-hyperpolarisation develops, this becomes more and more difficult, and more current is needed. Effectively, by this mechanism, the primary range of the motoneurone is extended over a large range of input currents (Figure 3.6). The duration of the after-hyperpolarisation is rèlated to the size of the cell; small cells have longer after-hyperpolarisation than large cells. It is also related to the minimal steady firing frequency of the cell. Cells with longer after-hyperpolarisations can discharge steadily at lower rates than those with shorter ones. Thus small cells can maintain slower firing frequencies than large cells, becuase of their long after-hyperpolarisation. Conversely, large cells with a shorter after-hyperpolarisation can fire more rapidly within the primary range than small cells. This is of some use since large motoneurones innervate large motor units which must discharge at higher frequencies to produce a fused contraction than their smaller neighbours. Similarly, rapid initial firing rates are used to produce rapid increases in muscle tension (see below), for example at the onset of a contraction (see Burke, 1981; Granit, 1970).

Control of Motor Units and their Recruitment Order

Most muscles contain a mixture of unit types. Since the contractile properties of each type of unit are different, it should come as no surprise to find that the control of the units is carefully regulated. Earlier this century, Denny-Brown (1929) had shown that slow pale muscles (that is, those with a preponderance of slow units) were activated preferentially in tonic contractions such as the stretch reflex, whereas red, fast muscles were activated only in rapid contractions such as the scratch reflex. More recently, this question of relative threshold, or *recruitment order* has been extended to the individual motor units of a muscle.

Henneman, Somjen and Carpenter (1965) performed most of the original experiments on the recruitment order of motor units. In order to examine the firing pattern of a small number of different units, they recorded the nerve action potentials from ventral root filaments during contraction. The method is superior to recording motor unit potentials from active muscle, since the recording electrodes do not get dislodged by the contraction. A variety of different types of contraction were examined: reflex contractions produced by stretch or by ear twisting, and contractions produced by electrical stimulation of the brain at various sites. For any ventral root filament that they examined, during any type of contraction, they always found that the axons were recruited in a particular order (Figure 3.7). In general, axons with small action potentials were recruited first in the contraction; only at higher levels of contraction did axons with larger potentials begin to discharge. Since the size of the potentials picked up by the recording electrode is related to the size of the active axon, this meant that small motor axons were recruited before large ones. The size of a motor axon is related to the number of muscle fibres that it innervates, and also to the size of the cell body within the spinal cord. Thus, at the level of the muscle, this meant that as force built up in the contraction, small motor units were recruited before large ones. At the level of the motoneurone, small neurones began firing before large ones. The rule that Henneman and his colleagues postulated was simple: the recruitment order of a motor unit depended, in all types of contraction, solely on the size of its motoneurone. This has become known as the *size principle*.

In addition to investigating the order of activation using facilitatory inputs, they also examined the *derecruitment* of units by inhibitory inputs. This followed the reverse order: large active units always were inhibited before small ones.

Much of this work was started before the physiological classification of motor

Figure 3.7: Recruitment Order of Two Motoneurones Recorded in a Ventral Root Filament During a Stretch-Reflex-Induced Contraction of the Triceps Surae Muscle. The amount of tension developed by the muscle is shown by the separation of the upper two beams in each frame. Stretch increases from frames 1 to 5 and then decreases back to zero in frames 6 to 9. A unit with a small action potential is recruited first (frame 2) and is followed by a larger unit (frame 4). Derecruitment proceeds in the opposite order: the large unit drops out first (frame 7), followed by the small unit (frame 9) (from Henneman, Somjen and Carpenter, 1965; Figure 1, with permission)

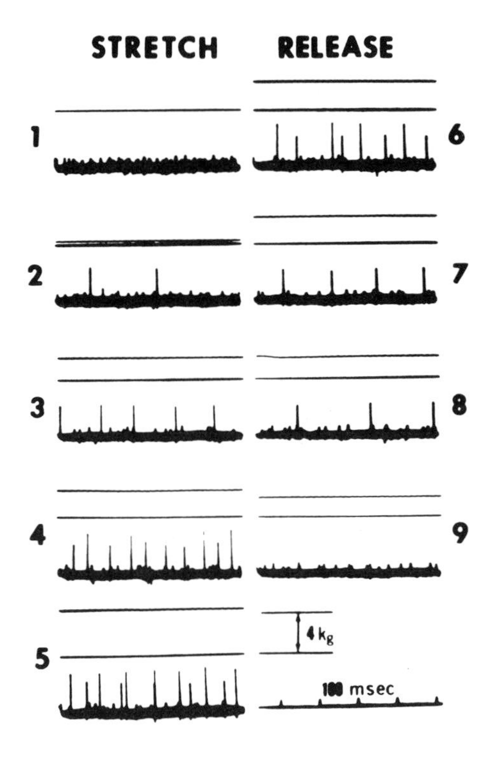

units into different types. Later studies have confirmed the supposition that in general, motor units are recruited in the order S, FR, FF. Thus, small, slow twitch units are activated first. They are suited to participate in long-lasting but relatively weak tonic contractions. The FR units are activated later, and the FF units last of all. The FF units generate large forces but are easily fatigued, and hence cannot sustain tension over long periods. The whole scheme seems beautifully adapted to allow the CNS to control body muscles in the most logical manner.

Mechanisms Responsible for the Recruitment Order of Motoneurones

The mechanisms underlying motoneurone recruitment order have been investigated most thoroughly during stretch reflex contractions produced by monosynaptic input from group Ia afferents. These are the largest fibres in peripheral nerve, and have the lowest threshold for electrical stimulation. They can therefore be excited independently of other smaller fibres in the nerve trunk. A large number of these fibres can be excited simultaneously using this method, and the size of the monosynaptic EPSP produced in different motoneurones may be recorded by a microelectrode. This EPSP is known as a composite or aggregate EPSP since it is produced by the activation of many separate Ia fibres. Eccles *et al.* (1957) initially showed that this EPSP was larger on average in motoneurones innervating grossly slow muscles, such as soleus, than it was in motoneurones innervating grossly fast muscles such as gastrocnemius. Burke and colleagues (1973) have since extended these findings to the individual units within the motoneurone pools of many different muscles. The composite EPSP usually is larger in motoneurones innervating identified type S units than it is in motoneurones innervating FF units. Type FR units have EPSPs more nearly related to the size of those in motoneurones of slow units (Figures 3.8 and 3.10).

It is thought that all the alpha motoneurones of the ventral horn of the spinal cord have approximately the same voltage threshold for impulse generation. Thus, the different EPSP sizes can readily explain the preferential recruitment of slow motor units by facilitatory inputs. The derecruitment of units by inhibitory inputs also follows the size principle. However, from this statement alone it is not possible to deduce anything about the size of the IPSPs in different motoneurones. Since large motoneurones receive only weak excitation, either a weak or a strong inhibition would be sufficient to reduce their excitation below threshold. In fact, it has been shown that, like the excitatory input, group Ia IPSPs are largest in those motoneurones innervating slow muscle units. See Figure 3.9 for a summary diagram.

Although the composite PSP sizes can account for the order of motoneurone recruitment, it is not clear why the PSPs are of different sizes in different motoneurones. The original idea of Henneman and his associates was that the order of recruitment was determined by properties dependent solely on the size of the motoneurone. It was assumed that the passive electrical properties of the cell membrane were the same in motoneurones of any size. In particular, the resistance of a unit area of membrane (the specific resistivity) was assumed to be constant. As described above, this means that the input resistance of a small motoneurone will be larger than that of a large motoneurone. Given this assumption, it is possible to explain recruitment order by postulating that a given input to a motoneuronal pool evokes the same synaptic current in all cells. If the current is the same, then small motoneurones with a large input resistance will be depolarised more than large motoneurones. If voltage threshold is the same in all cells, then the small cells will fire impulses before the large cells.

But why should the synaptic current be equal in all motoneurones? It might have been expected that because of their larger surface area, there would be more synaptic boutons on large than on small cells, and that the total synaptic current would be greater the larger the cell. Several possibilities can be put forward to explain the assumption of equal synaptic currents:

Figure 3.8: Relationship Between Motor Unit Type and Size of the Composite Group Ia EPSP in the Cat. The EPSPs were recorded in the ventral horn of the spinal cord from motoneurones innervating medial gastrocnemius. A synchronous Ia input to the motoneurones was achieved by stimulating the peripheral nerve innervating the muscle. The graph below shows the relation between size of group Ia EPSP and the tetanic tension produced in the muscle by each innervated muscle unit. Units which are innervated by motoneurones with large Ia EPSPs produce only small tetanic tensions and *vice versa*. The different symbols in the figure indicate motor units of different physiological types. The histograms opposite illustrate the number of units in each category with Ia EPSPs of a given size. The EPSPs are smaller in the FF units than in the type S units. Black dots indicate data collected from one animal. Arrows indicate mean values. (from Burke, Rymer and Walsh, 1976; Figures 2 and 3, with permission).

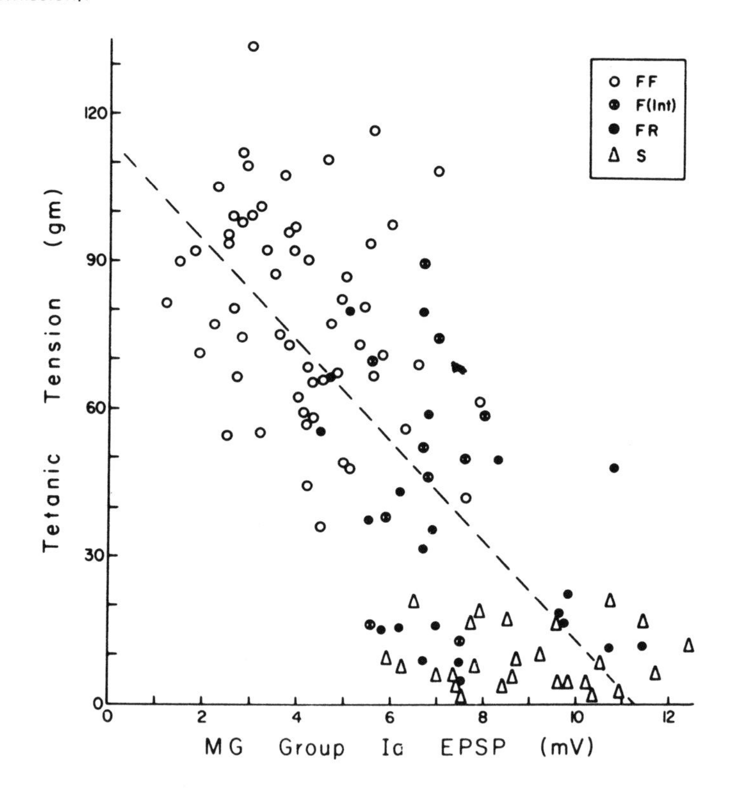

1. The synaptic efficiency of boutons on small cells may be greater than those on large cells. This could be because: (a) the input to small motoneurones is located near to the cell body and produces larger soma PSPs than those on large motoneurones which are located on distal dendrites; (b) the synapses on small cells could produce larger PSPs by being physically larger or produce larger conductance changes than those on large cells; or (c) if large neurones had a disproportionately large soma relative to dendritic area, then dendritic inputs would be less effective than those in small cells.

There is no strong experimental evidence for any of these possibilities.

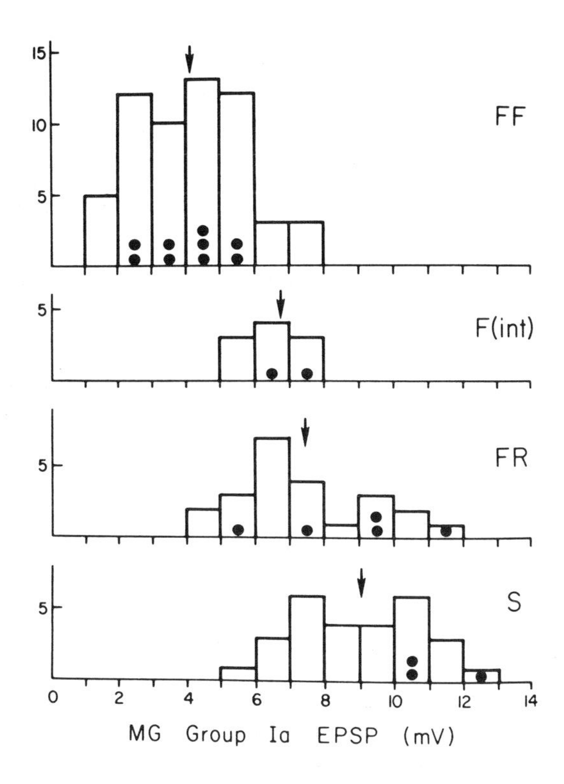

Recent anatomical studies using horseradish peroxidase (HRP) staining of Ia afferent terminals show little difference in anatomical location on motoneurones of different size. Electrophysiological techniques have confirmed this. Electrical models of the motoneurone have shown that if it is assumed that all the afferent inputs arrive synchronously at the cell, the composite PSP recorded in the soma from inputs located primarily on distal dendrites lasts longer in proportion to its size than the input produced by more proximal synapses (see above). There is no experimental evidence of this type to suggest that the terminals may be located slightly more distal on large than on small motoneurones. Similarly, anatomical studies have shown that there is no difference between the ratio of soma size to dendritic area in large and small motoneurones. At present there is no direct evidence concerning the conductance of single synapses in the membranes of large and small motoneurones. Analysis of the quantal potentials making up the EPSP in motoneurones of different sizes showed that the voltage changes were larger in small than in large cells. However, this was probably not due to any differences in the membrane conductance of single ion channels, but simply to the larger input resistance of small cells.

2. There may be the same *number* of synapses on motoneurones of all sizes. This would mean that the *density* of terminals would be lower on large cells.

Figure 3.9: This page: Three-dimensional Diagram Summarising the Relationships Between a Variety of Physiological, Morphological and Histochemical Profiles of the Motor Units of Cat Medial Gastrocnemius Muscle. The unit types FF, F(int), FR and S are shown by the shaded areas, together with percentage figures of their predominance in the muscle. The diagram emphasises that a great many properties of motoneurones and the motor units that they innervate are correlated one with another. Opposite: diagram summarising the recruitment of different proportions of the motor unit pool of cat medial gastrocnemius during different types of muscular activity from quiet stance, through walking, to jumping. During standing, only about 25 per cent of the total population of units is active. These are the type S units. Types FR and F(int) are recruited during walking and running. The type FF units (which are 45 per cent of the total population) are only rarely recruited for brief periods of time in very powerful contractions such as needed in jumping (from: this page, Burke, 1981 (after Burke, 1975); Figure 35, with permission; opposite, Walmsley, Hodgson and Burke, 1978; Figure 9, with permission)

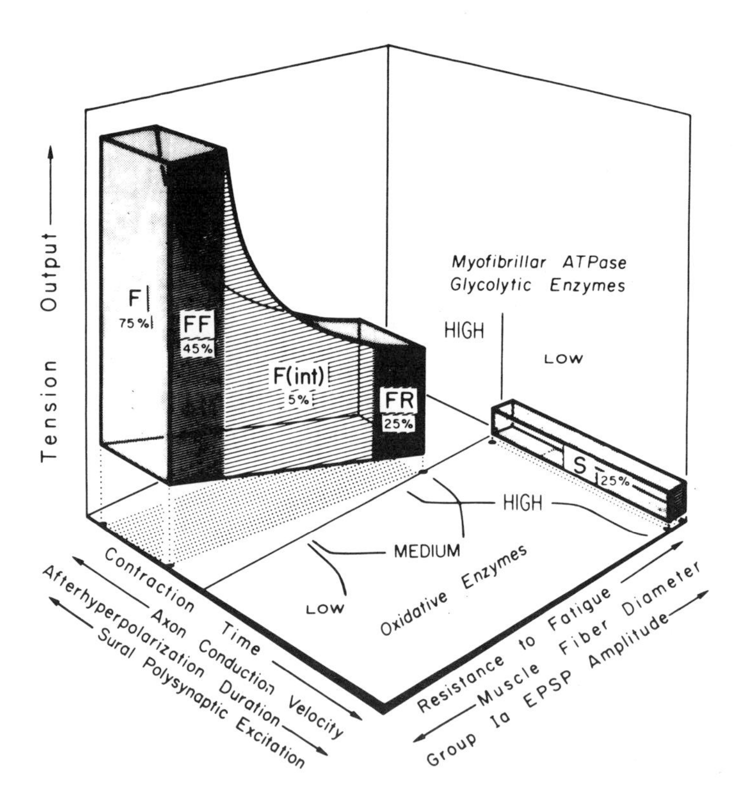

From an experimental viewpoint this hypothesis is very difficult to verify. It is never possible to stain selectively all terminals of a particular afferent class, or even a known proportion of them, so that they can be counted anatomically. However, some authors have favoured this possibility.

3. There may be an equal density of synapses on all motoneurones (and hence more synapses in total on large cells), but perhaps because the axons branch

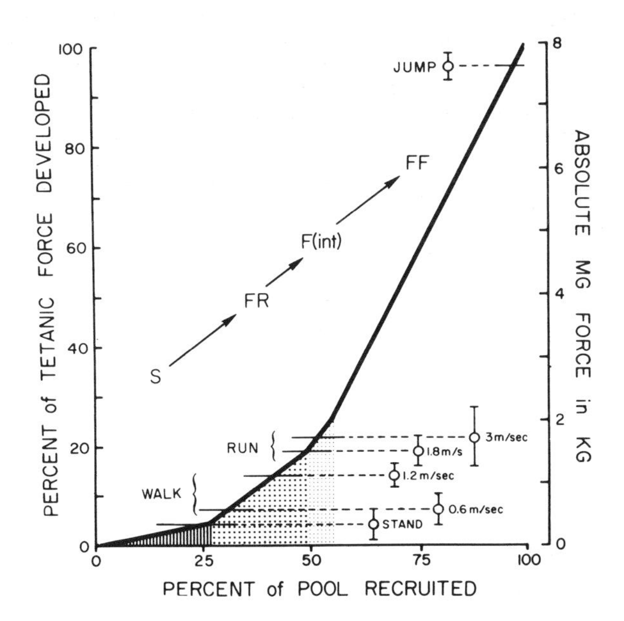

more often to innervate more terminals, the invasion of the presynaptic nerve impulse into the terminal arborisation is less complete for synapses on large than on small cells. This means that proportionately less synapses would be activated on large motoneurones. Recent studies of the quantal fluctuations in EPSP size in motoneurones of different size suggest that this is a possibility.

In the past five years, the idea that recruitment order depends solely on size-related properties of motoneurones has come under criticism. For example, direct estimates of total cell area using HRP techniques show that motoneurones innervating different types of motor units have rather small differences in average surface area. This makes it difficult to imagine that the size of a cell is the sole determinant of its recruitment order. Other possibilities put forward to account for the orderly recruitment of motoneurones include the idea that membrane properties may vary between cells. For example, if specific membrane resistance were to vary, this might be the reason why cells of apparently similar surface areas can have different recruitment thresholds. At the present, the matter is unresolved (see articles by Burke, 1981; Gustaffson and Pinter, 1984; Henneman and Mendell, 1981; Zengel *et al.*, 1985, for further information on the work covered in this section).

Exceptions to the Normal Order of Recruitment

The main reason for using Ia monosynaptic excitation and Ia disynaptic inhibition in the investigation of the mechanisms underlying the order of motoneurone recruitment, is that the methods are technically feasible, rather than being

Figure 3.10: Relationship Between Input Resistance and Size of Group Ia EPSP in Motoneurones Innervating Cat Tibialis Anterior Muscle. The larger the Ia EPSP, the larger the input resistance. Points obtained from experiments on a single cat are marked by arrows. Motoneurones innervating motor units of different physiological types can be distinguished by the symbols (from Dum and Kennedy, 1980; Figure 2, with permission)

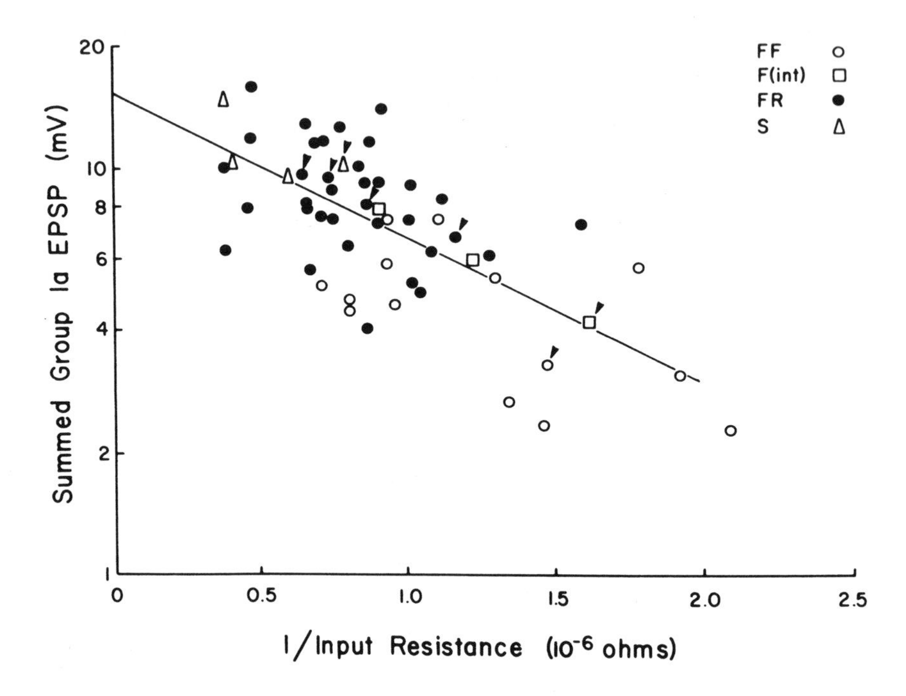

necessarily the most important. There are many other inputs to the motoneurones, but these use polysynaptic pathways which are not as easy to investigate directly. In their original experiments, Henneman and his colleagues (1965) studied the recruitment order of motoneurones elicited by stretch reflex contractions, crossed extensor reflexes, pinna reflexes and by direct stimulation of sites in the motor cortex and brainstem. Except for the stretch reflex, all these inputs use pathways other than the Ia afferent neurone. Despite this, from all the sites, motoneurones were recruited in order of size. One possible interpretation is that all the inputs tested had a synaptic weighting which favoured the small motoneurones just like the Ia connections. However, this is not necessarily true. For example, if there were any *tonic* activity in the Ia pathways, this would *bias* the motoneurone pool by bringing small neurones nearer to threshold than large neurones. Thus, even if other inputs projected with equal synaptic weight to the motoneurones, the tonic Ia firing would assure recruitment of motoneurones according to size.

Despite the generality of the size principle, there are exceptions to the rule which have been a source of debate for some time. These are the inputs from

Figure 3.11: Diagram to Illustrate Recruitment Order of Motoneurones with Different Weightings of Synaptic Input. Four types of motoneurone, labelled A, B, C and D received synaptic input with an efficacy proportional to the width of the arrows in the figure. The stippled arrows on the left represent the 'conventional' recruitment order of the motoneurones, with both excitatory (E) and inhibitory (I) inputs being most effective on motoneurones type A and least effective on type D. This type of organisation produces recruitment and derecruitment sequences like those shown in the upper spike traces. On the right is input from another source (for example, cutaneous afferents) in which the weighting of synaptic input is the reverse of that on the left of the figure. Combining inputs from left and right might lead to dislocations from 'expected' recruitment order as shown in lower set of spike diagrams (from Kanda, Burke and Walmsley, 1977; Figure 8, with permission)

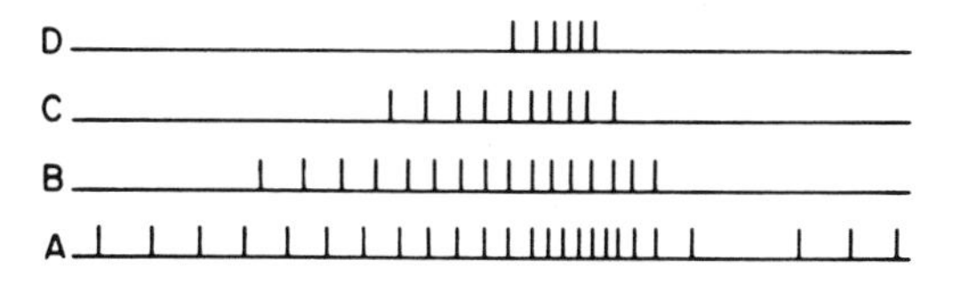

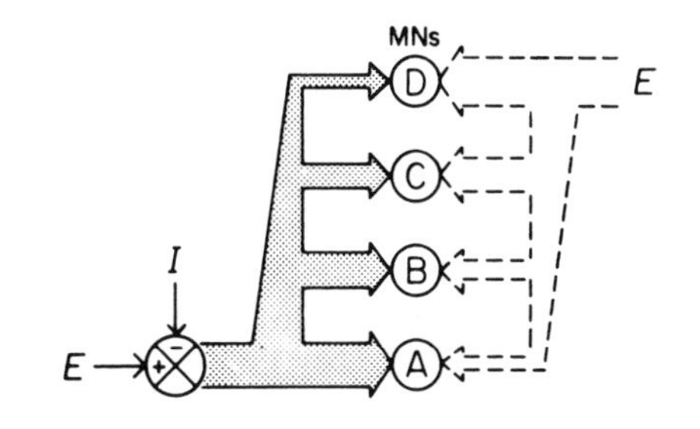

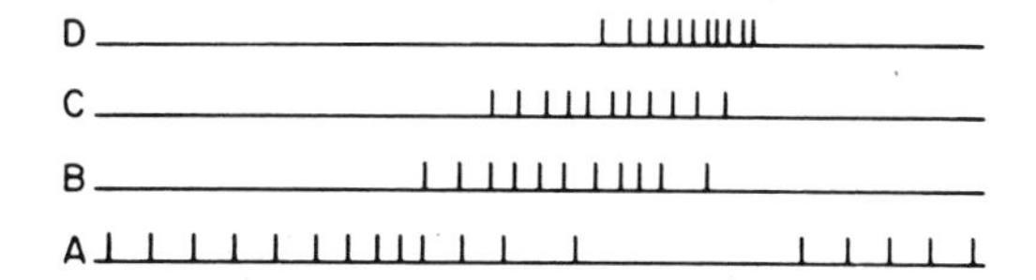

the rubrospinal tract and cutaneous afferents. Both these systems project via polysynaptic pathways so that the composite EPSP produced by stimulation of a large number of axons in a cutaneous nerve, or of a large number of cells in the red nucleus, is more diffuse than that from the monosynaptic pathway. Generally, a train of stimuli needs to be applied, which often produces a combination of excitation and inhibition in the cell. The most rapid effects from both the rubrospinal tract and cutaneous afferents probably traverse a trisynaptic pathway. In both cases, there is a much stronger excitation of fast motor units. Slow units often are inhibited by these inputs, which is reverse of the situation proposed by the size principle.

The implication of this is that the synaptic weighting to motoneurones of different size can vary with the input source. The arrows on the left in Figure 3.11 show the usual (size principle) weighting of inputs, whereas those on the left show the reverse pattern expected from cutaneous and rubrospinal input. One consequence of this hypothesis would be that unusual patterns of recruitment might be observed when both types of input are combined. Unless the exact synaptic weighting onto each neurone were known specifically, it would be impossible to predict the order of unit activation during combined stimulation (see articles in Desmedt, 1981 for further details).

Examples of Motoneurone Recruitment during Normal Movement (Figure 3.10)

In animal experiments, the pattern of activity of whole muscles rather than single units has been investigated. The two favourite muscles for investigation are the predominantly fast medial gastrocnemius and the slow soleus muscle.

In the cat hindlimb, Burke and his colleagues have classified three 'grades' of muscle activity: (1) postural maintenance, as in quiet quadrupedal standing; (2) sustained walking or running; and (3) movements made only in brief bursts, such as jumping, galloping and scratching. The soleus muscle produces almost the same level of force during all these activities, and even during walking, is generating a force near to its maximal output. In contrast, the medial gastrocnemius produces only 8 to 12 per cent of its maximal output during quiet stance, and only 25 per cent maximal force during walking or running. Only during the third category of movement does it generate maximal force. These are probably the only activities during which the 45 per cent FF units are recruited. Indeed, jumping often is achieved with a lengthening contraction of the muscle which may produce more force than that seen in a maximal tetanic contraction (see Chapter 2).

Such results can be interpreted as a direct consequence of the size principle of motor unit recruitment. If there are common inputs to the medial gastrocnemius and soleus motoneurone pools, then it is not surprising to find that the soleus, with its 90 per cent type S units, is always recruited before the gastrocnemius and may be fully activated at relatively low force levels. It is obviously a muscle which is well suited to tonic, sustained muscle contraction. However, it is generally not only active during tonic contractions. Even in jumping, the soleus is maximally active together with the gastrocnemius, only in this case the gastrocnemius produces most of the force.

There is one well documented exception to this orderly recruitment of motor units. That is the *preferential* activation of lateral gastrocnemius during rapid paw shakes in the cat. In the example cited above, the lateral head of gastrocnemius is activated together with the medial head, and hence is recruited after the soleus. However, during rapid paw shake, which is a vigorous automatic movement made by the animal to remove an object from its paw, the soleus muscle is not activated at all. The movement is made using only the fast ankle extensor. The practical reason for this may be that the movement is so fast

(10–12Hz) that it cannot be followed by the soleus muscle, so that if the soleus were active it would interfere with rather than assist the contraction. Interestingly, the movement is triggered preferentially by large low threshold cutaneous afferents from the plantar pads, which may, as noted above, be instrumental in inhibiting the slow soleus motor units (Walmsley *et al.*, 1978; Burke, 1981).

The Study of Motor Units in Human Physiology

The ease with which single motor units can be recorded in normal man, and the ability of human subjects to perform complex movements with little practice, has made the study of human motor units a particular rewarding field in the control of movement. At the same time, the neurophysiological investigation of motor units plays a very important role in the diagnosis of neurological disease (Buchthal and Schmalbruch, 1980; Desmedt, 1981; Freund, 1983).

Recording Motor Units in Man

Concentric needle electrodes were first used by Adrian and Bronk in 1929. Essentially, they consist of a hollow needle with a fine wire inserted down the centre to the tip and insulated from the shaft with resin. The tip of the wire is bared and acts as one electrode, while the barrel of the needle acts as the other. A differential amplifier is used to amplify the potential difference between core and shaft. The diameter of a typical needle usually is of the order of 0.5–1mm and, as such, will record from only a small portion of the whole muscle when it is inserted.

The recording lead of a conventional concentric needle electrode picks up activity of many individual muscle *fibres*. However, all the fibres of a given motor *unit* will discharge almost simultaneously, so that for each unit, only one aggregate spike is picked up by the electrode. Generally, the spike is of largest amplitude if the needle tip is at the centre of the motor unit territory, indicating a larger density of fibres at this point. Different motor units recorded at the same time from the same electrode usually have spikes of different shapes and sizes because their active fibres have a different spatial distribution about the recording tip. Motor units of human limb muscles have an average territory 5–10mm diameter, allowing space for the (intermixed) fibres of 15 to 30 motor units.

For even finer localisation, other types of electrode with two or more wires in the shaft can be used, and recordings made from the exposed tips with the needle barrel acting as a shielding ground. In such a situation, when the barrel is not used as one electrode, the wires usually will record only those motor unit potentials at a distance less than 1mm from the recording head.

Histochemistry of Human Motor Units

Human motor units are obviously more difficult to characterise histochemically than those in the cat. The earliest studies inferred the presence of different unit types in man by examining the properties of whole muscle, or strips of excised muscle which contain a preponderance of one or other type of histochemically identified fibre. When this was done, it was found that muscles which had high levels of ATPase and lower levels of SDH (that is, like type II fibres) tended to have fast contraction times. Those with slower contraction time had higher levels of SDH and less ATPase.

A small number of experiments have been performed on single motor units in man which are directly comparable with those performed in animals. For example, Garnett, O'Donovan, Stephens *et al.* (1978) inserted a small needle into the medial gastrocnemius muscle and used it to stimulate selectively a single motor unit. Using this technique it was possible to obtain the average time course of a twitch following a single stimulus. In addition, repeated tetani could be given in order to test the fatiguability of the unit as in cat experiments. With these data it was possible to classify the unit physiologically as type S, FR, or FF. The final procedure was to stimulate the unit tetanically for a continuous period of up to two hours to deplete its stores of glycogen. At this point the muscle was biopsied and the specimens frozen. Serial sections were cut to identify the glycogen depleted fibres and to characterise them for levels of SDH and myosin ATPase. As expected from cat experiments, physiologically identified FF units were composed of type IIB fibres, whereas the physiologically identified type S units were composed of type I fibres. Only a single FR unit was found, and had type I fibres.

From these studies it appears that motor units in man may be classified using the same terminology as that used in the cat.

Mechanical Properties of Human Motor Units

Electrophysiological recordings of motor units provide a less invasive method of identifying the mechanical properties of single units during a muscle contraction. The technique involves detection of the discharge from a single unit with needle electrodes. This signal is then used to average the force exerted by the unit from the total force produced by the muscle. The idea is that the random fluctuations in muscle force during contraction will cancel out and one will be left with the average twitch force of the unit being recorded (Figure 3.12).

There are three main restrictions. The first is that there should be no fusion of the unit contractions or else the average unit force will not represent that of a single twitch contraction. Thus, the unit should fire at a low frequency (see below). Second, there must be no synchrony between the firing of different motor units in the same muscle. If another unit were synchronised with the unit under study, this would contaminate the average twitch force which was recorded. Third, the unit must be able to be distinguished from all others throughout the time of contraction. This is relatively easy at low force levels. However, with conventional needle electrodes, it is impossible to identify a single unit during contractions of moderate to high force.

Despite this, a range of twitch sizes and contraction times has been found.

Figure 3.12: Method of Measuring the Twitch Force of a Single Motor Unit in the Human First Dorsal Interosseous Muscle. Discharges of a single motor unit are recorded by a needle electrode inserted into the muscle (top trace), while the subject exerts a fairly constant voluntary contraction of the muscle (second trace, MG). The spike of the selected motor unit then is arranged to trigger an electronic averager of the muscle force record. If all the active units in the muscle are firing asynchronously, the average force twitch time-locked to the unit discharge represents the twitch contraction of that unit (lower trace: note difference in force calibration in this part of the figure compared with that in the upper part) (from Desmedt, 1981; Figure 1, with permission, pp. 97–136)

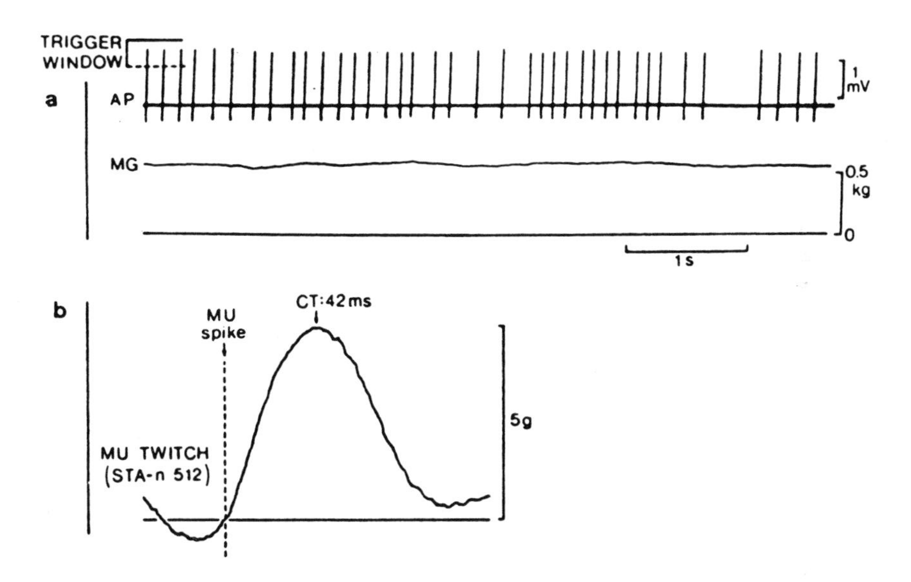

In experiments in which fatigue has been induced during prolonged contractions, it has been seen that units with high initial twitch tensions fatigued earlier than those with low twitch tensions.

Recruitment Order of Motor Units in Man

With this technique it is possible to examine the recruitment order of motor units during voluntary contractions (Figure 3.13). The order in which units are activated is studied first and then the twitch force of each unit is averaged out later. In almost all types of contraction, small slow twitch units are recruited before large units. However, the recruitment threshold of a motor unit, that is, the muscle force at which the unit first begins to fire, is dependent upon the speed of contraction. The faster the contraction, the earlier the unit is recruited, although the *order* of recruitment, with respect to the other units of the muscle, remains constant. The one exception to this occurs in very rapid contractions during which units are recruited so rapidly one after the other that the conduction velocity of the efferent axon becomes an important variable. Slow units, which

Figure 3.13: Twitch Force Profile and Recruitment Order of Two Motor Units from the First Dorsal Interosseous Muscle. Records a and b show the twitch forces of the two units and their respective action potentials (inserts to left of force record). Unit a was a small, slow twitch unit, and unit b was a larger, faster unit (note time to peak tension). The spikes of both these units could be recorded and distinguished through a single selective needle electrode. This is shown in c in which the subject was asked to produce a slowly increasing voluntary contraction of the muscle. As force rises, the first unit is recruited (it has a predominantly up-going spike potential), and later, at higher force levels, the second unit is brought to discharge (its action potentials can be distinguished by the down-going EMG deflections). In this slowly increasing contraction, unit a is recruited at a threshold force of 20g, while unit b is recruited at about 280g.

Records d to g are from rapid ('ballistic') voluntary contractions to different peak forces ranging from 10–110g. Unit a is recruited in the small contraction of trace e. Unit b is recruited only in the largest contraction (g). Note that this unit is recruited even though the peak tension of the contraction is only 100g. This is below the threshold force seen in the slow voluntary contraction (c). Thus, the force threshold of this unit depends on the type of contraction which is being made. Its recruitment order, however, remains unchanged (from Desmedt and Godaux, 1977; Figure 1, with permission, copyright 1977 Macmillan Journals Ltd)

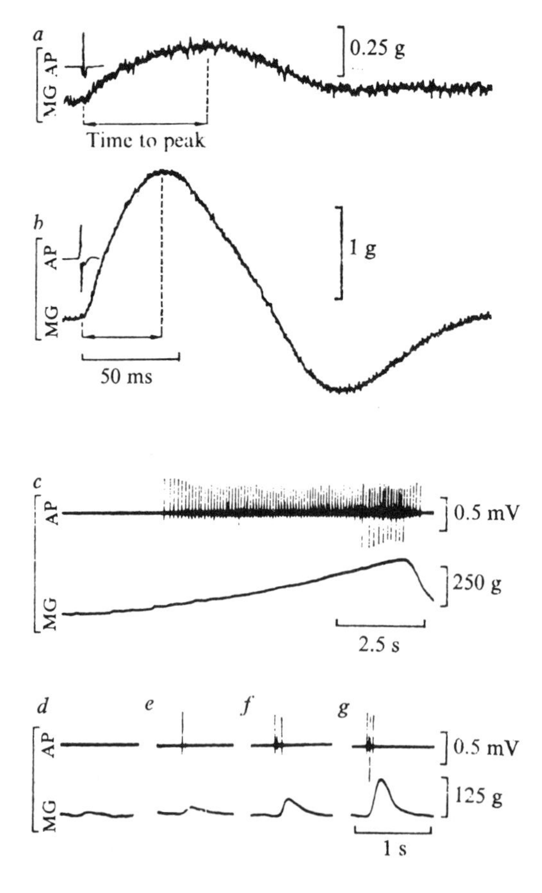

may have been recruited first at the level of the motoneurone actually may fire after some of their larger colleagues because of the time delay in conduction down their small axons.

One consequence of this change in recruitment threshold is that if very rapid adjustments are made from one level of force to another, there is large but brief increase in motoneurone drive which is needed to produce the high velocity of contraction. This reduces the threshold force at which new units are recruited. Because of this, discharge rates in all units will increase and large units will be recruited transiently during the change in force. When the steady level of force is reached, the motoneurone drive will fall and the largest units will stop firing.

Another way of looking at recruitment threshold, rather than examining the muscle force at which single units began to *discharge*, is to look at the level of muscle force at which the units actually begin to contribute *tension*. The two thresholds are different because there is a time delay of up to 100ms between electrical activation of a unit and the peak tension that it produces (see previous chapter). When this delay is taken into account, then the force level at which the unit's tension is recruited is the same in maintained (but not twitch) contractions at any speed.

Changes in Recruitment Order

As in the cat, there have been several reports of changes in the 'fixed' recruitment order of human motor units. Electrical stimulation of cutaneous nerves can reverse the usual pattern of motor unit recruitment (Figure 3.14). Studies examining the effect of stimulation on the firing rate of continuously active units have shown that, like the cat, cutaneous input has an overall inhibitory effect on small units and an excitatory effect on large units. In the first dorsal interosseous muscle, stimulation of the digital (cutaneous) nerves of the forefinger at three times the sensory threshold can reduce the force threshold for recruitment of large units and increase the threshold for recruitment of small units.

Changes in recruitment order also have been seen in certain bifunctional muscles. The abductor pollicis can be used either as prime mover in thumb abduction or as synergist in thumb flexion. Recruitment order of the units depends upon whether the muscle is being used as prime mover or synergist — that is, upon the direction of thumb movement. Such results suggest that the weighting of synaptic input to different motoneurones varies according to the motor command.

One hypothesis to explain this is that the prime mover command might be effected by synapses that are equally distributed to all units in the muscle. This would give rise to a 'normal' recruitment order of the units. However, the command to synergist muscles might use a different set of synapses which are preferentially distributed to certain of the units. Another possibility is that the pattern of afferent input produced by different movements, rather like the known effects of cutaneous inputs, could change the threshold for recruitment of motor units when the muscle is used as a synergist (Datta and Stephens, 1981; Desmedt, 1981).

Figure 3.14: Changes in Recruitment Threshold of Motor Units in the First Dorsal Interosseous Muscle Produced by Electrical Stimulation of the Digital (Cutaneous) Nerves of the Index Finger. The upper part of the figure shows nine panels of EMG and force recordings taken as the subject made a gradually increasing voluntary isometric contraction of the muscle. The subject was instructed to follow a target ramp increase in force (the straight, saw-toothed line) with his voluntary contraction (the irregular trace superimposed on the saw-tooth). Two units can be distinguished in the EMG traces: a unit with small action potentials producing the black smudge around the baseline and a unit with much larger potentials, seen as the more distinct large spike. The control state, with no stimulation is seen in row A. During three consecutive ramp contractions the large unit was recruited at a force of about 2.5N. Row B shows the behaviour of the same units 30s after switching on a continuous electrical stimulation of the digital nerves of the forefinger. The recruitment threshold of the large unit is now less than 1N. In contrast, the threshold for the small unit has risen. Row C is a further control series recorded 10s after the end of stimulation. The recruitment thresholds are similar to those in row A. The lower part of the figure summarises data from a number of experiments. Threshold forces at which units were recruited in the control state are plotted on the left; thresholds during skin stimulation are on the right. The trend is for high threshold units to be recruited at lower levels during stimulation. The opposite is true for low threshold units: they are recruited at higher force levels during cutaneous stimulation (from Garnett and Stephens, 1981; Figures 1 and 4A, with permission)

Firing Frequency of Human Motor Units

Unlike motor units in many sensory systems, which have firing rates which can vary continuously between 0–300Hz or more, the range of firing rate modulation of human motor units is very small. For the muscles of the forearm, units only discharge continuously at frequencies from 6–8Hz to 20–35Hz. This frequency range corresponds to the *primary* range of motoneurone firing described by Kernell (1965) (see above). It is well matched to the mechanical properties of the motor units, since the contractions of most units are fused at above 40Hz so that no extra force can be obtained by stimulating at higher frequencies. The only time it is necessary to exceed this limit is to increase the *speed* of a developing contraction. During very rapid contractions, firing frequencies within the secondary range are observed transiently. Such rates of 150Hz or more do not alter the maximum force of contraction, but they do shorten the time taken to reach peak tension because the individual twitches summate very rapidly.

The lower limit of 6–8Hz is more difficult to understand. The only way to discharge units at frequencies below 6–8Hz is for the subject to perform a series of small rapid muscle contractions at that frequency rather than maintain a constant level of contraction in one muscle.

The factors responsible for the low frequency limit are not clear. Animal experiments in which currents were injected via a microelectrode directly into the motoneurone have shown that even under these conditions, minimal firing rates are of the order of 5Hz. Thus, the low frequency limitation cannot be a result of the excitatory synaptic input to the motoneurone. Renshaw inhibition has been regarded as a possible candidate (see Chapter 4). The firing probability of Renshaw cells is very high at low motoneurone discharge rates, but gets progressively smaller at higher rates. It could be that the inhibition is so

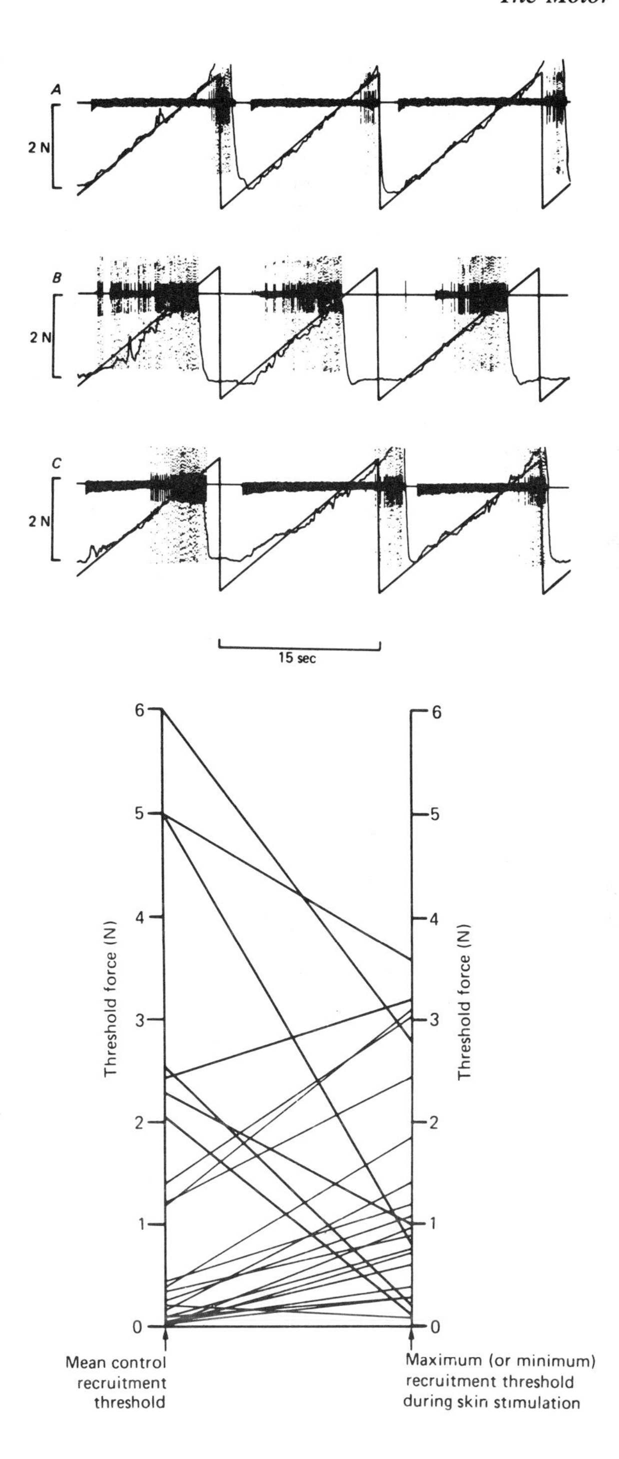
A
2 N
B
2 N
C
2 N
15 sec
Threshold force (N)
6
5
4
3
2
1
0
Mean control recruitment threshold
Maximum (or minimum) recruitment threshold during skin stimulation

profound as to limit the lowest frequencies of discharge.

Gradation of Force in Human Muscle Contraction

Ever since the first recordings of single unit activity by Adrian and Bronk (1929), one of the basic questions of muscle control has been whether muscle force is graded by the *number* of active units, or the *rate* at which each unit is stimulated. Either system of control seems feasible. For example, in the extreme case of recruitment (number) control, each unit, when recruited, would fire immediately at high frequency, and produce its maximum force. Under such circumstances, force would be graded by the number of active units. In the extreme case of rate control, every unit could be recruited at the same threshold force. In such a case, every unit would be firing at all levels of contraction and force could be graded only by changing the rate of unit activation. These extremes of recruitment and rate modulation are never seen; each mechanism contributes to the gradation of force. The relative contribution of each of these mechanisms has been analysed in some detail for various human muscles.

One of the problems in these experiments is to be able to follow the firing pattern of a single motor unit during large forces of contraction. Only recently have very selective microelectrodes been introduced with computer decomposition techniques which enable experimenters to distinguish a single unit among many during maximal effort. Some of the first experiments were performed on the small muscles of the hand. In these it was found that during a ramp isometric contraction against a slowly increasing force, small units were recruited first and began to fire at about 9Hz. As the force increased, these units increased their firing frequency and at the same time new, larger units were recruited which began firing at about the same initial rate of 9Hz. By the time the force reached 40 per cent maximum voluntary contraction (MVC), all the motor units in the muscle had been recruited and the rest of the force modulation was provided by changes in firing frequency up to a maximum of about 40Hz.

In the deltoid, a much larger muscle (about 1,000 units compared with the 120 of first dorsal interosseous) with a similar proportion of unit types, recruitment proved more important. During slowly increasing contractions new units were recruited up to force levels of 80 per cent MVC. Rate modulation played little part in increasing force since the firing frequency of units only rose to about 25Hz after initial recruitment at about 13Hz (De Luca, Lefever, McCue *et al.*, 1982).

This latter observation is of some interest, since at such frequencies, many of the motor unit contractions may still be unfused. If this is so, it means that the MVC was not, in fact, the maximum force of which the muscle was capable of exerting. There remains a large potential for increasing the force of contraction still further, a fact which might explain some of the feats performed by people under stress. This mismatch of MVC and the maximal force available in muscle is not seen in the small muscles of the hand. In the first dorsal interosseous and adductor pollicis, MVC is indeed the maximum force of muscle contraction and cannot be exceeded by an electrical tetanus to the nerve.

The reason for the different proportions of rate and recruitment modulation of force in these muscles can probably be related to the size and function of the muscle involved. For example, the first dorsal interosseous is a small muscle with only 120 units. Gradation of force solely by recruitment would perhaps prove far too coarse a mechanism in a muscle used in finely tuned tasks such as writing. In contrast, the deltoid is a large muscle used primarily for generating forceful contractions. Its 1,000 motor units may give a sufficiently finely graded contraction to suit its function.

Muscle Fatigue

During a maximum contraction of many mixed muscles, it is not uncommon to see a 50 per cent reduction of muscle force over a period of a few seconds. This is known as fatigue. There are three main reasons why this might occur: (1) failure in transmission at the nerve–muscle junction, or in conduction at the fine terminal branches of motor axons; (2) failure of the contractile machinery in muscle; and (3) reduction in the central drive to motoneurones below that necessary to sustain maximal muscle activity.

Failure of neuromuscular transmission was believed to be a major factor in fatigue for many years. Animal experiments using repetitive stimulation of muscle nerves had shown that transmission failure could occur. In humans, such failure can be demonstrated by supramaximal electrical stimulation of a motor nerve at frequencies of 50Hz or above. A single supramaximal shock produces a mass muscle action potential (M wave) due to the synchronous activation of all the units in the muscle. After 20s or so of 50Hz stimulation, the size of the action potential following each shock declines, indicating that action potentials are no longer being generated in all the units of the muscle. There has been a failure of electrical or neuromuscular transmission between the stimulating electrode and the muscle.

However, as noted above, such high rates of unit activity of 50Hz or more are never seen over prolonged periods in man. It turns out that the neuromuscular system is so designed that it operates at frequencies just below those at which neuromuscular transmission fails, even during muscle fatigue. This was demonstrated quite elegantly by Merton (1954), who recorded the size of the maximal motor potential (M-wave) in the adductor pollicis muscle at different times during the course of maximal voluntary contraction. In contrast to results during maximal electrical stimulation, during voluntary muscle fatigue there was no change in the size of the M-wave, indicating that electrical transmission had remained secure.

Fatigue probably is not due to central mechanisms either since throughout the course of an MVC in the small hand muscles, tetanic electrical stimulation of the muscle nerve cannot generate extra force over and above that produced by a well trained and motivated subject (Figure 3.15). Thus, fatigue is most probably caused by failure in the production of contractile force, although the precise nature of this effect is not known.

Figure 3.15: Effect of Fatigue on A, Force of a Sustained Maximal Voluntary Contraction, and B, Discharge Rates Recorded from Single Motor Units During the Contraction. The muscle used was adductor pollicis, a favourite for this type of experiment because it is the only muscle in the thenar eminence supplied totally by the ulnar nerve. In A are superimposed traces of muscle force and the surface rectified EMG. Both decline over the 40s period of contraction. Supramaximal ulnar nerve stimuli given at the times indicated by the arrows fail to produce any further increase in muscle force over and above MVC. Hence, the contraction must have remained maximal throughout this period. In B, highly selective needle electrodes were used to record short periods of activity from 21 identified single motor units. At the start of contraction, the firing rates are high (>30Hz) and slowly decline to less than 20Hz (from Bigland-Ritchie, Johansson, Lippold *et al.*, 1983; Figures 4A and B, with permission)

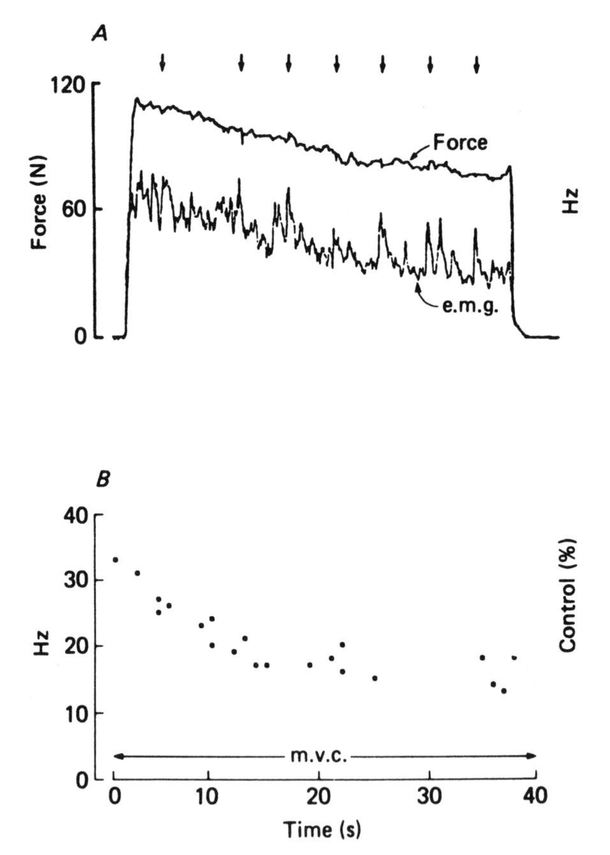

During the course of a maximal contraction, the firing rate of the motor units is modulated in a remarkable way. At the start of a rapid contraction they may fire at frequencies up to 150Hz. Indeed, the instantaneous firing frequency of the first two or three impulses of a train can be as high as 150 to 200Hz. Firing rate then declines slowly throughout the rest of the maximal contraction to rates as low as 20Hz (Figure 3.15). One consequence of this decline is that transmission failure does not occur at the neuromuscular junction. The surprising feature of this decline in motoneurone firing rates is that it is not accompanied by any additional drop in muscle force. Even after 40s MVC, when units are firing at less than 20Hz, the muscle force is maximal. The reason for this is that during fatigue there is a change in the mechanical properties of the motor units. The main effect is that the relaxation time of single unit twitches increases. Because of this increase in twitch duration, the stimulation frequency needed to produce a fused tetanus declines. The result is that the same unit can be activated maximally at reduced firing frequencies. Motoneurone firing rates are carefully matched to the changing properties of the motor units.

Figure 3.16 shows an example of the 'muscular wisdom' which matches firing frequencies to changing motor unit properties during fatigue. A and F are maximum voluntary fatiguing contractions of adductor pollicis. B,C,D and E are attempts to mimic the time course of the contraction using electrical stimulation of the muscle nerve. Initial force is high if very rapid trains are used (B,C and D), but declines more rapidly than MVC because of neuromuscular failure. In E, a good fit to MVC is achieved by changing the stimulus frequency during the course of stimulation, in much the same way as motor units have been found to change discharge frequency during a MVC (Merton, 1954; Bigland-Ritchie *et al.*, 1983).

Figure 3.16: Comparison of Voluntary and Electrically Excited Contractions of Adductor Pollicis. In each panel traces are force of contraction (top) and surface (bottom) EMG from the muscle. A and F represent a 95s maximal voluntary contraction. B,C,D, shows contractions produced by electrical stimulation of the muscle nerve at 100, 50 and 35Hz, respectively. E is an electrical contraction in which stimuli were given at 60Hz for 8s, 45Hz for 17s, 30Hz for 15s, and 20Hz for 55s. Only in this latter case does the force output from the muscle resemble that of an MVC. Calibrations are 5kg and 10mV (from Marsden, Meadows and Merton, 1983; Figure 3, with permission, pp. 169–211, copyright 1983, Raven Press, New York)

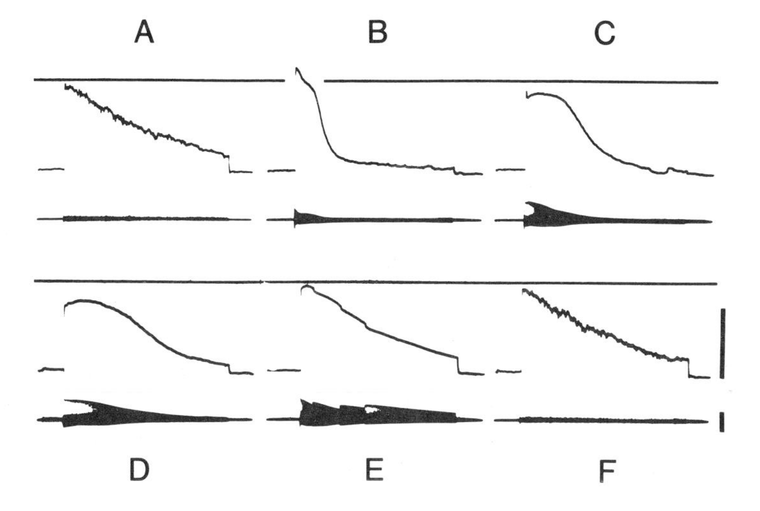

Pathophysiology of the Motor Unit

By definition, the motor unit has two components: the motoneurone and the muscle fibres which it innervates. Diseases of the motor unit are divided into two categories, depending upon which of these components is affected: (1) *neurogenic diseases* affect the motoneurone, either at its cell body (motor neurone diseases) or along its axon or myelin sheath (peripheral neuropathies); (2) *myopathic diseases* affect the muscle itself. Electrophysiological study of the motor unit using conventional needle electrodes plays an important part in diagnosis of these disorders (Kimura, 1983).

The Normal Electromyogram (Figure 3.17)

There is no electrical activity in a normal muscle when subjects are completely at rest. When a concentric needle recording electrode is inserted, there is a brief burst of activity (known as 'insertional activity') due to irritation of the muscle membrane with the needle. This dies away rapidly and the recording becomes silent. Voluntary effort recruits motor units which usually have di- or triphasic action potentials with a duration of 10ms or so. As voluntary effort increases, more units are recruited and it becomes more difficult to distinguish individual muscle action potentials. At maximum effort, single unit discharges are no longer visible and the EMG is said to show a full 'interference pattern'. This is because all the units of the muscle discharge asynchronously and their spikes interfere with each other in the electrical recording.

Figure 3.17: Typical EMG Findings Recorded with Conventional Needle Electrodes in Lower and Upper Motoneurone Disorders, and in Myopathic Lesions. See text for full explanation (from Kimura, 1983; Figure 13.4, with permission)

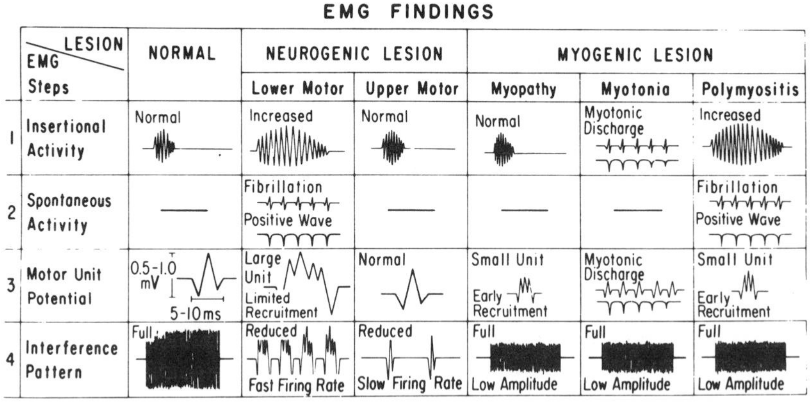

EMG FINDINGS

	LESION / EMG Steps	NORMAL	NEUROGENIC LESION		MYOGENIC LESION		
			Lower Motor	Upper Motor	Myopathy	Myotonia	Polymyositis
1	Insertional Activity	Normal	Increased	Normal	Normal	Myotonic Discharge	Increased
2	Spontaneous Activity	—	Fibrillation Positive Wave	—	—	—	Fibrillation Positive Wave
3	Motor Unit Potential	0.5-1.0 mV 5-10ms	Large Unit Limited Recruitment	Normal	Small Unit Early Recruitment	Myotonic Discharge	Small Unit Early Recruitment
4	Interference Pattern	Full	Reduced Fast Firing Rate	Reduced Slow Firing Rate	Full Low Amplitude	Full Low Amplitude	Full Low Amplitude

The Electromyogram in Neurogenic Diseases

Complete lesion of a motor nerve produces remarkable changes in the muscle which it innervates. There is, of course, no voluntary recruitment of motor units. However, several days after the lesion, there is an increase in the level of spontaneous activity in the muscle. In the EMG record this is seen as trains of very small action potentials firing at frequencies of about 15Hz. These are known as 'fibrillation potentials'. They are small in amplitude because they are due to the activity of single muscle fibres, rather than to the ensemble of muscle fibres which fire together when a motor unit is active.

This spontaneous activity is thought to be due to denervation supersensitivity. When the motor end-plates on a muscle are no longer active, the acetylcholine receptors begin to proliferate on regions of the muscle fibre membrane distant from the end-plate region. This proliferation can produce a 1,000-fold increase in sensitivity of the muscle fibre to acetylcholine. The muscle membrane also becomes more sensitive to mechanical deformation such as stretch or pressure. Fibrillation potentials may represent the response of some hypersensitive fibres to small amounts of circulating acetylcholine which normally are ineffective in exciting intact muscle.

If the lesion was not complete, or if regrowth occurs along the damaged segment of nerve, some of the denervated muscle fibres will become reinnervated by remaining or regrowing axons. Intact axons begin to sprout near their terminals on the muscle. These sprouts either form new motor end-plates on denervated fibres, or take over the original end-plates left by dying neurones. When a new neuromuscular connection is formed, the extra acetylcholine receptors which had proliferated over the entire surface of the muscle fibre membrane, are reabsorbed. The new end-plate again becomes the only point on the membrane where acetylcholine receptors are found. Fibre sensitivity to applied acetylcholine declines and spontaneous activity (fibrillation potentials) disappears.

It is not clear what induces the nerves to begin sprouting, nor why the sprouts form terminals only on denervated fibres. However, the nett result is that intact axons remaining after a partial nerve lesion expand the territory of their motor units by reinnervation of denervated muscle fibres. In such circumstances, there are fewer units in the EMG (since there are fewer intact axons innervating the muscle), but the units which remain are much bigger than those of normal muscle. The consequence is that action potentials recorded through needle electrodes may be up to ten times larger than average but the interference pattern during maximal contraction is reduced.

A final feature is that the motor unit action potentials, in addition to being large, also are far longer and more polyphasic than normal. This is because conduction along the axonal sprouts which reinnervate denervated muscle fibres is not secure. There is intermittent conduction block or slowing along the new part of the axon so that some fibres of the unit only fire intermittently or much later than others. The asynchronous discharge of a large number of fibres gives the motor unit action potential its polyphasic shape.

Such an EMG picture of denervation and reinnervation is typical of many neuropathic diseases which affect the lower motoneurone, such as chronic peripheral neuropathies and polio. The picture is quite different for diseases of the upper motoneurone. In these cases, there is usually no denervation of the muscle since the spinal motoneurones remain intact. However, loss of descending excitatory input to the motoneurones makes it more difficult to recruit motor units in a voluntary contraction. With an incomplete lesion of the upper motoneurone, motor units are normal, but cannot be discharged at high frequencies by voluntary effort. In addition, it is impossible to recruit all motor units into contraction, so that the interference pattern is reduced.

In some patients with chronic lesions of the upper motoneurone, there may be a small amount of trans-synaptic degeneration of spinal motoneurones. A proportion of cells deprived of their normal input from supraspinal sources may die. If this occurs, there may be typical signs of denervation (fibrillation potentials) and reinnervation (large, polyphasic units). Some diseases, such as amyotrophic lateral sclerosis (ALS, or motoneurone disease) affect both upper and lower motoneurones.

The Electromyogram in Myopathic Diseases

In most myopathic diseases such as the muscular dystrophies, the change in the EMG is not so marked as in neuropathic lesions. Death of muscle fibres usually is reflected only as a decrease in size of the motor unit action potentials. In extreme cases, only a single fibre of a unit may remain, so that voluntary activity produces a record indistinguishable from a fibrillation potential. Potentials may be more polyphasic than normal.

There are two types of muscle disease which affect the electrical properties of the muscle fibre membranes and produce characteristic changes in the motor unit potentials. These are the myotonias and periodic paralysis.

In *myotonia*, the muscle, once activated, tends to discharge repetitively. In EMG studies this is evident in two ways: (1) insertion of the needle provokes a repetitive 'myotonic' discharge of activated muscle fibres which far outlasts the normal period of insertional activity; and (2) there is delayed relaxation of muscle contraction and electrical activity following voluntary contraction, percussion or electrical stimulation. This extra activity persists even after nerve block or curarisation, implying that the site of the disorder is in the muscle membrane. It is thought that in some forms of myotonia, this could be due to a decrease in Cl^- conductance of the membrane.

Following an action potential, the muscle fibre membrane repolarises and there is a movement of K^+ out of the cell. Because of the complex infolding of the fibre membrane, much of this is into the transverse tubular system, where the ions are effectively trapped. Such an accumulation of K^+ depolarises the membrane. Under normal circumstances, the amount of depolarisation is reduced by movement of Cl^- across the membrane. In situations where Cl^-

movement is decreased, this depolarisation can become large enough to initiate a further action potential and so on.

Periodic paralysis is the result of reversible inexcitability of the muscle membrane. Clinically, it is a condition in which patients have episodes of paralysis, lasting an hour or more and during which the muscle membrane cannot be driven to discharge an action potential. Three major subtypes are distinguished according to the level of serum K^+ during an attack: hypo-, hyper- and normo-kalaemic. However, serum K^+ levels play no causal part in producing an attack. A possible clue as to the abnormality comes from the finding that the muscle membrane is substantially depolarised. This may lead to depolarisation block and fibre inexcitability.

Changes in Firing Rate and Recruitment Order of Motor Units

The firing rate of a motoneurone depends on the amount of excitatory input which it receives; its recruitment order depends in most instances upon its size relative to the other members of the same motoneurone pool. What happens to these properties when some of the inputs to the motoneurone are lost, as in patients with lesions of the upper motoneurone, or when the motor units are changed by sprouting and reinnervation?

Interruption of descending supraspinal pathways, as in patients with hemiplegia due to motor stroke, decreases the firing rate of spinal motoneurones. Motor unit discharge rates are abnormally low and more variable than usual in spastic muscles (that is, less than 8Hz). Abnormal patterns of discharge also have been seen in patients with Parkinson's disease (low firing rates and difficulty in recruiting units) and Huntington's chorea (irregular discharges). These changes are probably caused by a decrease in size and increase in variability of the descending motor command onto the spinal motoneurones, rather than a change in the properties of the motoneurones themselves.

In spastic patients, such low firing rates may contribute to muscular weakness. If patients cannot achieve firing frequencies high enough to produce a fused contraction of motor units, then maximum force will not be available from the muscle. Unfused contractions do not produce maximum force. This effect can be seen in the relation between force and EMG in spastic muscles. Figure 3.18 shows an experiment in which a hemiparetic patient was asked to make an isometric flexion at the elbow while the surface EMG was recorded from the biceps muscle. On the paretic (weak) side, far more EMG was needed to produce the contraction than on the normal side. The most probable reason for this discrepancy (taking into account difficulties in electrode placement, muscle size, and so on) is that the contraction of the motor units on the paretic side was not fused. Thus, more units have to be recruited to produce a given level of force, and more EMG is recorded from the surface.

Recruitment order and mechanical properties of motor units usually are unaffected by upper motoneurone lesions. However, changes do appear in

Figure 3.18: Dissociation Between the Amount of Electrical Activity and the Force Exerted in Paretic Muscle as Compared with Normal. A patient with hemiparesis due to a capsular stroke was asked to perform isometric contractions of the biceps brachii, while the EMG was recorded from surface electrodes placed over the belly of the muscle. The upper diagram shows some raw data from the paretic (weak) arm (left), and the normal arm (right). In order to exert the same force (bottom traces), the subject produced much larger amounts of EMG activity (rectified signals; top traces) on the paretic side. The lower graph shows the relationship between force exerted and average EMG activity in the same subject for a range of different contractions. Closed circles and dotted line are for the normal side; open circles and solid line are for the paretic side (from Tang and Rymer, 1981; Figures 2 and 3, with permission)

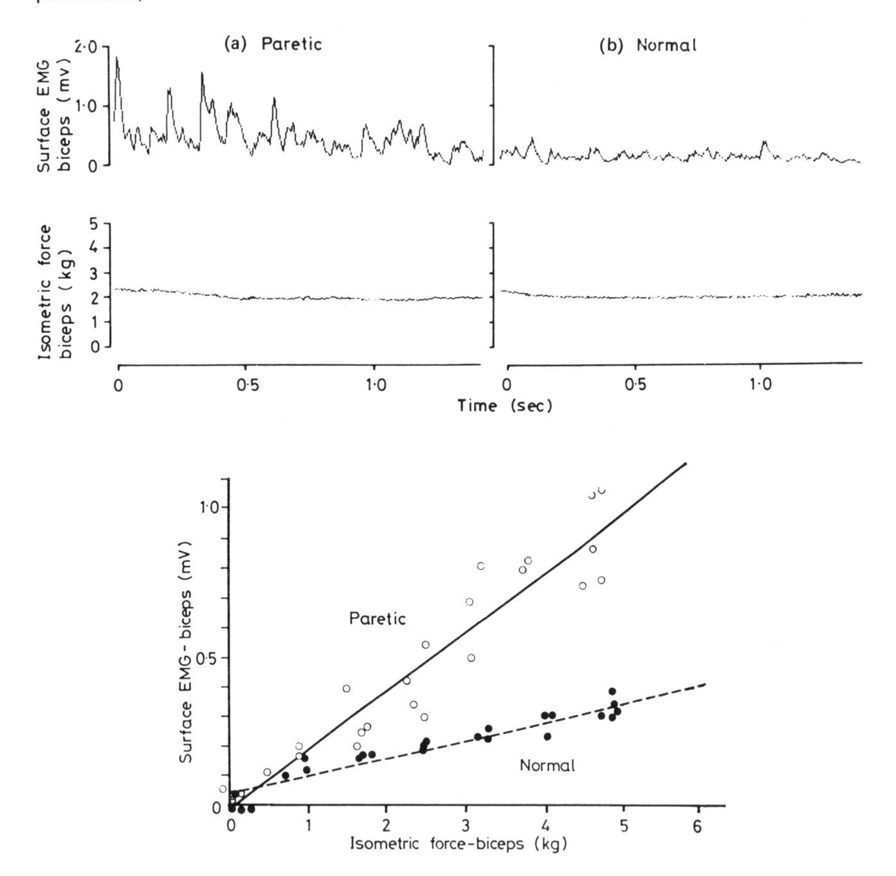

diseases affecting the lower motoneurone, which produce signs of denervation and reinnervation in the muscle under study. As noted, reinnervated motor units have potentials which are larger than normal, consistent with the increased number of muscle fibres within each unit. However, averaging the force produced by each enlarged unit in patients with peripheral neuropathy or motoneurone disease gives a rather unexpected result: the twitch tensions are no bigger than in normal muscle, and may in many cases be even smaller, on average, than normal. In other words, the large unit appears to be a less efficient contractile machine than normal. The reason for this change is not known. It could be due to changes in the contractile mechanisms of muscle fibres or to fibrosis affecting the linkage between motor unit contraction and externally monitored force.

The recruitment order is not affected in most diseases of the motor unit. Units with small twitch tensions are recruited earlier than those with large twitch tensions. This is true even when motor unit territory has been expanded through reinnervation. This may be because the number of new fibres reinnervated by existing units is proportional to the size of the original motor unit. Thus, large units may remain large, and small units remain relatively small.

An exception to this has been described in a patient with surgical repair of a completely severed nerve. Regrowth occurred over two years and reinnervation produced a normal spectrum of unit sizes. However, recruitment order was abnormal: there was no consistent size order. It may be that recruitment order was not preserved in this case because large and small motoneurones regrow and reinnervate at the same rate. The situation may then arise that a large axon captures only a small number of fibres, while a small axon captures a large number of fibres. Assuming that no retrograde changes in size of the motoneurone occur, this means that even if the motoneurones retain their recruitment order, the size of recruited motor units in muscle would not be orderly (Rosenfalck and Andreassen, 1980; Tang and Rymer, 1981; Milner-Brown, Stein and Lee, 1974).

References and Further Reading

Review Articles

Buchthal, F. and Schmalbruch, M. (1980) 'Motor Unit of Mammalian Muscle', *Physiol. Rev.*, *60*, pp. 90–142

Burke, R.E. (1981) 'Motor Units: Anatomy, Physiology and Functional Organisation' in V.B. Brooks (ed.) *Handbook of Physiology*, sect. 1, vol. 2, part 1, Williams and Wilkins, Baltimore, pp. 345–411

Desmedt, J.E. (ed.) (1981) *Motor Unit Types, Recruitment and Plasticity in Health and Disease. Progress in Clinical Neurophysiology*, vol. 9, Karger, Basel

Dum, R.P. and Kennedy, T.T. (1980) 'Synaptic Organisation of Defined Motor-unit Types in Cat Tibialis Anterior', *J. Neurophysiol.*, *43*, pp. 1631–44

Freund, H.-J. (1983) 'Motor Unit and Muscle Activity in Voluntary Motor Control', *Physiol. Rev.*, *63*, pp. 387–436

Garnett, R. and Stephens, J.A. (1981) 'Changes in the Recruitment Threshold of Motor Units Produced by Cutaneous Stimulation in Man', *J. Physiol.*, *311*, pp. 463–73

Granit, R. (1970) *The Basis of Motor Control*, Academic Press, New York

Henneman, E. and Mendell, L.M. (1981) 'Functional Organisation of Motorneuron Pool and Its Inputs' in V.B. Brooks (ed.), *Handbook of Physiology*, *Sect. 1*, vol. 2, part 1, Williams and Wilkins, Baltimore, pp. 423–508

Kanda, K., Burke, R.E. and Walmsley, B. (1977) 'Differential Control of Fast and Slow Twitch Motor Units in the Decerebrate Cat', *Exp. Brain Res.*, *19*, pp. 57–74

Kernell, D. (1965) 'High-frequency Repetitive Firing of Cat Lumbosacral Motoneurones Stimulated by Long-lasting Injected Currents', *Acta Physiol. Scand.*, *65*, pp. 74–86

Kimura, J. (1983) *Electrodiagnosis in Diseases of Nerve and Muscle: Principles and Practice,* F.A. Davis, Philadelphia

Lewis, D.M. (1981) 'The Physiology of Motor Units in Mammalian Skeletal Muscle' in A.L. Towe and E.S. Luschei (eds.), *Handbook of Behavioural Neurobiology,* vol. 5, Plenum, New York, pp. 1–67

Original Papers

Adrian E.D. and Bronk, D.W. (1929) 'The Discharge of Impulses in Motor Nerve Fibres. Part II. The Frequency of Discharge in Reflex and Voluntary Contractions', *J. Physiol., 67,* pp. 119–51

Bigland-Ritchie, B., Johansson, R., Lippold, O.C.J. *et al.* (1983) 'Changes in Motoneurone Firing Rates During Sustained Maximal Voluntary Contractions', *J. Physiol., 340,* pp. 335–46

Burke, R.E. (1975) 'A Comment on the Existence of Motor Unit "Types" ' in D.B. Tower (ed.) *The Nervous System. The Basic Neurosciences,* vol. 1, Raven, New York, pp. 611–19

Burke, R.E., Levine, D.N., Tsairis, P. *et al.* (1973) 'Physiological Types and Histochemical Profiles in Motor Units of the Cat Gastrocnemius', *J. Physiol., 234,* pp. 723–48

——— Rymer, W.Z. and Walsh, J.V. (1976) 'Relative Strength of Synaptic Input from Short-latency Pathways to Motor Units of Defined Type in Cat Medial Gastrocnemius', *J. Neurophysiol., 39,* pp. 447–85

——— and Tsairis, P. (1973) 'Anatomy and Innervation Ratios in Motor Units of Cat Gastrocnemius', *J. Physiol., 234,* pp. 749–65

Datta, A.K. and Stephens, J.A. (1981) 'The Effects of Digital Nerve Stimulation on the Firing of Motor Units in Human First Dorsal Interosseous Muscle', *J. Physiol., 318,* pp. 501–10

De Luca, C.J., Lefever, R.S., McCue, M.P., *et al.* (1982) 'Behaviour of Human Motor Units in Different Muscles During Linearly Varying Contractions', *J. Physiol., 329,* pp. 113–28

Denny-Brown, D. (1929) 'On the Nature of Postural Reflexes', *Proc. Roy. Soc. B., 104,* pp. 252–301

Desmedt, E. and Godaux, E. (1977) 'Fast Motor Units are not Preferentially Activated in Rapid Voluntary Contractions in Man', *Nature, 267,* pp. 717–9

Eccles, J.C., Eccles, R.M. and Lundberg, A. (1957) 'The Convergence of Monosynaptic Excitatory Afferents onto Many Different Species of Alpha-motoneurone', *J. Physiol., 137,* pp. 22–50

Edstrom, L. and Kugelberg, E. (1968) 'Histochemical Composition, Distribution of Fibres and Fatiguability of Single Motor Units', *J. Neurol. Neurosurg. Psychiat., 31,* pp. 424–33

Garnett, R.A.F., O'Donovan, M.J., Stephens, J.A. *et al.* (1978) 'Motor Unit Organisation of Human Medial Gastrocnemius', *J. Physiol., 287,* pp. 33–43

Granit, R., Kernell, D. and Smith, R.S. (1963) 'Delayed Depolarisation and the Repetitive Response to Intracellular Stimulation of Mammalian Motoneurones', *J. Physiol., 168,* pp. 890–910

Gustaffson, B. and Pinter, M.J. (1984) 'Relations among passive electrical properties of lumbar α-motoneurones of the cat'. *J. Physiol., 356,* pp. 401–32 and 433–42

Henneman, E., Somjen, G. and Carpenter, D.O. (1965) 'Functional Significance of Cell Size in Spinal Motoneurons', *J. Neurophysiol., 28,* pp. 560–80

Jami, L. Murthy, K.S.K., Petit, J. *et al.* (1982) 'Distribution of Physiological Types of Motor Unit in the Cat Peroneus Tertius Muscle', *Exp. Brain Res., 48,* pp. 177–84

Kugelberg, E. and Edstrom, L. (1968) 'Differential Histochemical Effects of Muscle Contractions on Phosphorylase and Glycogen in Various Types of Fibres: Relation to Fatigue', *J. Neurol. Neurosurg. Psychiat., 31,* pp. 415–23

Luscher, H.-R., Ruenzel, P.W. and Henneman, E. (1983) 'Effects of Impulse Frequency, PTP, and Temperature on Responses Elicited in Large Populations of Motoneurones by Impulses in Single Ia Fibres', *J. Neurophysiol., 50,* pp. 1045–58

Marsden, C.D., Meadows, J.C. and Merton, P.A. (1983) ' "Muscular Wisdom" that Minimises Fatigue During Prolonged Effort in Man: Peak Rates of Motoneurone Discharge and Slowing of Discharge During Fatigue' in J.E. Desmedt (ed.), *Adv. in Neurol.,* vol. 39, Raven Press, New York

Merton, P.A. (1954) 'Voluntary Strength and Fatigue', *J. Physiol., 123,* pp. 553–64

Milner-Brown, H.S., Stein, R.B. and Lee, R.G. (1974) 'Contractile and Electrical Properties of Human Motor Units in Neuropathies and Motor Neurone Disease', *J. Neurol. Neurosurg. Psychiat., 37,* pp. 670–6 (and pp. 665–9)

Rall, W. (1967) 'Distinguishing Theoretical Synaptic Potentials Computed for Different Somadendritic Distributions of Synaptic Input', *J. Neurophysiol.*, *30*, pp. 1138–68

Rosenfalck, A. and Andreassen, S. (1980) 'Impaired Regulation of Force and Firing Pattern of Single Motor Units in Patients with Spasticity', *J. Neurol. Neurosurg. Psychiat.*, *43*, pp. 907–16

Tang, A. and Rymer, W.Z. (1981) 'Abnormal Force-EMG Relations in Paretic Limbs of Hemiparetic Human Subjects', *J. Neurol. Neurosurg. Psychiat.*, *44*, pp. 690–8

Walmsley, B., Hodgson, J.A. and Burke, R.E. (1978) 'Forces Produced by Medial Gastrocnemius and Soleus Muscles During Locomotion in Freely Moving Cats', *J. Neurophysiol.*, *41*, pp. 1203–16

Zengel, J.E., Reid, S.A., Sypert, G.W. *et al.* (1985) 'Membrane Electrical Properties and Prediction of Motor-unit Types of Medial Gastrocnemius Motoneurons in the Cat', *J. Neurophysiol.*, *53*, pp. 1323–44

4 PROPRIOCEPTORS IN MUSCLE, JOINT AND SKIN

Proprioceptive organs signal to the CNS information about the relative positions of the body parts. Apart from pressure receptors in the soles of the feet, they do not supply any information as to the orientation of the body with respect to gravity; they only signal the position of one part of the body with respect to another. The receptors involved lie in the muscles (spindles and Golgi tendon organs), the joints and the skin. In this chapter, only the structure and characteristics of the afferent discharge will be summarised for each type of receptor. The possible roles of these receptors in the control of movement will be discussed in subsequent chapters. Despite its age, Matthews's (1972) book on muscle receptors is still one of the best complete reviews on these topics.

Muscle Receptors: I. The Muscle Spindle

Anatomy

The muscle spindle is one of the most well studied and sensitive sensory receptors in the body. It rivals the eye and the ear in its complexity and yet, in comparison, relatively little is known of its role in sensation or even in the control of movement.

Spindles were first identified in muscle about 120 years ago and were seen to consist of both nervous and muscular elements. At first, there was some speculation as to whether the spindles might be 'muscle buds' (the site of formation of new muscle fibres), an inflammatory focus or perhaps a sensory ending. The matter was resolved in 1894 by Sherrington, who cut the ventral root supply to a number of muscles. After allowing some weeks for degeneration of the motor fibres, he found a dense, surviving, afferent supply to the spindles. The spindle was proved to be a 'sensorial end organ'. At the same time, Ruffini (1898) published the first histological diagrams of the nervous innervation of the spindle. He classified, on morphological grounds, three types of nerve ending on the spindle fibres which are still known by the same names today — primary, secondary (both now known to be afferent) and plate (now known to be motor) endings.

The presently accepted structure of the muscle spindle is shown in Figure 4.1. This illustrates the structure of a typical spindle from the tenuissimus muscle of the cat hindlimb, a favourite muscle for such studies. The spindle consists of a bundle of specialised muscle fibres which lie in parallel with the fibres of the main extrafusal muscle. The intrafusal fibres are about 10mm long, which is much shorter than fibres in the main muscle. At each end they may attach to extrafusal fibres or to tendinous insertions. For about half its length

Figure 4.1: Diagram of a Typical Muscle Spindle from a Cat Hindlimb Muscle. The spindle consists of two bag fibres (DB_1, SB_2: dynamic bag_1, and static bag_2) and a number of chain fibres (C), encapsulated in a fluid-filled space. Sensory endings are found in the p and s_1 regions of the fibres. Primary spiral endings are seen on all fibres of the spindle in the equational, p, region. These endings give rise to the large group Ia afferent fibres. Secondary endings consist of spiral terminations on the chain fibres and less extensive spray-like terminations on the primary endings in the s_1 region. Secondary endings are rarely seen on the DB_1 fibre. They give rise to the group II afferent fibres from the spindle. The motor (γ) supply to the spindle terminates outside the equatorial region of the fibres. The endings sometimes are described as 'plate endings' on the bag fibres and 'trail endings' on the chain fibres. The γ_d efferents innervate the DB_1 fibre, while the γ_s efferents innervate the DB_2 and chain fibres. The βsupply to the spindle is not shown (from Boyd, 1980; Figure 1, with permission)

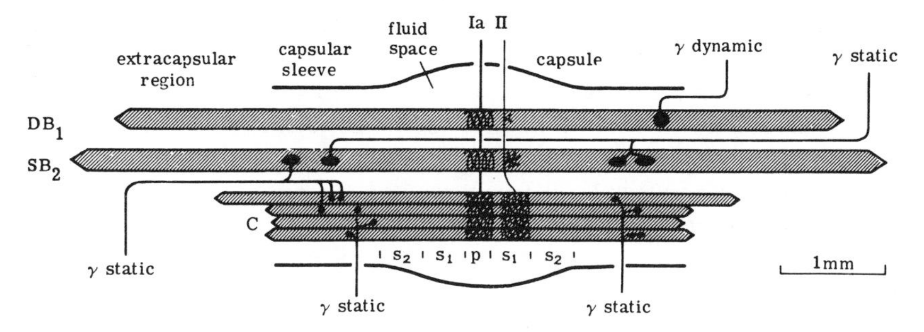

the spindle is contained within a thick connective tissue capsule which expands in its central 2mm to form a fluid-filled space, giving the structure its characteristic 'fusiform' appearance.

There are two types of muscle fibre in the spindle: generally, there may be two (or three) bags and three to five chain fibres. Like all muscle fibres, they are formed from a syncitium of cells whose cell walls break down during development to form a continuous structure. Up to 100 or so of the nuclei of the bag fibres are collected together at the spindle equator, where they may be three or four deep when seen in cross-section. In contrast, the chain fibres only have a single row of nuclei in the same region. On the basis of other features (see below), the bag fibres have been subdivided into two separate categories: bag_1 and bag_2.

The sensory innervation of the spindle is of two types, as described by Ruffini (1898). The large diameter Ia neurones distribute primary endings to all the muscle fibres of a spindle. These endings occupy the most central region of each fibre. Either side of them may be the secondary endings of the group II afferent neurones. There are few secondary endings on the bag_1 fibres. Primary endings consist of spiral terminations on all types of spindle fibre, whereas secondary endings usually only have spiral terminations on the bag_2 and chain fibres. The secondary innervation of bag fibres consists of less extensive spray-like terminations, which may also be seen at times on the chain fibres.

The motor supply to the spindle fibres consists mainly of the small diameter γ neurones, which distribute their endings to the poles of the fibres, usually within the connective tissue capsule. The neurones were first subdivided using physiological techniques (see below) into γ_s and γ_d categories. This physiological difference is reflected in differences in their anatomical termination on the intrafusal muscle fibres. The γ_d neurones innervate only the bag_1 fibres, whereas the γ_s neurones innervate bag_2 and chain fibres. There is a possibility that there are sub-groups of γ_s fibres which innervate just bag_2 or just chain fibres.

Recently, more attention has been directed towards another source of efferent neurones to the intrafusal fibres. In some cases, up to 20 per cent of the motor supply has been shown to arise from branches of alpha-motoneurones innervating the extrafusal muscle. Such shared axons are known as β-axons, a confusing term since they have the same axon diameter as the unshared α-axons. Attempts have been made to subdivide them into static β and dynamic β. The former tend to have a higher conduction velocity than the latter and appear to have endings only on chain fibres. However, the relative roles of the γ and β systems in spindle control have yet to be understood. It is worth mentioning, though, that β-innervation is the only type of innervation seen in amphibian spindles. The additional system found in mammals is therefore thought to be an evolutionary 'improvement' in design. It allows for the possibility of independent control of extrafusal and intrafusal motor systems.

Figure 4.1 is the 'typical' spindle structure characteristic of many limb muscles in both cat and man. Other types of spindle are known, some of the most complex being found in the small paravertebral muscles of the cervical cord. These muscles, many of which are so small that they have never been given specific names, have some of the highest densities of spindles in the body. In many cases, the spindles are 'atypical'; they may lack the bag_1 fibre or may be found in large arrays of up to 10 spindles, sharing intrafusal fibres and capsules (Boyd, 1962, 1976, 1980, 1985; Boyd and Ward, 1975; Boyd *et al.*, 1981; Hulliger, 1984; Matthews, 1981).

Physiology of Spindle Afferent Responses

The first systematic study of muscle spindle physiology was made by B.H.C. Matthews in the early 1930s. He subdivided the nerve to a variety of muscles in the cat until only one afferent fibre remained intact. Recording from such filaments, he classified three types of response — A, B, C — which he presumed came from three different receptors. The C endings were rare and fired impulses both during stretch and release of the muscle, but not during sustained contraction produced by electrical stimulation of the muscle nerve. The responses disappeared when the fascia was removed completely from the muscle and it is possible that they arose from free nerve endings (see below).

The other two classes of response were studied in much greater detail (Figure 4.2). Both endings increased their firing rate in response to muscle stretch, but the A endings had a lower threshold than the B endings and frequently discharged spontaneously at a low rate. The crucial difference between them was

Figure 4.2: Contrasting Responses of a Spindle Ending and a Tendon Organ During a Twitch Contraction of the Muscle. The spindle ending pauses during the twitch, while the tendon organ discharges (from Matthews, 1972; Figure 3.2, with permission

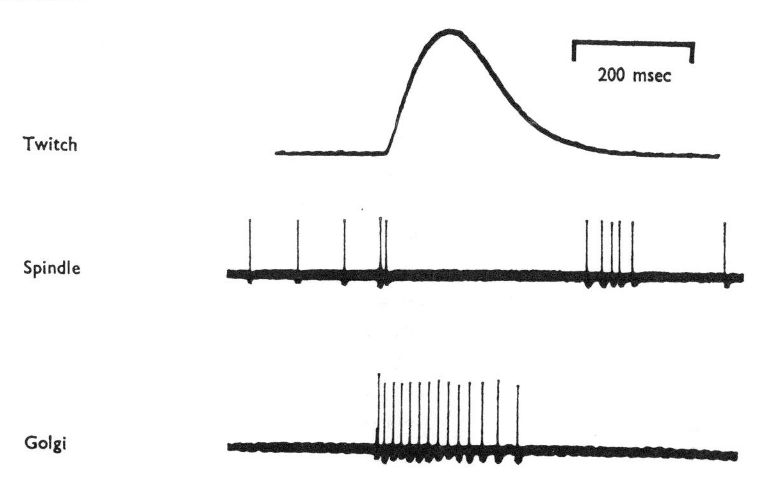

seen in their responses to an electrical muscle twitch. The A endings silenced at the onset of the twitch, whereas the B endings increased their rate of firing. This is exactly the behaviour predicted for muscle spindle receptors and Golgi tendon organs. Spindles, lying in parallel with the main muscle, would be unloaded by the extrafusal contraction and reduce their discharge, while tendon organs, in series with the extrafusal muscle, would be excited by the twitch. This 'twitch test' is still used today in both human and animal work to differentiate spindle and tendon organ afferent fibres. Matthews also found that the A endings (spindle) were sensitive to both the rate of stretch of the muscle as well as its extent. The receptors were said to have both dynamic and static sensitivity. Later studies, some 30 years later, improved on this description and distinguished between the responses of the primary and secondary ending.

If recordings are taken from spindle afferent fibres during a maintained muscle contraction or stretch, it is not possible to distinguish between the discharges from primary or secondary endings since they both fire at approximately the same rate. Under these circumstances they may be differentiated only on the basis of the conduction velocity of their afferent fibres. The Ia fibres in the cat conduct at about 80–120ms^{-1} and the group II fibres from 20–70ms^{-1} (values are slightly less in man; maximum sensory nerve conduction velocity is never greater than 80ms^{-1}). During muscle stretch, their responses are considerably different (Figure 4.3). The primary ending is very sensitive during the dynamic phase of stretch. At the onset of a ramp stretch the discharge frequency suddenly jumps to a new value and then continues to increase throughout the remainder of the stretch (see Figure 4.3, top). The end of stretch produces a decline in discharge over a period of 0.5s or so to a new

Figure 4.3: This page: Contrasting Responses of a Spindle Primary and Secondary Ending to a Rapidly Applied Stretch Shown in the Presence (VR intact) and Absence (VR cut) of Fusimotor Activity. The endings were in the soleus muscle of a decerebrate cat, and the afferent fibres had a conduction velocity of $85ms^{-1}$ (primary) and $44ms^{-1}$ (secondary). Note the dynamic sensitivity of the primary ending during the course of stretch (from Matthews, 1972; Figure 4.2, with permission).
Opposite: Comparison of the Responses of De-efferented Primary and Secondary Endings to Ramp Stretches of Different Velocity. The receptors were located in the tibialis posterior muscle of a decerebrate cat. The firing rate is plotted as dots in an instantaneous frequencygram. Each spot represents the occurrence of an action potential, and its height above the baseline is proportional to the time interval since the immediately preceding spike. Measurement of the 'dynamic index' is shown in record b. It is the difference in frequency between the discharge near the end of the dynamic phase of stretch and that occurring when the muscle has been maintained at its final length for 0.5s. The time marker only applies while the muscle is at a constant length; during the phase of dynamic stretching, it has been somewhat expanded (from Brown, Crowe and Matthews, 1965; Figure 1, with permission)

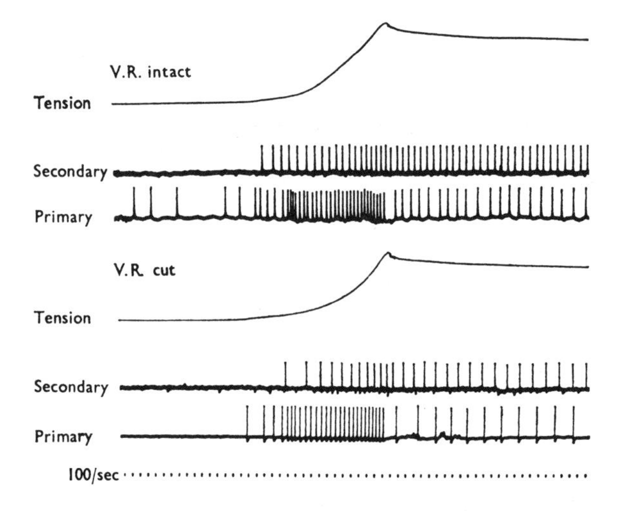

steady, or static level of activity. The decline in discharge during the first 0.5s after the end of the dynamic phase is termed the dynamic index, and is a convenient way of separating primary and secondary responses (Figure 4.4, bottom). In contrast, the final discharge level, measured some 0.5s after the end of stretch (the static or position sensitivity of the ending) is approximately the same for both types of ending.

The behaviour of the primary ending is thought to be due to the mechanical response of the bag_1 fibre (which is not shared with the secondary ending) to stretch. The poles of the bag_1 fibre are stiffer than the equatorial, sensory region. Cine films of single intrafusal fibres show that when a stretch is imposed on a bag_1 fibre, a much larger proportion of the extension is taken up by the sensory region than the poles, and at this stage the afferent discharge can reach very high peak frequencies. When the fibre is held in the stretched position, the poles 'creep' back towards the sensory spiral. This movement reduces the

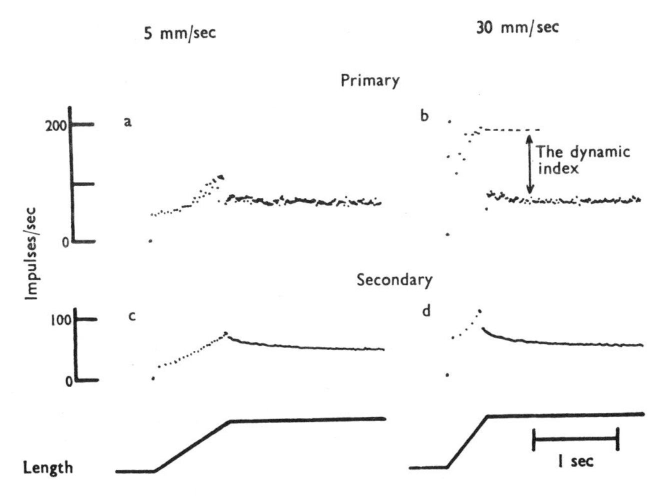

degree of extension of the equatorial region and results in the decay of primary afferent discharge (Figure 4.5).

It is sometimes said that the primary endings are *velocity*-sensitive. Indeed, this is true, but one should be quite clear what part of the discharge profile in Figure 4.3 represents velocity sensitivity. Ramp stretches were applied to the spindle, which is the same as giving a *step* change in velocity: the velocity of stretch suddenly increased from zero to, say, 30mm/s. If the primary ending were only sensitive to velocity, then a ramp stretch should produce a step increase in firing, from one value at rest to another steady value for the duration of the stretch. What actually happens is that there is a step change in firing frequency at onset of stretch, but superimposed on this is a slower ramp increase in firing. The primary ending is not only velocity-sensitive, but also is length-sensitive, and this ramp increase represents the length sensitivity of the ending during stretch.

Sometimes, an 'initial burst' of firing is seen in the primary response at onset of stretch (Figure 4.3, part b): the initial discharge of the first two or three impulses overshoots the rate seen during the remainder of stretch. This initial burst may be referred to as an 'acceleration' response. It varies considerably under different conditions. It is most prominent in spindles which are stretched from a resting position; changes in acceleration during an ongoing stretch do not produce an 'acceleration' response. If stretch is applied some seconds after a period of intense fusimotor activity, the initial burst is enhanced. Further stretches applied without intervening fusimotor activity then fail to evoke the response. These features are believed to be a consequence of the nonlinear characteristics of the ending discussed in detail below.

Some other differences between the responses of primary and secondary spindle endings to different types of muscle stretch are shown in Figure 4.6. In general, primary endings are much more sensitive to rapidly changing stimuli; they are particularly responsive to tendon taps and to sinusoidal stretching at

Figure 4.4: Top: The Relationship Between Dynamic Index and Velocity of Stretching for a Primary and Secondary Spindle Ending in the Soleus Muscle of a Decerebrate Cat. De-efferented preparation (from Matthews, 1972; Figure 4.4, with permission). Bottom: The Relationship Between Frequency of Static Firing and the Degree of Muscle Extension With (VR Intact) and Without (VR Cut) Spontaneous Fusimotor Activity. Measurements were made from receptors in the soleus muscle of a decerebrate cat, 0.5s after completion of a ramp stretch to the appropriate muscle length. Note the increased sensitivity of both types of ending in the presence of spontaneous fusimotor drive (from Jansen and Matthews, 1962; Figure 4, with permission)

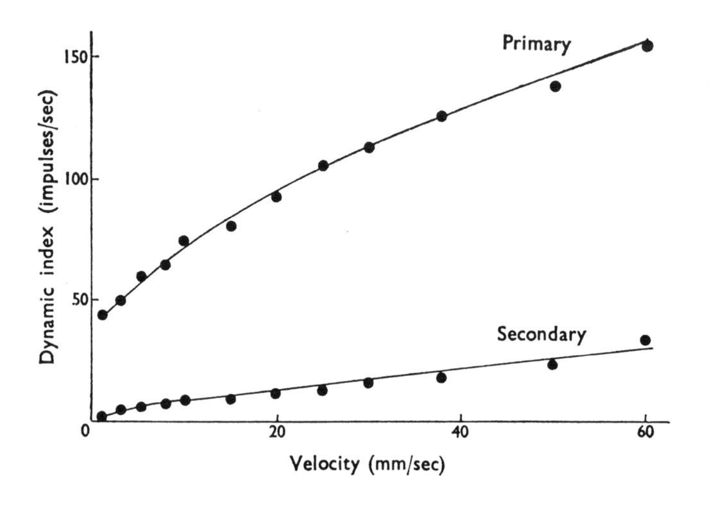

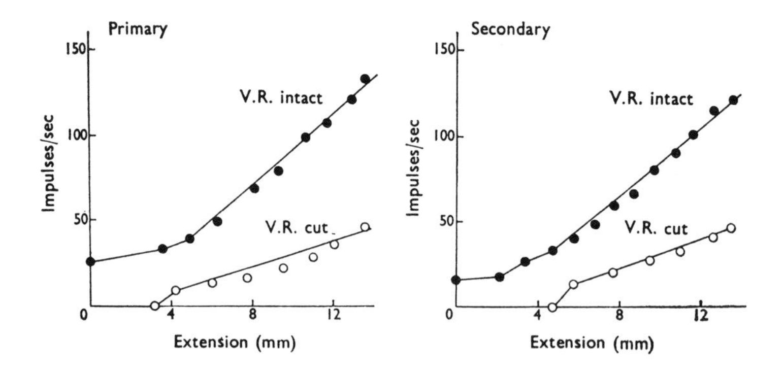

Figure 4.5: Top: Intrafusal 'Creep' in the Primary Sensory Spiral of a Dynamic Bag_1 Fibre in an Inactive Spindle Following a Ramp and Hold Stretch. During stretch, the dynamic bag_1 (Db_1) spiral extends a great deal; at the end of stretch the spiral creeps back towards the spindle equator. There is no creep in the static bag_2 (Sb_2) or the chain (Ch) fibres.
Bottom: Comparison of Intrafusal 'Creep' in the Sensory Spiral of a Dynamic Bag_1 Fibre Following a Ramp and Hold Stretch With (Filled Circles) and Without (Open Circles) γ_d Stimulation at 100Hz. Note how the spiral is extended more, and shows greater creep during γ_d stimulation (top from Boyd, 1976; Figure 4, with permission; bottom from Boyd, Gladden and Ward, 1981; Figure 1, with permission)

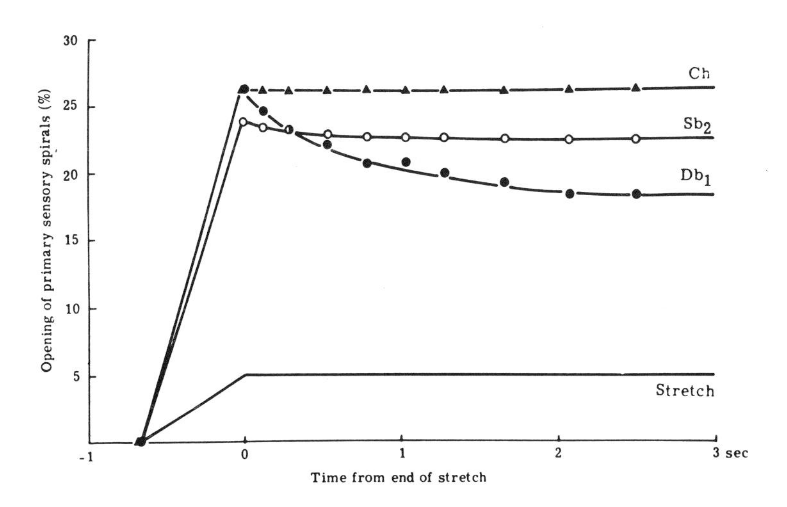

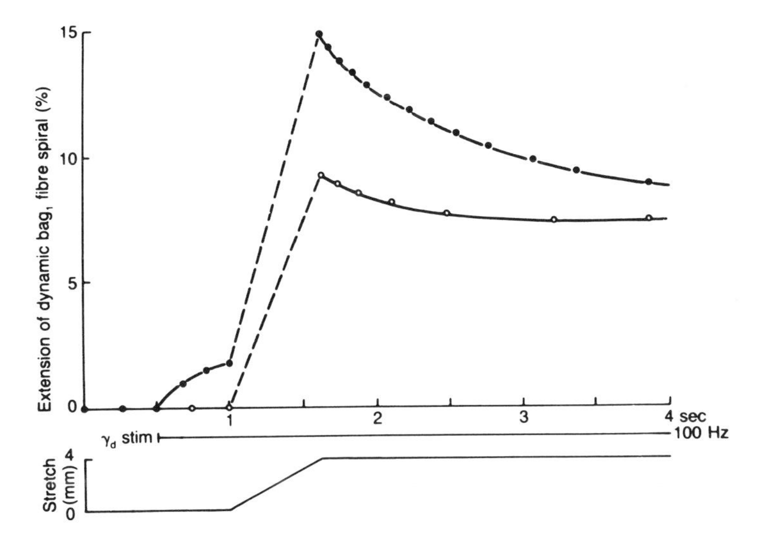

Figure 4.6: Diagrammatic Comparison of the Typical Responses of Primary and Secondary Spindle Endings to Large Stretches Applied in the Absence of Fusimotor Activity (from Matthews, 1964; Figure 2, with permission)

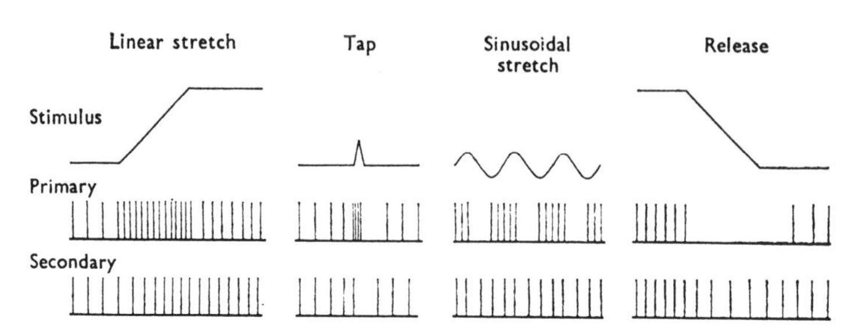

relatively high frequencies. Secondary endings show little response to either type of input (see Matthews, 1933; Boyd, 1980; articles in Barnes and Gladden, 1985).

The Effect of γ Activity on Spindle Afferent Responses

Following the description of the two types of spindle afferent response, two types of γ-efferent action soon were described. The technique used was to record the response to stretch of single afferent fibres in dissected dorsal root filaments, while stimulating single γ-efferent fibres within ventral root filaments. Primary endings were found to respond in two different ways to stimulation of γ neurones (Figure 4.7). Stimulation of some filaments (γ_s) during ramp and hold stretches was found to increase the resting level of discharge and the static sensitivity of the ending, while decreasing the dynamic sensitivity of the ending. Stimulation of the other category of γ fibres (γ_d) also increased the resting discharge, but not so much as the γ_s stimulation. Its main effect was to produce a much greater dynamic response of the primary ending. One way of indicating this difference is to plot the dynamic index (see above) of the ending during stimulation of γ_s and γ_d fibres at various frequencies (Figure 4.7, bottom): γ_d stimulation increases the dynamic index whereas γ_s stimulation decreases it.

The response of secondary endings usually is unaffected by γ_d stimulation, supporting the idea that γ_d fibres terminate only on the bag_1 fibres. Stimulation of γ_s fibres has a similar effect to that seen on the primary endings — an increase in resting discharge and a slightly greater sensitivity to static length changes.

The mechanism of action of the γ-efferents on the discharge from the sensory endings is thought to be a result of the contractile properties of each of the types of intrafusal muscle fibre. Contraction of all three types of fibre is limited to the capsular sleeve or extracapsular region. Activation of any fibre will therefore result in stretch of the central region and a consequent increase in the resting discharge rate.

Neither of the bag fibres can conduct action potentials like the extrafusal muscle. When they are activated by γ axons, non-propagated depolarisations are

Figure 4.7: Top: Different Effects of Stimulation of a Single Static or Dynamic Fusimotor Fibre on the Response of a Primary Ending to Ramp and Hold Stretching. Soleus muscle of a decerebrate cat. In C and D, stimulation started before the onset of the traces and was given at 70Hz in each case.
Bottom: Different Effects of Static and Dynamic Fusimotor Stimulation on the Dynamic Index of a Primary Spindle Ending: γ_d stimulation increases the dynamic sensitivity, while γ_s stimulation decreases it (from Crowe and Matthews, 1964; Figures 1 and 3, with permission)

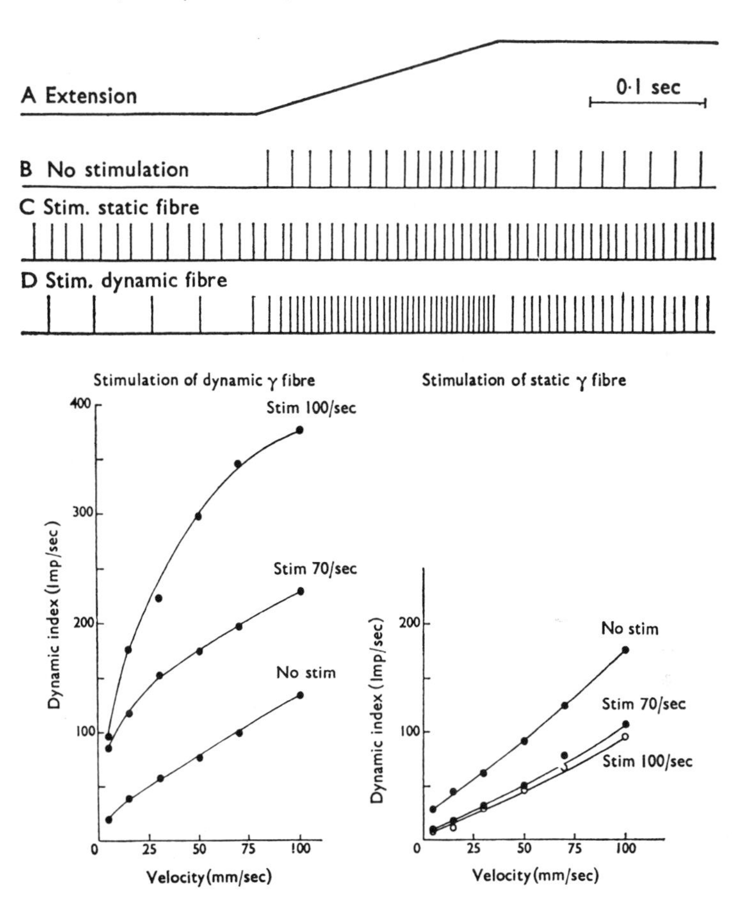

produced at the endplates. In the absence of the amplifying capacity of an action potential such stimuli produce relatively small mechanical contractions of the fibre. This is particularly true of the bag_1 fibre, which contracts very little even when stimulated tetanically (70Hz) by a γ_d axon. There may be only a 5 per cent extension of the equatorial, sensory region. This explains why γ_d stimulation has relatively little effect on the resting discharge of primary endings. Although γ_d stimulation produces a rather small contraction of the bag_1 fibre, during muscle stretch it increases the stiffness of the poles of the fibre quite considerably. When an active bag_1 fibre is stretched, a much greater proportion of the extension is taken up by the primary sensory spiral than if the fibre is inactive. At the end of stretch, the 'creep' back to static length also takes much longer (Figure 4.5). The result is that both primary afferent discharge during stretch, and the dynamic index at the end of stretch are increased by γ_d stimulation.

In contrast to the behaviour of the bag_1 fibre, the bag_2 fibre, which is innervated by γ_s axons, may contract strongly and stretch the sensory region by 20 per cent or so. This can increase resting discharge by 20–30Hz in a primary ending. However, the effect of γ_s stimulation on the bag_2 fibre seems limited simply to increasing the basal firing level of its sensory endings (both primary, and to a lesser extent, secondary), with little effect on either dynamic or static responses to stretch.

Chain fibres are capable of conducting action potentials, yet even when activated by a γ_s axon, the deplorisation seems confined to the outer regions since the sensory, equatorial region is stretched by up to 20 per cent. Single stimuli produce twitch contractions, which may give small, transient extensions of the sensory spiral. Such stimuli can 'drive' the primary ending (with one impulse per twitch) to fire at up to 75Hz. Secondary endings do not show this effect. Sustained stimulation increases the basal rate of sensory discharge in both primary and secondary endings and also (by an unknown mechanism) increases the static length sensitivity of the chain fibre endings (Banks, Barker, Bessou *et al.*, 1978; Boyd and Ward, 1975; Boyd, 1980, 1985; Crowe and Matthews, 1964; Hulliger, 1984; Matthews, 1981).

Response of Spindle Endings to Very Small Displacements

In the description of spindle responses given above, both primary and secondary endings emerge as relatively insensitive endings, giving an increase in firing frequency of only about 5Hz for every millimetre change in length. However, if much smaller stretches are given, the response of the primary ending becomes relatively much greater and may reach up to 500imp/s/mm. In general terms, the primary ending is exquisitely sensitive to very small displacements and much less sensitive to large ones. It is, in engineering terms, a non-linear receptor. In contrast, the secondary ending is fairly linear throughout the physiological range (Figure 4.8).

The analysis of the response to small amplitude stretches was first performed by Matthews and Stein (1969) who used sinusoidal stretch analysis. This is an engineering technique in which the sensitivity of a receptor is plotted in terms of its response to different frequences of sinusoidal inputs. The reason for using this approach is that it provides a complete description of spindle response to all types of input. Plots of dynamic and static sensitivity are useful, but do not describe, for example, how the spindle will discharge to continuously changing inputs. The theory of sinusoidal analysis is that all inputs to a spindle can be broken down by Fourier analysis into sinusoidal components of different frequencies and amplitudes. Knowing the response to such sinusoidal inputs when given alone then allows one to predict the response to complex inputs by

Figure 4.8: High Sensitivity of Primary Endings to Very Small Changes in Length and Their Non-linear Behaviour as Larger Stretches are Given. Secondary endings are less sensitive and more linear over a wide range of stretch amplitudes. This graph was constructed by giving continuous sinusoidal stretches at 1Hz to the soleus muscle of a decerebrate cat with intact ventral roots and spontaneous fusimotor activity. The muscle receptors had a steady background level of discharge which was modulated by the sinusoidal stretch input. The change in firing frequency is plotted on the y-axis, and the amplitude of the sinusoidal stretch (that is, half the peak-peak extent) on the x-axis (from Matthews and Stein, 1969; Figure 4, with permission)

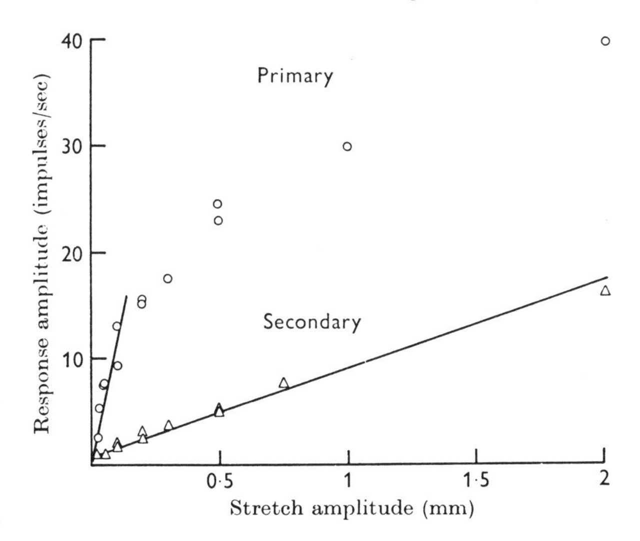

adding up the contributions from the various component frequencies.

There is, however, a rather serious limitation to this form of analysis: that the behaviour of the receptor is linear. In sinusoidal analysis, a system is tested with an input of constant amplitude but of varying frequencies. It is assumed that if the amplitude were halved, then the response would also be halved. Unfortunately, this is not true for large stretches applied to a spindle. However, it is true when very small stretches are applied.

Figure 4.8 shows the region of linearity for primary and secondary endings. Secondary endings are linear over most of the range tested, but primary endings are linear only over a very small range. Within this range, the primaries are very sensitive indeed, equivalent to 150imp/s/mm at 1Hz.

Within their linear range, the sensitivity of the primary ending increases as the frequency of the input is increased. This is shown in Figure 4.9, in which sensitivity is plotted against frequency of stretching for a sinsusoidal input of constant amplitude. The horizontal part of the curve is equivalent to the static sensitivity of the ending (that is, there is no dependence on frequency of stretch). The sloping part indicates a dependence on stretch velocity (that is, frequency), beginning at about 1.5Hz. When analysed in this way over their linear range, primary and secondary endings seem remarkably similar. Two points are quite unexpected: (1) secondary endings are sensitive to velocity: the reason this was

Figure 4.9: Comparison of Sensitivity to Sinusoidal Stretching Within the Linear Range of a Primary and Secondary Ending in the Soleus Muscle of a Decerebrate Cat with Intact Ventral Roots. The sensitivity (y-axis) is the amplitude of response divided by the amplitude of stretching (that is, the slope of the lines in Figure 4.8). Continuous lines are theoretical vector sums of a response to velocity (sloping region) and length (horizontal region) components of stretch. Deviation of points from the primary ending above this line at high frequencies of stretch may be taken as indicating an 'acceleration sensitivity' of the primary ending (from Matthews and Stein, 1969; Figure 5, with permission)

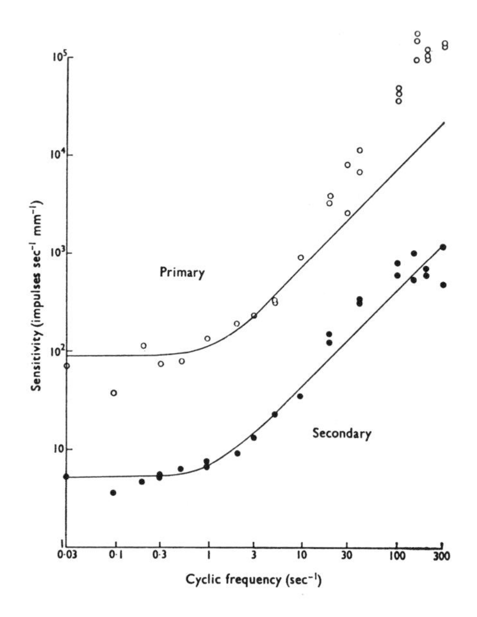

not appreciated before is that (at least in the linear range) it is some ten times smaller than the sensitivity of the primary ending (see shift of the curves in Figure 4.9); and (2) over their linear range, the primary endings are remarkably sensitive to static stretch. Like the velocity component, there is a ten-fold difference between primary and secondary response.

Such studies emphasise how differently endings can behave, particularly the primary ending, for stretches of small amplitude. The initial high sensitivity of the primary ending is believed to be due to the presence of cross-bridges between actin and myosin in the poles of the intrafusal fibre. These resist any extension with a large short-range stiffness, until some critical deformation when they break and reform. Up to this critical point, most of the stretch will therefore be taken up by the central, poorly striated, portion of the fibre, resulting in a larger proportional deformation of the primary ending than during larger displacements.

The functional significance of this behaviour is that the CNS may be informed of very small changes in muscle length. In the decerebrate cat, a change in length of the soleus muscle of only 50μm (that is, 0.1 per cent of the muscle length) may produce a change in discharge of some 20imp/s from an ending which gave only a 50imp/s discharge when the muscle was fully extended. On being given a large stretch, the sensitivity of the ending falls, but when the new ('operating') length is reached, the sensitivity is reset rapidly within a few seconds to a high value once again. This prevents saturation of the response, and effectively increases the working range of a highly sensitive device.

Short-range stiffness of the muscle spindle also has been invoked to explain the existence of the initial burst or 'acceleration' response of spindle primary endings. If a spindle is stretched to a constant length, cross-bridges detach and reattach at the new length over a period of several seconds. If fusimotor stimulation is given, this process is speeded up by increasing the rate at which cross-bridges recycle. When the cross-bridges have formed at the new length, their short-range stiffness changes the mechanical properties of the spindle. The initial portion of a stretch is taken up predominantly by the equatorial (sensory) region of the spindle and produces the initial burst of afferent activity. As stretch continues, the short-range bonds break, and the initial burst declines (Emonet-Dénand and Laporte, 1981; Hulliger, 1984; Matthews, 1981; Matthews and Stein, 1969; Morgan, Prochazka and Proske, 1984).

Muscle Receptors: II. Golgi Tendon Organs

Anatomy

Tendon organs were first described properly by Golgi in 1880, although attention had been focused upon them some five years previously by the discovery of tendon reflexes. They rarely lie in the tendon itself, but usually are found at the musculo-tendinous junction. The mammalian tendon organ is composed of a spindle-shaped connective tissue capsule enclosing a number of collagen strands. These strands are derived from the main tendon at one end, and at the other attach to about ten to 20 individual muscle fibres. Within the spindle sheath, the collagen strands are less densely packed than those outside and are divided into separate fascicles by transverse septae. The afferent nerve fibre is of large diameter (group Ib; 8–12μm; 70–120ms^{-1} in the cat) and breaks up after entering the capsule into a series of fine non-medullated branches which are closely applied to the collagen fascicles.

For many years, it was believed that when tension was applied to the organ the collagen fibres within would be straightened out and crowded together, leading to compression of the nerve terminals between them. Squeezing the nerve endings was therefore believed to be the active stimulus for the afferent nerve. More recently, recordings from individual tendon organs during the contraction of single motor units has cast doubt on this interpretation. The response of a Golgi tendon organ is *not* related to the amount of tension developed by a motor unit: small units may be able to produce as large a response as large units. Proske (1981) has suggested that the critical stimulus to a tendon organ is not compression, but *stretch* of the ending caused by elongation of collagen strands within the capsule. In this model, the size of the response elicited by a motor unit would depend on the size of the collagen strands to which its fibres were attached. If their cross-sectional area was small, then they could be extended a large amount by small motor units, and hence their discharge would be high.

Physiology of Golgi Tendon Organ Responses

As noted above, B.H.C. Matthews was the first to distinguish the responses of muscle spindles and tendon organs in recordings from single afferent fibres. In his hands, the tendon organs of the soleus muscle had a high threshold both to passively applied stretch or active muscle contraction. For the next 30 years, Golgi tendon organs were believed to be high threshold receptors with a function limited to signalling dangerously high muscle tensions.

Since then, that view has changed entirely, and tendon organs have become accepted as very sensitive muscle receptors with a more widespread role in movement control. Part of the reason for this change has been the finding that, particularly in certain muscles such as the cat soleus, much of the passively applied force on a muscle is not transmitted through the tendon organs but via the surrounding connective tissue. When this is dissected away, tendon organ sensitivity can be increased considerably. In fact, tendon organs are, despite Matthews's original assertion, very sensitive to actively generated muscle force. This is because in a contracting muscle, all the force produced by the muscle fibres is exerted through the tendon organs at the musculo-tendinous junctions.

As an example of the sensitivity of tendon organs to active force, Houk and Henneman (1967) recorded the responses of single tendon organs to tetanic stimulation of single α-motor fibres. They found that each individual organ could be excited by the contraction of just four to 15 motor units (Figure 4.10). Since only a maximum of 25 muscle fibres insert into each tendon, then each motor unit must have been capable of exciting the tendon organ with the contraction of only one or two of its single muscle fibres. Houk and Henneman (1967) estimated that if the contraction of a single fibre can excite a tendon organ, the threshold of sensitivity would be about 0.1g. In the same experiments, if the tendon organ had a resting discharge, then contraction of a small number of other units sometimes was found to silence the ending. Such an effect arises from the unloading of a tendon organ by a parallel contraction of neighbouring motor units that do not insert into the tendon organ. Later work by Binder, Kroin, Moore, *et al.* (1977) has suggested that Houk and Henneman (1967) underestimated the sensitivity of tendon organs. Tendon organs actually discharge their impulses during the rising phase of a motor unit twitch, and not at the peak of twitch tension. Their sensitivity to such dynamically changing forces is probably as low as 5–10mg.

Although the threshold stimulus for a tendon organ may be very low, its sensitivity as a device for measuring muscle tension (that is, in terms of the increase in firing rate per Newton increase in tension) is lower than one might

Figure 4.10: Sensitivity of Tendon Organ Discharge to the Contraction of Particular Groups of Motor Units Within a Muscle. The two sets of record are from a tendon organ in the soleus muscle of an anaesthetised cat. In each record the upper trace represents the force exerted by the muscle through its tendon and the lower trace the discharge of the tendon organ recorded in a dorsal root filament. The ventral roots were divided into filaments each containing motor axons to several motor units in the soleus. Stimulation of one filament (bottom record) elicits a vigorous tendon organ discharge even though the contraction was very small (18g). Stimulation of another filament produces a larger force (off scale), but the receptor, which initially was discharging due to positive stretch, drops below threshold throughout the contraction. This is due to unloading of the receptor by parallel contraction of neighbouring motorunits. Sweep duration is 1s (from Houk and Henneman, 1967; Figure 1, with permission)

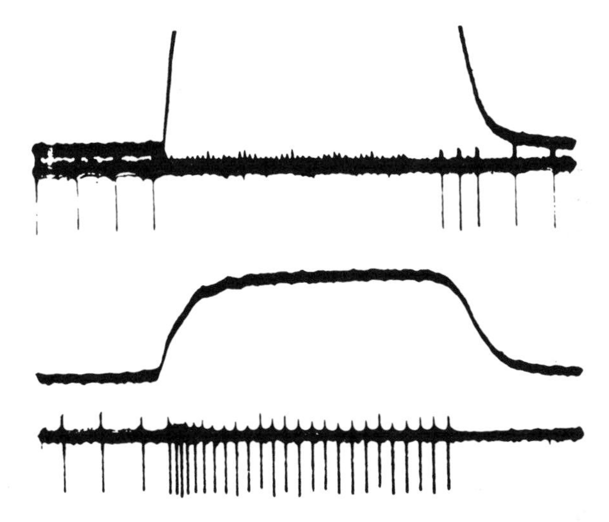

expect. There are two main reasons for this. The first, referred to above, is that, particularly in a stretched muscle, the total force recorded externally at the tendon is not the same as the force exerted by the active contraction of the muscle fibres: some of the total force is produced by stretch of connective tissue around the muscle. The second is that the firing of the tendon organ does not summate linearly to equal increments of force produced by the firing of different motor units. For example, if an individual tendon organ were caused to fire at 5imp/s by the separate tetanic contraction of two different motor units, then it would fire at less than 10imp/s if both units were activated simultaneously.

Non-linear summation of responses has certain implications for models of receptor transduction. It may mean that the generator potentials produced by two separate twitches do not summate linearly. Alternatively, it may be a consequence of the microanatomy of the receptor. If, for example, each muscle fibre inserted onto a single collagen strand within the tendon organ, and each strand was innervated by branches of the same sensory axon, then stretch of any one strand would activate the afferent fibre. If each sensory terminal was capable of producing not only a generator potential, but also an action potential,

Figure 4.11: Static Sensitivity of a Single Tendon Organ to Development of Active Force by a Whole Muscle. Data are from the soleus muscle of a decerebrate cat, in which the crossed extensor reflex was used to provoke a natural contraction of soleus. The points represent the average discharge rate over the last half of a 1s period of maintained force. Except for the rapid rise in discharge at low force levels (shown by squares), the receptor follows the total muscle force very well (from Crago, Houk and Rymer, 1982; Figure 3A, with permission)

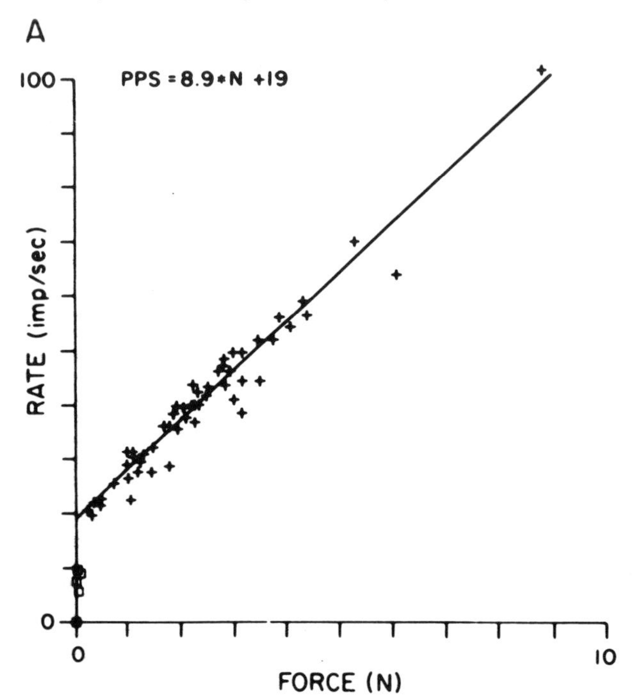

then the terminal that was firing at the highest frequency would predominate. Its spikes would antidromically invade the terminals of other, more slowly firing branches of the parent axon, and prevent summation of responses. The fact that combined stimulation of two or more motor units gives a larger response than stimulation of any of the individual motor units on their own is explained by cross-linkages between receptor strands, so that a single fibre, although mainly exerting tension through a single strand, also exerts a small fraction of its force through other poorly linked adjacent strands.

The type of stepwise motor unit recruitment necessary to demonstrate non-linear behaviour of Golgi tendon organs is not seen during normal contraction of muscle. Although one unit is recruited first, it usually does not reach its maximum tetanic tension before the next unit begins to fire. Its firing rate continues to increase, as does its tension, during the recruitment of higher threshold units. Under these circumstances, during normal whole muscle (isometric and isotonic) contractions, the most recent work suggests that tendon organs sample whole muscle force very well, with a sensitivity of 3.4 to

13imp/s/N (Figure 4.11). The sensitivity is fairly linear over a wide range, except for a small region of high sensitivity at low force levels, when only one or two of its muscle fibres are active (see Crago, Houk and Rymer, 1982; Houk and Henneman, 1967; Proske, 1981; for further details and references on tendon organs).

Muscle Receptors: III. Other Types of Ending

Two other types of nerve ending found in muscle deserve mention. These are Paciniform corpuscles and free nerve endings. Little is known about the physiological responses or the central effects of either type of ending, although they form an appreciable proportion of the total sensory innervation of the muscle. There are twice as many free nerve endings in muscle as in any other type of sensory receptor, and the number of Paciniform corpuscles may be anything up to 30 per cent of the number of muscle spindles.

Paciniform corpuscles are similar in structure, although rather smaller than the Pacinian corpuscles found in the skin. They are most frequently observed at the musculo-tendinous junction in the vicinity of tendon organs and are supplied by a large diameter (group II) medullated fibre which may innervate several separate corpuscles. Very little work seems to have been performed specifically on the Paciniform corpuscles in muscle. It is generally accepted that they are rapidly adapting end organs with a sensitivity to high frequency vibration like true Pacinian corpuscles in skin. Their central connections have not been studied.

Free nerve endings are found in close association with almost every structure in muscle (spindles, tendon organs, muscle fibres, fascia, fat and blood vessels, excluding capillaries). It is not known whether their function depends on their location, but the fact that a single fibre may innervate receptors in several different regions argues against this possibility. All non-medullated fibres and the majority of small (less than 5μm) group III medullated axons terminate as free nerve endings. Recording from slowly conducting muscle afferents (less than 1m/s for non-medullated, less than 24m/s for group III medullated) has shown that these endings are rarely excited by classical proprioceptive stimuli such as muscle stretch or contraction or muscle vibration. Instead, they are principally activated by high threshold mechanical stimulation of the muscle, such as pinching or pricking, and have been termed 'pressure-pain' receptors. In addition, they also respond to other nociceptive stimuli like ischaemia and injection of hypertonic saline. It is possible, however, that some of these fine endings may be more specifically responsive to humoral or metabolic stimuli and hence play a role in mediating local cardiovascular or respiratory reflexes.

Joint Receptors

Anatomy

Three main types of ending are associated with the synovial joints of the body: (1) free nerve endings, which are the most numerous type of joint receptor, are found throughout the connective tissue; (2) Golgi endings similar to tendon organs are found in the joint ligaments; and (3) Ruffini endings (see below) are found in the joint capsule. Paciniform-like corpuscles also have been described in the joint capsule: there are no nerve endings on the cartilaginous surfaces of the joint or in the synovial membranes. Golgi endings are innervated by large diameter (10–17μm) group I medullated fibres; Ruffini endings by slightly smaller (5–10μm) group II fibres and free endings by group III or non-medullated nerves.

The Ruffini ending is similar to the same ending found in the skin. It consists of a small capsule which encloses a number of spray arborisations from a single afferent fibre (Figure 4.12).

Physiology

Initial experiments on the responses of joint receptors to passive movement of the joint were performed in the 1950s. The results were quite promising. When single group I-II afferent fibres were recorded from the articular nerves of the cat knee joint, it appeared that individual fibres had a distinct 'turning curve'. That is, the response of a fibre was limited to a small range of joint angles, with

Figure 4.12: Diagram of Location and Type of Receptors Commonly Found in Joints. Type I endings are Ruffini endings; type II, Paciniform endings; type III, Golgi endings; type IV, free nerve endings (from Brodal, 1981; Figure 2.2, with permission)

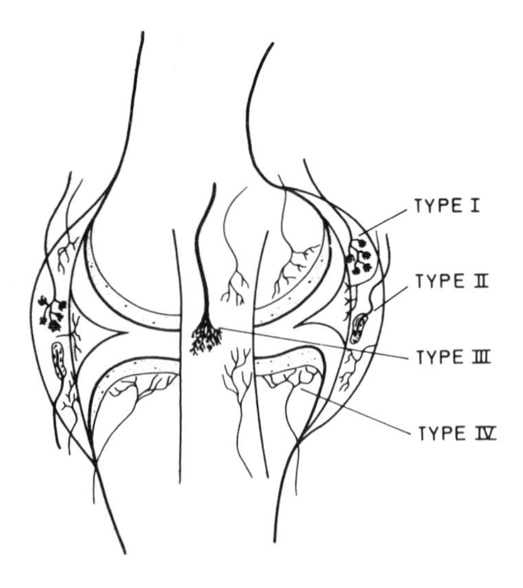

Figure 4.13: Adapted Firing Frequencies of a Number of Knee Joint Afferents at Different Positions of the Knee. Points joined by lines are from a single receptor (solid lines from one cat, dotted lines from another cat). Other experiments in the same series showed a predominance of fibres which fired at the two extremes of joint position rather than in the mid-range (from Skoglund, 1956; Figure 19, with permission)

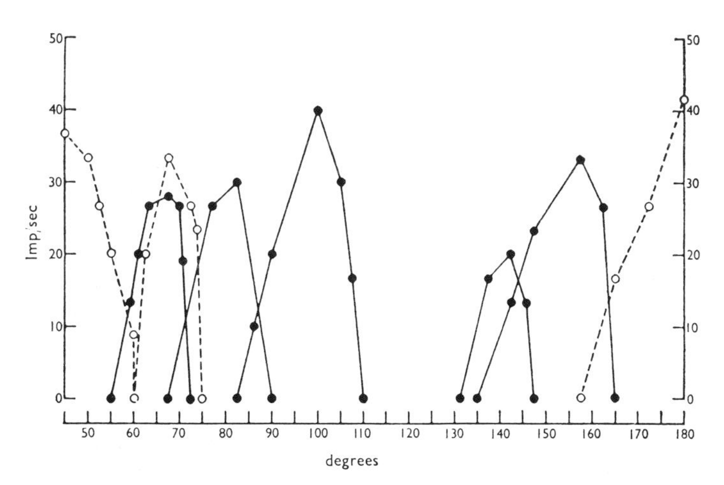

maximum firing occurring at a particular point within that range (Figure 4.13). The discharge rate was very slow to adapt and hence this population of receptors was thought to be particularly suited to provide information to the CNS about static joint position.

However, this view has been challenged. In a re-investigation of knee joint receptor sensitivity, Clark and Burgess showed that only a small proportion of the joint afferents actually had responses like those described above. The majority had no response to joint angles in the mid range, but only fired at the extremes of joint rotation. Moreover, many of these receptors did not even distinguish between extremes of flexion and extension, so that it became difficult to see what role, if any, they could play in signalling static joint position. Later work has extended these findings to other joints with the same result. The generally accepted conclusion at present is that joint receptors with group I-II afferent fibres (that is, Golgi and Ruffini endings) are responsive only to deformation of the joint capsule or ligaments. At joint positions where no stress is placed on the capsule, the receptors remain virtually silent. The temptation therefore, has been to conclude that joint receptors serve only to indicate extremes of joint rotation, presumably with some protective function. However, in the living animal muscle, activation may place different stresses on the joint capsule and ligaments and change the firing pattern of joint afferents from that seen in the experimental preparation. Thus, the precise role of these receptors is still subject to much discussion.

The firing pattern of non-medullated and group III joint afferents has not been investigated in such detail. In the cat knee joint, many of the fibres are activated by movements within the normal limits, although a large proportion only fire in response to noxious mechanical stimuli. The firing of such afferents may be a 'warning' signal that the joint is about to leave its normal working range (see Clark and Burgess, 1975; Schiable and Schmidt, 1983; for further work on joint receptors).

Cutaneous Mechanoreceptors

Three types of receptor are found in the skin: thermoreceptors, nociceptors and mechanoreceptors. Much of the interest in the study of these receptors obviously lies in the field of sensory physiology and, particularly with reference to man, in the psychophysics of sensation and sensory discrimination. Nevertheless, it is quite clear that mechanoreceptors also play a very important role in control of movement. This is particularly true of the densely innervated regions of the hand and foot. For example, signals from the pressure sensors in the sole of the foot are used in balance. Peripheral neuropathies which cause loss of sensation in the extremities may be accompanied by Romberg's sign: patients are unable to stand unassisted with their feet together when they close their eyes. Although the neuropathy may involve muscle receptor input from the intrinsic muscles of the foot, it is believed that the major contribution to pressure sensation comes from cutaneous receptors. Without information on the differential distribution of pressure on the soles of the feet, the reflexes from the intact vestibular system of these patients are unable to maintain balance alone. In the same way, any complex manipulation of objects in the hand is grossly impaired by cutaneous anaesthesia: writing with a pen, or picking up coins from a flat surface or doing up the buttons of one's shirt, all become difficult, if not impossible, even when vision of the hand and object is allowed. Visual input cannot substitute for the loss of cutaneous mechanoreceptor signals.

Despite the importance of cutaneous input, it is still not clear precisely how these signals are used in the control of movement. In this section some of the anatomical features and physiological properties of the four main types of mechanoreceptor will be detailed. Much of the work has been done on the receptors of the monkey hand: recent work on the physiological responses of human cutaneous mechanoreceptors will be summarised in the next section.

Of the four types of receptor, two are found close to the epidermal–dermal junction within a few hundred micrometres from the surface of the body. These are the Merkel discs and the Meissner corpuscles. In the glabrous skin of the human palm, the Merkel discs are found at the base of the epidermal infoldings in the dermis, whereas the Meissner corpuscles are found at the tip of the dermal protrusions. In the monkey, Merkel discs are located just underneath dome-like projections in the stretched skin known as dome corpuscles. The two other types of receptor, Ruffini endings and Pacinian corpuscles, are found in the deeper layers of the skin (Figure 4.14).

Figure 4.14: Location of the Four Main Types of Cutaneous Mechanoreceptors in the Glabrous Skin of the Hand. Ruffini endings are in the dermis; Pacinian corpuscles are located deeper, in the subcutaneous tissues. Merkel discs and Meissner corpuscles are found nearer the surface at the dermal–epidermal junction (from Brodal, 1981; Figure 2.1A, with permission)

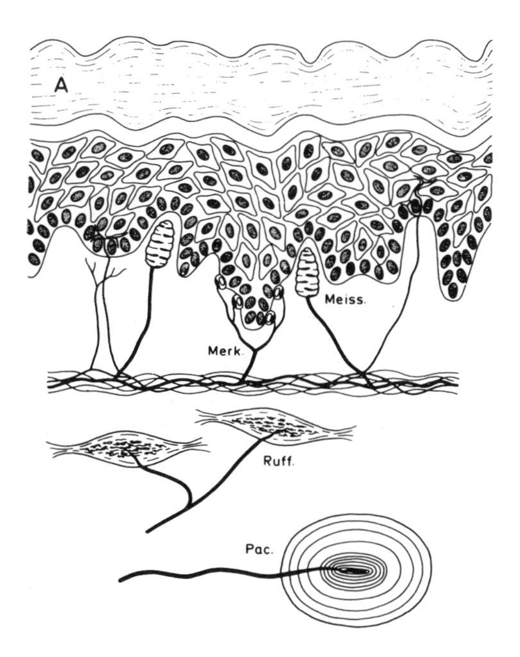

Merkel discs are concave, flattened, disc-like formations of cells within the stratum granulosum of the epidermis. They do not contain keratin and hence are differentiated from the rest of the epidermal cells. A single nerve axon innervates the whole group of cells. It emerges from the subdermal plexus and loses its myelin sheath at the dermal—epidermal junction, branching many times to each Merkel disc. In the monkey, records from single axons show that at rest there is little or no discharge. The receptor is sensitive only to localised vertical pressure on the surface of the 'touch dome', and does not respond to any lateral stretch of the skin. The response to a maintained indentation has a dynamic and static phase. In the dynamic phase, very high rates of discharge may be reached, but these slow over about half a second or so to reach their final steady rate, which may be constant for up to 10 minutes. It is not known to what extent the initial adaptation is due to tissue movement under the stimulus probe, or whether this represents a fundamental receptor property.

Meissner corpuscles are ovoid and are found with their long axis oriented perpendicular to the surface of the skin within a dermal papillum. From two to six separate axons innervate each Meissner corpuscle and form a complex network within the structure. Each axon may innervate more than one corpuscle. Krause end bulbs, found in the lips, tongue, conjunctiva and so on, are similar in structure and are thought to be analogous with the Meissner corpuscles of glabrous skin. The receptive field of single axons is small, and the discharge to maintained pressure adapts extremely rapidly. The discharge lasts only a few seconds.

Ruffini endings are found within the dermal layer, and consist of nerve terminals from a single axon intimately associated with collagen fibrils which merge with dermal collagen. The whole structure is surrounded by a fluid-filled space which is enclosed by a thinly lamellated capsule. These receptors are responsive to stretch of skin over a wide area (up to $25cm^2$), and have a slowly adapting response to continuous stimulation. Stretch along the long axis of the structure stimulates, and stretch at right angles inhibits, the discharge. This gives the receptor some directional sensitivity, which is usually only ascribed to joint receptors. However, Ruffini endings in the skin also seem likely candidates to signal some aspects of limb position to the nervous system via their response to skin stretch.

Pacinian corpuscles are found wtihin the subdermal fasciae. They are the largest receptors in the skin, and may be 1–4mm in size. A single axon terminates within a large number of concentric cytoplasmic lamellae which are separated by fluid-filled spaces. The bare nerve terminal is sensitive to the mechanical deformation which it receives through the 'onion skin' capsule. This filters any slow frequency components of the signal, leaving the Pacinian corpuscle sensitive only to very rapidly changing stimuli. Such rapidly changing pressure stimuli frequently travel in a wave of vibration through the tissues of the skin and the bones, and are picked up by the Pacinian corpuscle from over a wide area.

Recordings from Human Afferent Nerve Fibres

One of the most important advances made in technique over the past 20 years has been the ability to record single unit activity from nerve fibres in awake, co-operative human subjects. The technique, which was pioneered by Vallbo and Hagbarth (1968) in Sweden involves the production of fine tungsten microelectrodes with a tip diameter of 1–15μm which can be inserted through the skin and into underlying nerve trunks. The electrodes are insulated, apart from a length of 10–15μm at the tip. Usually, the procedure is not painful, although paraesthesias are elicited when manipulating the electrode. Permanent nerve damage is extremely rare.

Activity has been recorded from all classes of myelinated fibres, both afferent and efferent, except for γ-fibres to muscle spindles. The largest potentials are seen with activity in the largest afferents: the Aα class from skin receptors and group Ia spindle afferents. Vallbo has suggested that the shape and polarity of the potentials from these large fibres indicates that the impulse is recorded through a low-impedance pathway between the electrode tip and the inside of the fibre, and hence that the fibre has been impaled by the electrode. The potentials are believed to be recorded extracellularly with smaller fibres.

The major problem with the technique is the stability of recording. Single units may only be held for 15 minutes or so before the electrode moves within the nerve and the unit is lost. Set against this disadvantage is the ability of the human subject to co-operate with the experimenter, either by reporting sensations or by producing voluntary muscle movements. In the former case, for example, it has been possible to show that at the limit of detection a human subject can distinguish the firing of a single impulse in an afferent fibre, caused by minute mechanical stimuli at the finger tips. This section will outline only some of the experiments relating to the discharges of proprioceptive sensory endings in muscle and skin.

Muscle Spindles

Much of the work performed on human muscle spindles has been devoted to analysing their discharge during movement and in pathological disease states in man. This will be summarised in the next chapter. However, there are several features of their passive behaviour which will be mentioned here.

Because the activity in large nerve fibres tends to be picked up more easily by microelectrodes, most recordings have been made from group 1a axons. The responses of what have been presumed to be group II secondary endings are not so well documented. In general, human spindle endings have been shown to behave very similarly to those in the cat. In a subject at rest, the primary endings have an impressive sensitivity to small dynamic changes in muscle length. They may discharge even in response to the arterial pulse or to respiratory movements. However, their static position sensitivity is, in contrast, extremely low. In the finger–wrist flexor muscles, only 10 per cent of spindles show any discharge at all at a comfortable rest position of the hand. During imposed static wrist movements, they only change their firing rate at the order of 1imp/s/mm increase in muscle length. Such behaviour is very similar to the behaviour of de-efferented spindle endings in the cat. Without any γ-bias, any resting discharge disappears and length sensitivity, in the hip and knee extensor muscles, becomes about 4imp/s/mm extension (as opposed to about 10imp/s/mm in the decerebrate cat).

At first sight, 4imp/s/mm in the cat still seems higher than 1imp/s/mm in man. However, if the human figure is scaled up by a factor of four, to allow for the length difference between the cat extensor muscles and human forearm muscles, then the sensitivities are about the same. The rationale behind this is that the same external stretch applied to muscles of different sizes causes smaller internal length changes per unit length in larger muscles. Thus, much of the stretch applied to a large muscle will be taken up by the muscle fibres in series with a spindle rather than being transmitted directly to the spindle as in a small muscle. The main conclusion of such observations is that in man, at rest, there is no background fusimotor activity to muscle spindles.

A difference between the behaviour of primary and secondary spindle endings in cat and in man has been found. In the decerebrate cat, vibration at about 200Hz is known to be a selective stimulus for the primary spindle endings. Secondary endings are insensitive to such a rapidly changing stimulus. However, in the small number of secondary endings which have been recorded so far in man, such a dramatic difference does not seem to occur. Primary endings can follow vibration, giving one impulse per cycle at frequencies up to 200Hz (they may discharge at frequencies of up to 800Hz during muscle stretch). Secondaries only follow vibration up to 100Hz, but thereafter are still excited potently by vibration. The reason for this difference is not known. It may be due in part to the methods used to vibrate cat spindles *in vitro* and human spindles *in vivo*. In cat experiments, vibration is applied longitudinally to the muscle or isolated spindle, whereas in man, vibration is usually applied transversely across the tendon or muscle belly. In this case, vibration may produce a less potent mechanical displacement in man than in the cat.

Golgi Tendon Organs

As with the muscle spindle, human observations are in accord with previous findings in the cat. Tendon organs are distinguished from spindle endings by their difference in response to active muscle contraction. In response to a tendon tap, they fire during the active phase of contraction, whereas spindles fire in response to the initial stretch. During electrically induced muscle twitches, they respond during the twitch, in the period when spindles are usually silent. Their threshold to active contractions is very low and they appear to monitor the active tension in muscle. As in the cat, there is evidence to suggest that they may sample preferentially the contraction of particular motor units in series with them. During a steadily rising isometric contraction there may sometimes be discontinuities in the relationship between instantaneous impulse frequency versus muscle force attributable to the recruitment of additional motor units into the contraction.

Cutaneous Mechanoreceptors

Corresponding to the four anatomical types of receptor, four physiological types of afferent response can be distinguished in recordings from cutaneous nerves innervating the glabrous skin of the hand. There still is some uncertainty as to the exact correspondence between physiology and anatomy in human studies. In general, the assumptions have been based on parallel detailed studies of identified end organs in animal experiments.

The four types of response are subdivided on the basis of their rate of adaption to novel, sustained, stimuli and the size of the receptive field (Table 4.1). RA (rapidly adapting) units and SAI (slowly adapting; type I) units have small receptive fields of about 13mm^2 area. There are several points of maximal sensitivity within the field, suggesting that the axon terminates with a number of end organs spread within a small area of skin. These two types of receptors are very sensitive to mechanical stimuli which cause indentation of the skin: RA units have a threshold of about 10μm, whereas SAI units have a threshold of about 50μm. They are thought to correspond to Meissner corpuscles (RA units) and Merkel discs (SAI units). Each RA unit fires impulses only in response to rapidly changing or moving stimuli and can follow sinusoidal displacements up to about 100Hz. Such stimuli give rise to a localised sensation of 'flutter' in man. SAI units have a maintained discharge to constant pressure. Both types of unit are extremely dense on the tips of the fingers, declining substantially towards the proximal part of the palm.

Table 4.1: Receptive Field Size and Adaptation Rates of Cutaneous Mechanoreceptors

		Adaptation Rate	
		Fast	Slow
Receptive Field	Small	RA	SAI
Size	Large	PC	SAII

PC (Pacinian corpuscles) and SAII units (slowly adapting; type II) respond to stimuli applied over a much wider area, which can be almost as large as the whole palm in some cases. However, within this field there is only a single point of maximal sensitivity, suggesting a single terminal end organ. The SAII units are believed to correspond to Ruffini endings. They respond particularly well to stretch of the skin and have a directional sensitivity, increasing their discharge to stretch in one direction and decreasing (or not responding) to stretch in the opposite direction. The end organ is less sensitive to direct mechanical indentation of skin (threshold of about 250μm). Its response to stretch may be an important factor in joint position sense in view of the deformation produced in skin by movement at joints and by contraction of underlying muscles. The PC units are exquisitely sensitive to rapidly changing stimuli (threshold skin indentation of about 10μm), particularly high frequency vibration at 300–400Hz. A given unit may be driven to respond, for example, by a single tap with a pencil applied anywhere on the skin of the hand and fingers. Both PC and SAII units are less common in the glabrous skin of the palm than the RA and SAI units. They are distributed with a fairly even density over the whole area (Vallbo and Hagbarth, 1968; Vallbo *et al.*, 1979).

Contribution of Muscle Afferents to Sensation

The role of spindle and Golgi tendon organ afferents in sensation has been debated for many years. *A priori* it would seem most likely that spindles might contribute to one's sense of static limb position or joint movement (kinaesthesia), and that tendon organs would contribute to one's sense of the force of muscular contraction. However, conventional physiology has long held that muscles are insentient. An important experiment that appeared to confirm this was performed by Gelfan and Carter in 1967.

During routine operations at the wrist or ankle under local anaesthetic, these investigators pulled on the long tendons of the digits in such a way as to cause stretch of muscle, without any distal movements occurring. They found that, in all cases, the patients did not experience any sensation referrable to the muscle and suggested that no conscious sensation arose from activity in afferent fibres from muscle. Such observations seemed irrefutable at the time, but over the next few years evidence from several experiments began to accumulate and point to the fact that the muscle receptors might, on the contrary, produce conscious sensation.

Joint position sense and sense of movement up until the late 1960s were usually attributed to activity of joint receptors within the joint capsule. As noted above, this view had to be changed considerably in the light of experiments performed by Burgess and Clark. They showed that most joint receptors did not fire over the whole range of joint movement. On the contrary, most joint receptors respond preferentially to extremes of joint rotation, with many of them not distinguishing between extreme flexion or extension. Such receptor characteristics were clearly unsuitable to mediate joint position sense. In addition, it was shown that the position sense of patients with total hip joint replacement was almost as good as normals whose joint receptors were, presumably, intact. In the absence of accurate input from joint receptors, it was assumed that cutaneous or spindle afferent inputs played the most important role in position sense.

In the early 1970s, P.B.C. Matthews and his colleagues (see Goodwin, McCloskey and Matthews, 1972) conducted some psychophysical experiments which suggested very strongly that muscle spindle inputs could give rise to conscious sensation in man. Blindfolded subjects were seated at a table with their elbows resting and forearms vertical. One forearm was moved passively by the experimenter and the subjects were asked to indicate the position of that arm by matching it with their other, free arm (Figure 4.15).

This is performed very accurately under normal circumstances. However, if vibration was applied over the belly or tendon of biceps while the arm was being moved by the experimenter, subjects would consistently over-estimate the angle of extension at the elbow. That is, they perceived the vibrated arm to be further extended than it actually was. Since vibration is known to be a powerful stimulus for spindle receptors, it was concluded that the increased spindle input from biceps during vibration was interpreted by the CNS as being due to increased stretch of the biceps muscle. This led the subjects to believe that the elbow was extended.

Therefore, the weight of evidence was clearly becoming incompatible with the classical view that muscles are insentient. The tendon-pulling experiments were then repeated by Matthews and Simmons (1974). This time, patients reported that pulling the flexor tendons of the fingers at the wrist so as to stretch the muscle, produced a sensation of movements in the digit to which the tendon

Figure 4.15: Effect of Muscle Vibration on the Sense of Joint Position at the Elbow. Top: posed picture to illustrate the magnitude of the difference in position of the vibrated and tracking arm that can occur while the subject believes that he is managing to keep them aligned. The scale is marked in 10° divisions. The right arm was vibrated to produce a tonic vibration reflex in biceps and flexion at the elbow. With eyes closed, the subject was required to match the position of arm by moving his left arm. He perceives the right elbow to be more extended than it is.
Bottom: Experiment in Which One Arm was Moved Passively by the Experimenter While the Subject Tracked its Position With the Other Arm. Shortly after tracking began, vibration was applied to the tracking arm. This had the effect that the subject underestimated the degree of extension of the moved arm. That is, he perceived the tracking arm to be more extended than it actually was. On turning off the vibrator, the subject matched his elbow angles accurately once again. The process was repeated while the experimenter moved the passive arm in the opposite direction (from Goodwin, McCloskey and Matthews, 1972; Figures 2 and 8, with permission)

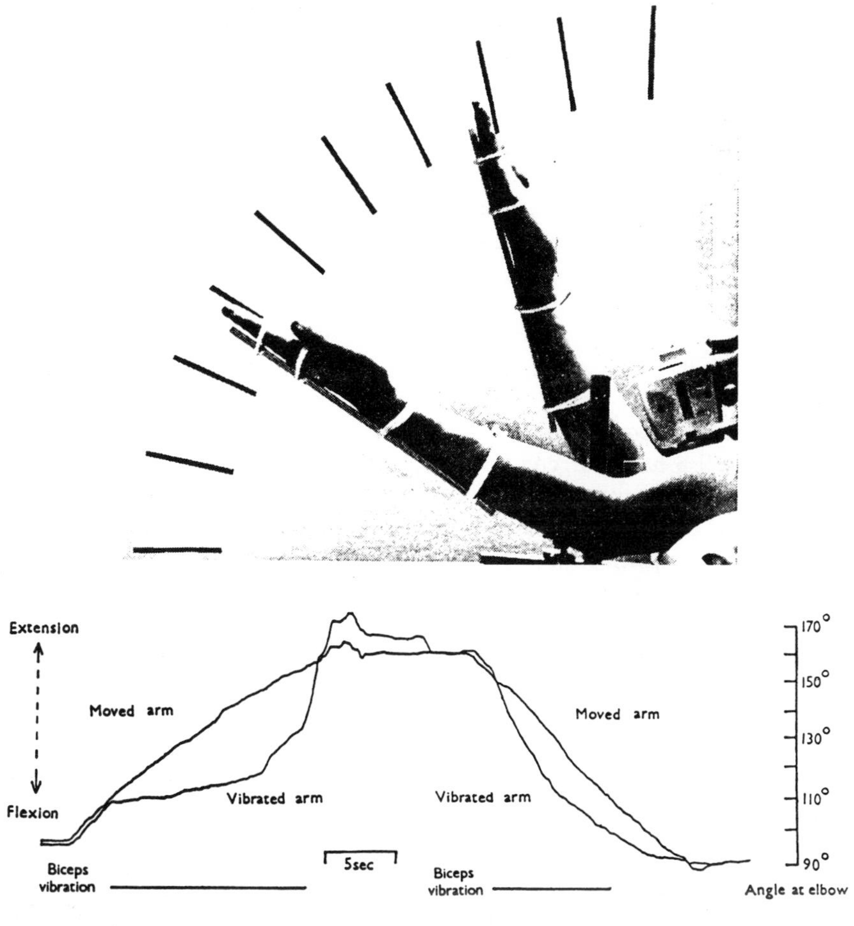

was attached. The experiments have been repeated by McCloskey, Cross, Honner *et al.* (1983) with the same conclusion. McCloskey is one of the firmest advocates of muscle sensibility; he even underwent local surgery at the ankle to be able to report for himself, under controlled laboratory conditions, the sensations of the movement which accompany pulls on the tendon of extensor hallucis longus. The evidence is now quite strongly in favour of a muscular sense, probably contributing to one's sense of joint position, which arises from muscle spindle receptors.

There are two caveats to this conclusion. One is that the relationship between joint position and spindle discharge varies with the state of contraction of the muscle under study. At rest, there is a linear relationship between spindle discharge and joint position, which could subserve a sense of joint position during passive movements. However, this relationship no longer holds during voluntary contractions. During mild isotonic muscle contractions, spindle discharge, as recorded in microneurographic experiments in man, hardly varies as a function of joint angle. This is because the α-γ co-activation in voluntary movement produces spindle shortening which maintains spindle discharge constant.

However, we still know where our joints are during voluntary contraction. Joint position could perhaps be extracted from spindle input, but it would mean that the CNS would need access to a copy of the fusimotor command sent to the spindles so that the result of active spindle shortening could be subtracted from the total spindle input to give the true shortening of the muscle. That this may be possible is shown by the fact that the vibratory illusions are absent in strongly contracting muscles and are less during moderate contraction. It is believed that spindle firing is still entrained by vibration in these conditions as in relaxed muscle. If so, then the CNS produces an illusion 'appropriate' to the *excess* level of spindle activity.

The other problem of muscle sense has been raised by Moberg (1983), a well-known Swedish hand surgeon. His argument is that, in view of the extreme sensitivity of some cutaneous afferents, not enough weight has been given to the possibility of cutaneous contributions to joint position sense. In a critique of the tendon-pulling experiments, he points out that in his own experience, if the skin overlying the muscle bellies is completely anaesthetised, then *no* sensation arises from stretch of muscle in conscious man. The argument still goes on!

A final question is whether tendon organ input may contribute to conscious sensation. Little work has been performed specifically on this since the illusions above may have involved discharges from both spindle and tendon organs. There is one experiment which suggests that it may: during matching of forces exerted, say, by the arms on each side of the body, it seems that the predominant factor is a matching of the size of the efferent motor command (sense of effort).

For example, when subjects match forces while one arm is being vibrated, they usually underestimate, since the vibrated arm is producing extra, reflex, force by way of the tonic vibration reflex. They are assumed to be matching the central command, rather than the force exerted. However, if instructed carefully to match the force applied, rather than the subjective 'effort', some subjects perform accurately even when the skin is anaesthetised under the point of application of the force. This may be due to an ability to monitor true muscle force via the tendon organs (see Goodwin *et al.*, 1972; Matthews, 1977; McCloskey, 1981; Vallbo, Olsson, Westberg, *et al.*, 1984; for further information on work covered).

References and Further Reading

Review Articles and Books

Barnes, W.J.P. and Gladden, M.H. (eds.) (1985) *Feedback and Motor Control in Invertebrates and Vertebrates,* Croom Helm, London
Boyd, I.A. (1980) 'The Isolated Mammalian Muscle Spindle', *TINS 3,* pp. 258–65
Boyd, I.A. (ed.) (1985) *The Muscle Spindle,* Macmillan, London
Brodal, A. (1981) *Neurological Anatomy in Relation to Clinical Medicine,* Oxford University Press, Oxford
Hulliger, M. (1984) 'The Mammalian Muscle Spindle and Its Central Control', *Rev. Physiol. Biochem, Pharmacol., 101,* pp. 1–110
Matthews, P.B.C. (1972) *Mammalian Muscle Receptors and Their Central Actions,* Arnold, London
——— (1977) 'Muscle Afferents and Kinaesthesia', *Br. Med. Bull., 33,* pp. 137–42
——— (1981) 'Review Lecture: Evolving Views on the Internal Operation and Functional Role of the Muscle Spindle', *J. Physiol., 320,* pp. 1–30
McCloskey, D.I. (1981) 'Corollary Discharges: Motor Commands and Perception' in V.B. Brooks (ed.), *Handbook of Physiology,* Sect. 1, Vol. II, part 2, Williams and Wilkins, Baltimore, pp. 1415–48
Proske, U. (1981) 'The Golgi Tendon Organ: Properties of the Receptor and Reflex Action of Impulses Arising from Tendon Organs' in R. Porter (ed.), *Int. Rev. Physiol., Neurophysiol., IV,* University Park Press, Baltimore, pp. 127–72
Vallbo, A.B., Hagbarth, K.-E., Torebjork, H.E. *et al.* (1979) 'Somatosensory, Proprioceptive and Sympathetic Activity in Human Peripheral Nerves', *Physiol. Rev., 59,* pp. 919–57

Original Papers

Banks, R.W., Barker, D., Bessou, P., *et al.* (1978) 'Histological Analysis of Muscle Spindles Following Direct Observation of Effects of Stimulating Dynamic and Static Motor Axons,' *J. Physiol., 283,* pp. 605–19
Binder, M.D., Kroin, J.S., Moore, G.P., *et al.* (1977) 'The Response of Golgi Tendon Organs to Single Motor Unit Contractions', *J. Physiol., 271,* pp. 337–49
Boyd, I.A. (1962) 'The Structure and Innervation of the Nuclear Chain Muscle Fibre System in Mammalian Muscle Spindles, *Phil. Trans. Roy. Soc. B., 245,* pp. 81–136
——— (1976) 'The Mechanical Properties of Dynamic Nuclear Bag Fibres, Static Nuclear Bag Fibres and Nuclear Chain Fibres in Isolated Cat Muscle Spindles', *Prog. Brain Res., 44,* pp. 33–50
——— Gladden, M.H. and Ward, J. (1981) 'The Contribution of Mechanical Events in the Dynamic Bag_1 Intrafusal Fibre in Isolated Cat Muscle Spindles to the Form of the Ia Afferent Axon Discharge', *J. Physiol., 317,* pp. 80–81P
Boyd, I.A. and Ward, J. (1975) 'Motor Control of Nuclear Bag and Nuclear Chain Intrafusal Fibres in Isolated Living Muscle Spindles from the Cat', *J. Physiol., 224,* pp. 83–112
Brown, M.C., Crowe, A. and Matthews, P.B.C. (1965) 'Observations on the Fusimotor Fibres of the Tibialis Posterior Muscle of the Cat', *J. Physiol., 177,* pp. 140–59
Clark, F.J. and Burgess, P.R. (1975) 'Slowly Adapting Receptors in Cat Knee Joint: Can They Signal Joint Angle?' *J. Neurophysiol., 38,* pp. 1448–63
Crago, P.E., Houk, J.C. and Rymer, W.Z. (1982) 'Sampling of Total Muscle Force by Tendon Organs', *J. Neurophysiol., 47,* pp. 1069–83
Crowe, A. and Matthews, P.B.C. (1964) 'The Effects of Simulation of Static and Dynamic Fusimotor Fibres on the Response to Stretching of the Primary Endings of Muscle Spindles',

J. Physiol., *174*, pp. 109–31 (see also *175*, pp. 132–51)
Gelfan, S. and Carter, S. (1967) 'Muscle Sense in Man', *Exp. Neurol.*, *18*, pp. 469–73
Goodwin, G.M., McCloskey, D.J. and Matthews, P.B.C. (1972) 'The Contribution of Muscle Afferents to Kinaesthesia Shown by Vibration-induced Illusions of Movement and by the Effects of Paralysing Joint Afferents', *Brain*, *95*, pp. 705–48
Jansen, J.K.S. and Matthews, P.B.C. (1962) 'The Effects of Fusimotor Activity on the Static Responsiveness of Primary and Secondary Endings of Muscle Spindles in the Decerebrate Cat', *Acta Physiol. Scand.*, *55*, pp. 376–86
Houk, J.C. and Henneman, E. (1967) 'Responses of Golgi Tendon Organs to Active Contractions of the Soleus Muscle of the Cat', *J. Neurophysiol.*, *30*, pp. 466–81
Matthews, B.H.C. (1933) 'Nerve Endings in Mammalian Muscle', *J. Physiol.*, *78*, pp. 1–53
Mathews, P.B.C. (1964) 'Muscle Spindles and Their Motor Control', *Physiol. Rev.*, *44*, pp. 219–88
——— and Simmons (1974) 'Sensations of Finger Movement Elicited by Pulling Upon Flexor Tendons in Man', *J. Physiol.*, *239*, pp. 27–28P
——— and Stein, R.B. (1969) 'The Sensitivity of Muscle Spindle Afferents to Small Sinusoidal Changes of Length', *J. Physiol.*, 200, pp. 723–43
Ruffini, A. (1898) 'On the Minute Anatomy of the Neuromuscular Spindles of the Cat, and on their Physiological Significance', *J. Physiol.*, *23*, pp. 190–208
Skoglund, S. (1956) 'Anatomical and Physiological Studies of Knee Joint Innervation in the Cat', *Acta Physiol. Scand.*, *36*, suppl. 124, pp. 1–101
Vallbo, A.B. and Hagbarth, K.-E. (1968) 'Activity from Skin Mechanoreceptors Recorded Percutaneously in Awake Human Subjects, *Exp. Neurol.*, *21*, pp. 270–89

Recent Papers

Emonet-Denand, F. and Laporte, Y. (1981) 'Muscle Stretch as a Way of Detecting Brief Activation of Bag_1 Fibres by Dynamic Axons' in A. Taylor and A. Prochazka (eds.), *Muscle Receptors and Movement*, Macmillan, London
McCloskey, D.I., Cross, M.J., Honner, R. *et al.* (1983) 'Sensory Effects of Pulling or Vibrating Exposed Tendons in Man', *Brain*, *106*, pp. 21–37
Moberg, E. (1983) 'The Role of Cutaneous Afferents in Position Sense, Kinaesthesia and Motor Function of the Hand', *Brain*, *106*, pp. 1–19
Morgan, D.L., Prochazka, A. and Proske, U. (1984) 'The After-Effects of Stretch and Fusimotor Stimulation on the Responses of Primary Endings of Cat Muscle Spindles', *J. Physiol.*, *356*, pp. 465–77
Schaible, H.-G. and Schmidt, R.E. (1983) 'Responses of Fine Medial Articular Nerve Afferents to Passive Movements of Knee Joint', *J. Neurophysiol.*, *49*, pp. 1118–26
Vallbo, A.B., Olsson, K.A., Westberg, K.-G., *et al.* (1984) 'Microstimulation of Single Tactile Afferents from the Human Hand', *Brain*, *107*, pp. 727–49

5 REFLEX PATHWAYS IN THE SPINAL CORD

Classification of Nerve Fibres

Reflexes are most easily observed and analysed when the spinal cord receives a synchronous volley of afferent input. Because of this, the afferent volley usually has been provoked by electrical stimulation of nerves, rather than by natural stimulation of peripheral receptors. The result has been that most reflex stimuli, especially in animal experiments, are described in terms of the intensity of electrical stimulation of the nerve, rather than in terms of which sensory receptors have been activated. Fortunately, in muscle nerves there is a fairly close relationship between the electrical stimulation threshold of a fibre and the sensory receptor which it innervates.

In 1943, Lloyd proposed a system of classification of muscle afferent fibres on the basis of fibre diameter (which is inversely related to electrical threshold). The diameters fall into four main groups, which Lloyd labelled I-IV. The largest are group I (about 15μm diameter), which also have the lowest threshold to electrical stimulation. They are followed by group II (8μm), group III (3μm) and the unmyelinated group IV. The largest group I fibres (Ia) arise from primary spindle endings. Golgi tendon organs have a slightly smaller average diameter (Ib), although there is considerable overlap with the Ia spectrum. Spindle secondary endings have fibres of the group II class.

Unfortunately, the afferents in cutaneous nerves are not classified by the same system. The corresponding myelinated fibres are called Aα, Aβ, Aδ. Unmyelinated fibres are designated C fibres. There are many more Aβ fibres in skin nerves than there are group II fibres in muscle nerve. The Aα fibres mediate sensations of light touch, pressure, flutter and vibration, and also contain the majority of joint afferents. Pain and temperature are mediated by Aδ and C fibres.

Reflex Pathways from Ia Muscle Spindle Afferents

Anatomy

Using the horseradish peroxidase (HRP) method it has become possible to visualise directly the localisation of Ia afferent terminals within the spinal cord. Single Ia afferent fibres are identified electrophysiologically in the dorsal root of anaesthetised cats by their rapid conduction velocity (> 80 ms^{-1}) and their responses to muscle stretch and muscle contraction. The fibres are then injected with a small amount of HRP, and the animal maintained for the next 12 hours or so.

Figure 5.1: Reconstructions of the Spinal Pattern of Termination of Group Ia (A and B), Group Ib (C and D) and Group II (E and F) Muscle Spindle Afferent Fibres. A, C and E show transverse sections of the cord with course and branching pattern of a single HRP-stained axon. B, D and F are three-dimensional representations showing the overall pattern of projection. A single fibre is shown entering the dorsal root and bifurcating into an ascending and descending axon with terminals in specific laminae of the spinal grey matter (from Fyffe, 1984; Figure 7, with permission)

Afferent Fibres

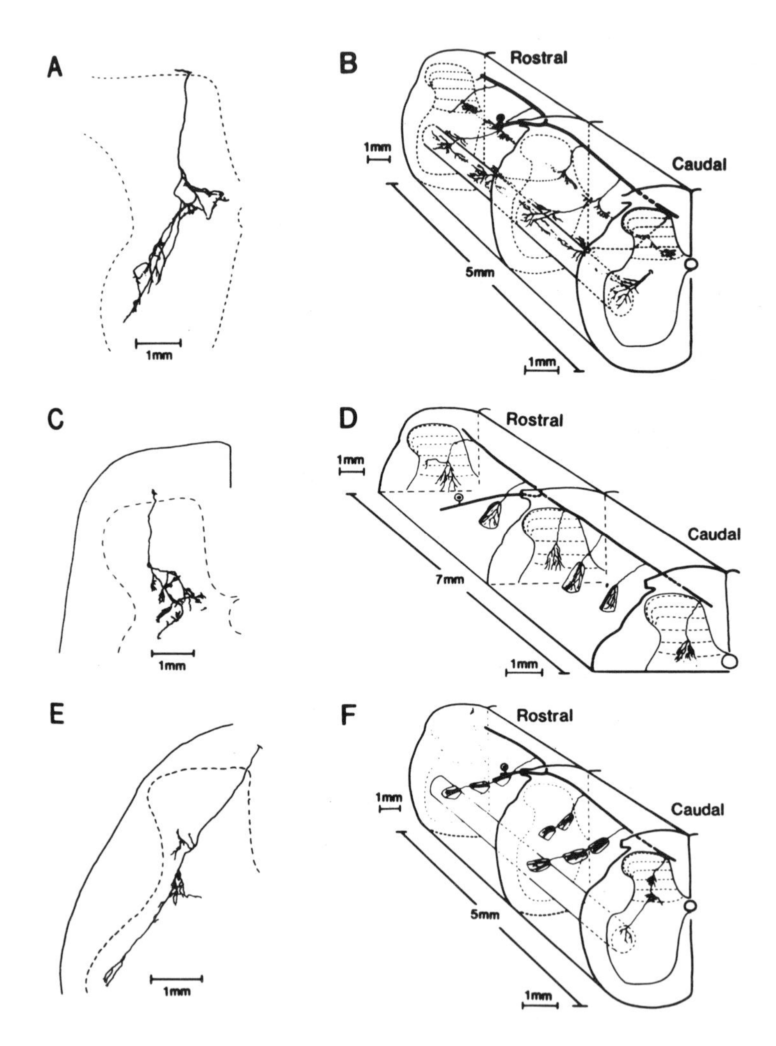

Over this period, the HRP is transported both anterogradely (towards the axon terminals) and retrogradely (towards the cell body in the dorsal root ganglion). The cat is then killed and the HRP is located by means of its reaction product with diaminobenzidene. This appears as a dark granule in the light microscope, and allows identification of all the synaptic terminals of that axon. HRP stain is visible at the electron microscope level, allowing even finer localisation of the sites of synaptic termination.

Figure 5.1 shows the pattern of termination of single Ia fibres from the triceps surae muscle of the cat, reconstructed from serial sections of spinal cord. Soon after entry into the spinal cord, the axon bifurcates into an ascending and descending branch which travel in the dorsal columns. Every millimetre or so, over one or two segments either side of the point of entry, the axon gives off collaterals which descend into the dorsal horn. There are three main areas of termination: Rexed's lamina VI, the intermediate region (lamina VII), and lamina IX. Lamina VII is just dorsal and medial to the motor nuclei, whereas lamina IX comprises the motor nuclei themselves. In the latter, the synaptic contacts appear to be made on to the soma and proximal dendrites of identified alpha-motoneurones (see Brown, 1981).

Figure 5.2: Schematic Diagram Showing Principal Connections of Ia Afferents and Renshaw Cells (RC). Inhibitory interneurones are in black. Ia afferents project monosynaptically to the homonymous α-motoneurones and disynaptically, via the Ia inhibitory interneurone (IaIN) to the α-motoneurones of the antagonistic muscle. Renshaw cells are excited by axon collaterals of α-motoneurones and project back to the same motoneurones, the γ-motoneurones and the associated Ia inhibitory interneurones. The box around the Ia inhibitory interneurones, and α and γ-motoneurones emphasises that these neurones receive very many of the same input connections. Dotted lines represent input from higher centres (from Hultborn, Lindstrom and Wigstrom, 1979; Figure 1, with permission)

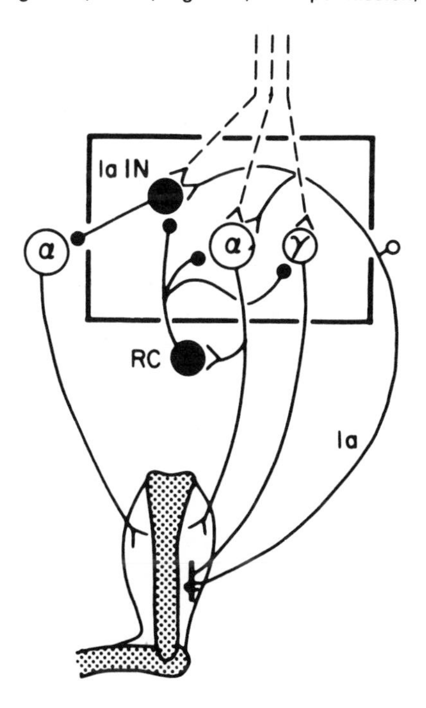

Electrophysiology

Monosynaptic Excitation. The monosynaptic reflex from Ia afferent fibres to motoneurones is probably the best known of all spinal reflexes (Figure 5.2). However, it was only in the 1930s and 1940s that precise measurements were made of the time taken for an afferent volley in the dorsal root to emerge from the ventral root. These revealed that there was a minimum delay within the cord which was consistent with monosynaptic transmission.

The earliest experiments simply demonstrated that the minimal delay between dorsal root stimulation and the emerging ventral root volley was very short. Stimulation of the dorsal root at low intensities produced an efferent volley in the ventral root with a delay of about 1.5ms. Increasing the stimulus intensity increased the size of the response and recruited later (probably polysynaptic) events, but did not change the latency of the earliest volley. This figure of 1.5ms includes both the synaptic delay and the conduction time of the fibres within the cord.

In order to remove the latter factor from estimation of the synaptic delay, Renshaw (1940) performed an experiment in which he inserted an electrode into the grey matter of the cord just dorsal to the motor nuclei of the ventral horn. Stimulating here, he found that it was possible to activate the motoneurones at two different latencies (Figure 5.3). At low intensities, there was a delay of about 1ms before impulses emerged through the ventral root, whereas the delay was reduced to 0.2ms with much higher intensities. Renshaw interpreted this experiment as follows. At low intensities, only those axons near the stimulating electrode were excited. These fibres then activated motoneurones synaptically, and produced the ventral root volley with a delay of 1ms. At high intensities, the motoneurones probably were stimulated directly by current spread and produced the ventral root volley with a delay of 0.2ms. Thus, the difference in latencies (that is, 0.8ms) between high and low intensity stimulation represents the synaptic delay time. Such a short delay is consistent with monosynaptic transmission.

During the course of such experiments, the conduction velocity of the responsible afferent fibres was measured and found to be of the order (in the cat) of 120ms^{-1}, which is compatible with the largest muscle afferent (Ia) fibres. (In man, the Ia fibres conduct more slowly at up to 80ms^{-1}.) However, the final proof that it was muscle spindle receptors which had monosynaptic access to the motoneurones, rather than any other type of receptors with large afferent fibres, only came some 15 years later, after the advent of intracellular recording. Natural muscle stretch, small enough to activate only primary spindle endings, finally was shown to produce monosynaptic excitatory postsynaptic potentials (EPSPs) in spinal motoneurones. The shape of these EPSPs, in terms of rapid rise time and short decay is consistent with the anatomical location of the Ia terminals on the soma and proximal dendrites of the motoneurones, as shown by the HRP method.

Intracellular recording has been used to investigate the distribution of Ia monosynaptic effects within the spinal cord motor nuclei. Consistent with the

Figure 5.3: Renshaw's Method of Estimating Synaptic Delay in the Spinal Cord. A shows arrangement of stimulating (S) and recording (R) electrodes in the grey matter and ventral roots. B shows spread of stimulating current at low and high intensities (dotted circles a and b, respectively). At low intensities, the axons of dorsal root fibres (DRF) or local spinal interneurones (i) might be activated in the vicinity of the stimulating electrodes. Motoneurones (m) could only be excited trans-synaptically. At high intensities, the motoneurones might be excited directly. C shows the records obtained from the ventral root at different stimulus intensities from 10–100ms. At high intensities, two responses can be distinguished labelled as m (direct motoneurone stimulation) and s (trans-synaptic motoneurone activity). A very small third wave is present which might have been produced by disynaptic excitation of motoneurones (from Patton, 1982; Figure 8.4, with permission (after Renshaw, 1940))

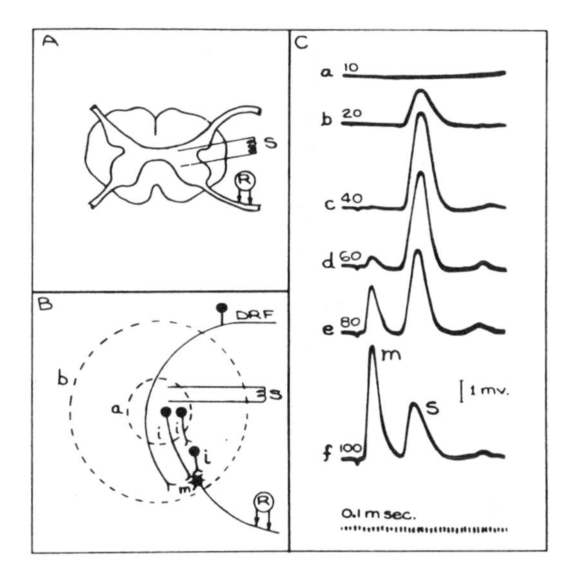

wide anatomical distribution of terminals within the cord, Ia fibres from one muscle may supply monosynaptic excitation to many other muscles (see original work by Eccles, Eccles and Lundberg, 1957). Conversely, the motor nucleus of one muscle may receive Ia inputs from many other muscles. Connections between different muscles are known as *heteronymous*, whereas those between a motoneurone and Ia fibres from the same muscle are known as *homonymous*. Each Ia fibre forms synaptic connections with up to 90 per cent of all the motoneurones within the homonymous motor pool. Heteronymous Ia excitatory projections also are provided, usually to mechanical synergists at the same joint or joints. For example, at the ankle joint, soleus motoneurones receive a substantial heteronymous excitation from Ia afferents in medial and lateral gastrocnemius. This group of muscles is sometimes termed the Ia synergists. Although the homonymous Ia excitation of a motoneurone is usually larger than that from other sources, the efficacy of heteronymous Ia inputs can vary considerably. In the muscles of the baboon's forearm, the heteronymous Ia EPSPs evoked in flexor digitorum communis are considerably larger than those produced by homonymous stimulation.

Disynaptic Inhibition. The other major reflex action of Ia afferents is the disynaptic inhibition of antagonist motoneurones. This was investigated extensively by Lloyd in the 1940s using the technique of monosynaptic testing. Single, low-intensity electrical stimuli were given to the muscle nerve supplying gastrocnemius in order to produce monosynaptic excitation of the gastrocnemius motoneurones in the spinal cord. The size of the reflex was monitored by recording the size of the ventral root discharge at various times before and after a conditioning stimulus was applied to the peroneal nerve, which supplies the antagonist anterior tibial muscles. Peroneal nerve stimulation produced inhibition of the gastrocnemius monosynaptic reflex, with maximum inhibition occurring if the peroneal nerve was stimulated some 0.5ms before the gastrocnemius nerve. Thus, for inhibition to occur, the antagonist nerve volley had to arrive at the spinal cord before the agonist test volley.

Araki, Eccles and Ito (1960) later used intracellular microelectrodes to show that the latency of the Ia inhibitory postsynaptic potential (IPSP) produced by stimulation of antagonist nerves was some 0.8ms longer than the onset of an EPSP evoked by stimulation of homonymous afferents in nerve filaments from the same dorsal root. This delay was attributed to the presence of an interposed interneurone in the inhibitory pathway, and the effect became known as *disynaptic* inhibition (see Figure 5.2). The extra time delay caused by an extra synapse is due to: (1) slowing of nerve impulses in the fine terminals of the presynaptic afferents; (2) the time taken for the postsynaptic neurone to reach firing threshold; and (3) the actual time taken for transmitter to diffuse across the synaptic cleft (which is of the order of 0.2ms).

Intracellular recording has revealed an extensive distribution of Ia inhibitory effects. A given motor nucleus draws its Ia inhibition mainly from its mechanical antagonists and also from other muscles connected to them in Ia synergism. This is known as *reciprocal Ia inhibition* because, in many instances, the inhibition to antagonist muscles is reciprocated by a similar inhibition from the antagonist Ia afferent fibres. Interestingly, there is no reciprocal inhibition between adductors and abductors.

The spinal interneurone which mediates disynaptic Ia inhibition has been termed the *Ia inhibitory interneurone*. It is one of the few interneurones whose input and output connections have been analysed in detail. Many other types of afferent fibres, besides the Ia afferent fibres converge onto it and will be discussed below. In electrophysiological experiments in which cells are impaled by microelectrodes in the spinal cord, the Ia inhibitory interneurone is identified by the fact that it is the only interneurone which (1) produces monosynaptic inhibition of motoneurones, and (2) receives monosynaptic Renshaw inhibition (see section on Renshaw cells below).

The anatomical location of these interneurones was demonstrated in an elegant series of experiments by Jankowska and Lindstrom (1971). After considerable effort, they succeeded in introducing a microelectrode into neurones in Rexed's lamina VII and identified some cells which received monosynaptic excitation from Ia fibres, and disynaptic inhibition from motoneurones (via Renshaw cells). These cells, in turn, could be stimulated and shown to evoke a monosynaptic inhibition of spinal motoneurones. Hence they are prime anatomical candidates for the physiologically identified Ia inhibitory interneurones. They lie in the second main region of Ia afferent fibre termination, dorsal and medial to the motor nuclei.

Polysynaptic Actions. Besides the well-studied monosynaptic excitation and disynaptic inhibition, there are, of course, many other routes by which Ia afferent fibres can influence the activity of spinal motoneurones. Weak trisynaptic inhibitory effects have been found on homonymous motor nuclei, with excitatory effects on heteronymous pools. Both probably are mediated by Ib interneurones, described in the next section. Much longer latency excitation of motoneurones also has been documented by Hultborn and Pierrot-Deseilligny (1979).

The more complex the pathway which such effects traverse, the more susceptible they are to changes in the state of the animal or to changes in the levels of other afferent inputs. Because of this, they are difficult to investigate and there is little information as to their effectiveness in normal movement. Future investigation will undoubtedly reveal them to be more important than they are considered at present (see Baldissera, Hultborn and Illert, 1981, for a complete review and further references to electrophysiological work on Ia afferents).

Reflex Pathways from Ib Tendon Organ Afferents

Anatomy

Using the HRP method to label physiologically identified Ib axons, the pattern of afferent termination has been found to be slightly different from that of the Ia fibres. The axons bifurcate on entering the cord, giving off a rostral and caudal branch running in the dorsal columns, but the main area of termination is in the intermediate laminae V-VII, as fan-shaped arborisations (Figure 5.1) (Brown, 1981).

Electrophysiology

Laporte and Lloyd (1952) made the first systematic study of Ib effects by using graded electrical stimulation of single muscle nerves. Using monosynaptic testing, they evoked reflexes in motoneurones of agonist and synergist muscles. They then measured the effect on the size of the evoked ventral root volley of single graded stimuli applied to a synergist muscle nerve at different times

Figure 5.4: Laporte and Lloyd's Demonstration of Ib Inhibition from Synergist Muscles. Monosynaptic reflex volleys were recorded in the ventral root after low intensity stimulation of the nerve to the plantaris muscle of a cat. The ventral roots were cut distal to the recording site to prevent antidromic invasion of the spinal motoneurones from activity in motor fibres. The test stimulus was preceded by a conditioning stimulus to the synergist muscle, flexor digitorum longus. Curve A shows the effect of a very weak conditioning stimulus. This was assumed to be mainly Ia in origin, and gave a pure monosynaptic facilitation. Curve B was obtained with slightly higher stimulation intensities. It was presumed to activate Ib fibres as well as Ia and inhibited the monosynaptic reflex with a latency some 0.5ms longer than that needed to produce monosynaptic excitation. The curve represents the effect of Ib disynaptic inhibition combined with Ia monosynaptic facilitation. Curve C was obtained with yet stronger stimuli which probably activated group II fibres (from Laporte and Lloyd, 1952; Figure 3, with permission)

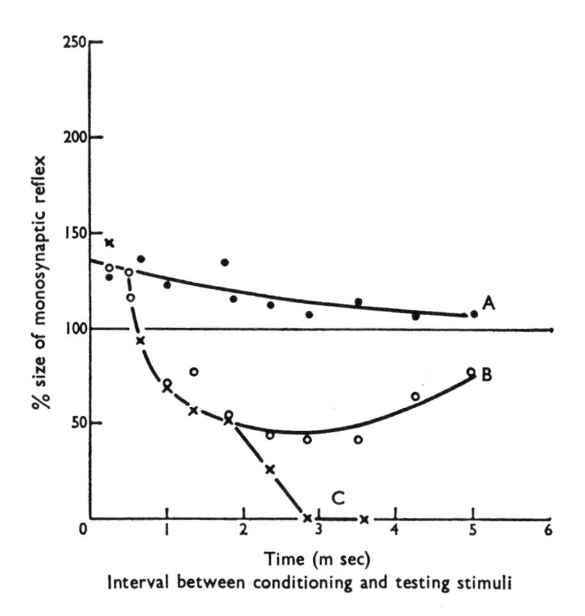

before and after the test stimulus. Weak stimuli to a synergist muscle nerve produced only facilitatory effects on the monosynaptic reflex. These were interpreted to be due to heteronymous, monosynaptic, Ia excitatory projections. However, when the intensity of the conditioning stimulus was increased slightly, an inhibition was revealed superimposed on the time course of facilitation. This inhibition began some 0.5–1ms later than the facilitation, and was attributed to a disynaptic inhibition produced by Ib fibres (Figure 5.4).

Stimulation of antagonist muscle nerves at these intensities was found to produce a di- or trisynaptic facilitation of the agonist. They termed such effects the *inverse myotatic reflex*, since the Ib actions seemed to be the exact opposite of those of the Ia fibres. Laporte and Lloyd also investigated the distribution of these effects, and found them to be particularly powerful from the Ib fibres of

Figure 5.5: Homonymous Ib Inhibitory Effects Shown by Intracellular Recordings From a Motoneurone Supplying the Gastrocnemius Muscle of a Cat. The ventral roots were cut in this preparation to prevent antidromic invasion of the motoneurone. Stimulation at low intensities (1.4 times group I threshold) produces a fairly pure EPSP in the records on the left. At higher intensities (1.8 times threshold activates almost all the group I fibres), the effect of the Ib IPSP manifests itself as a more rapid return of the initial EPSP towards baseline. In the records on the right, the cell was held hyperpolarised by passing a steady current through the microelectrode. The amount of hyperpolarisation was greater than the reversal potential of the IPSP. In these records, the Ib IPSP can be seen as a second depolarising wavelet superimposed on the EPSP (from Eccles, Eccles and Lundberg, 1957b; Figure 1, with permission)

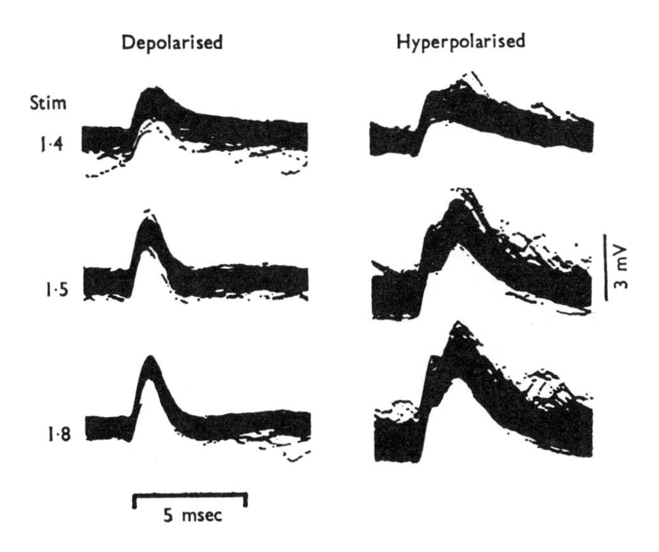

extensor muscles. Their actions may be summarised as: (1) disynaptic inhibition of motoneurones projecting to synergists, and (2) di- or trisynaptic excitation of motoneurones to antagonists.

Microelectrode studies confirmed that Ib fibres could produce inhibitory effects on homonymous motoneurones (Figure 5.5). On the basis of latency measurements, these effects were confirmed to be mediated by di- (or sometimes tri-) synaptic actions. However, such studies also revealed that the pattern of Ib projections was more widespread than originally envisaged by Laporte and Lloyd. The most recent work has shown that the Ib projections may be much more diverse than those of the Ia fibres. They may span more than one joint and often can be found to have actions opposite to those described as the inverse myotatic response. Such physiological findings of widespread action pose considerable problems in the interpretation of the role of Ib afferents in movement control (see Chapter 6).

The source of some of the interneurones within the Ib pathways has been localised to within Rexed's lamina V and VII, the area of the main Ib termination in the spinal cord. Many of these cells are monosynaptically excited by Ib fibres and project into the motor nuclei of the ventral horn. They are a different set of neurones to those mediating the Ia reciprocal inhibition and have been termed *Ib interneurones* (see Brink, Jankowska, McCrea, *et al.*, 1983 for further recent work, and Baldissera *et al.*, 1981, for a complete electrophysiological review).

Reflex Pathways from Group II Muscle Afferents and the 'Flexor Reflex Afferents'

Group II afferent fibres, even in muscle nerve, do not come exclusively from one type of receptor. Muscle spindle secondaries, joint receptors, Pacinian corpuscles and other receptors all have afferent fibres in the group II range. As shall be seen, this complicates the interpretation of any experiment in which group II fibres have been investigated purely by electrical stimulation.

Anatomy

The anatomy of group II spindle afferents has been studied in detail. Other categories of group II fibre have not been investigated so well. The spindle secondary afferents have a more variable morphology of termination than those of the Ia and Ib fibres. In general, they bifurcate into a rostral and caudal branch on entry to the cord. Collaterals then descend to terminate in three main regions: laminae IV-dorsal VI, lamina VI and dorsal VII and, in the ventral horn, lamina VII and IX (Figure 5.1). Figure 5.6 compares the regions of termination of group Ia, Ib and II fibres (Brown, 1981).

Electrophysiology

The earliest work on the group II projection was performed, like that of other afferent systems, using graded electrical stimulation of muscle nerves in combination with monosynaptic testing of the motoneurones of other muscles. In contrast with the group I effects, the group II effects from flexor and extensor muscles of the cat hindlimb were found to be the same. Stimulation of group II fibres from either set of muscles produced excitation of flexor muscles and inhibition of extensor muscles. Intracellular recording showed that the excitatory effects were produced by a disynaptic pathway and the inhibitory effects by a trisynaptic pathway. Some of the excitatory interneurones appeared to be located in the ventral part of Rexed's lamina VII, in a region of dense termination of group II afferents.

Later, it was found that electrical stimulation of a large class of both groups II and III muscle, cutaneous and joint afferents all produced the same action: excitation of ipsilateral flexors and inhibition of ipsilateral extensors. Weaker opposite effects were observed in muscles on the contralateral side of the body, that is, excitation of extensors and inhibition of flexors. This is sometimes described as ipsilateral flexion accompanied by crossed extension. In the intact animal, these reflexes would withdraw the limb from the stimulus, while at the

Figure 5.6: Summary Diagram of Main Areas of Termination of Muscle Afferent Fibres from Cat Lateral Gastrocnemius Near their Point of Entry into the Spinal Cord (from Brown, 1981; Figure 13.7, with permission)

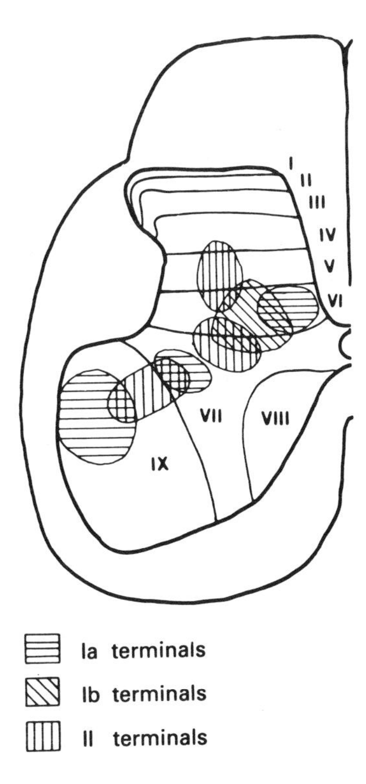

same time increasing the support provided by the contralateral limb to bear the weight of the animal. The umbrella term *flexor reflex afferents* (FRA) was used to describe these different classes of afferent because they all appeared to produce the same actions, presumably via common interneuronal projections (Eccles and Lundberg, 1959).

Although useful, the term FRA can sometimes be a little misleading. The precise action of FRAs depends on the state of the animal. In the mid-collicular decerebrate, the flexion reflex is virtually suppressed, whereas it is facilitated in the spinal preparation. There are also many reflexes mediated by cutaneous afferents, which do not fit into the general category of FRA actions. For example, gentle pressure on the plantar cushion of the cat hindlimb will activate muscles which plantar flex the toes, whereas a pin-prick applied to the same area would produce withdrawal of the foot by dorsiflexion. Thus the quality of the stimulus (and, in other cases, the point of application) is an important feature in determining the type of reflex response.

Similar arguments apply to the specific action of the group II secondary spindle afferents. In most muscle nerves it is not possible to stimulate spindle secondaries in isolation using electrical techniques. Other fibres from different classes of receptor inevitably are stimulated, and such fibres may have very different actions on the spinal alpha-motoneurones than the secondaries themselves.

A critical objection to the notion that spindle secondaries have actions which may be classed together with those of the 'flexor reflex afferents' was presented by Matthews (1969). According to the results of electrical stimulation techniques, only primary muscle spindle endings were supposed to contribute to the stretch reflex of extensor muscles. The secondaries, like all group II afferents, were expected to produce only inhibition of the extensor muscles. In order to test this hypothesis using more natural stimulation, Matthews examined the effect of two different types of muscle spindle stimulation on the size of the stretch reflex recorded in the cat soleus muscle.

In one experiment, the muscle was held slightly stretched, producing a weak stretch reflex contraction. Vibration at 200Hz was then applied to the tendon. Since this is an extremely effective stimulus for primary endings (especially in the cat), it would be expected to make all the primaries in the muscle fire at the same frequency. The effect was a marked increase in the reflex muscle tension. Next, without vibration, the muscle was stretched to a point where the reflex tension equalled that seen with vibration in the previous experiment. On the basis that only group Ia fibres contribute to extensor muscle excitation, it was argued that with this new degree of muscle stretch, they should all be discharging at 200Hz. However, when the vibrator was applied, the tension rose even further above this level (Figure 5.7). Matthews concluded from this that the spindle secondary endings must have been contributing to the stretch-imposed reflex contraction. This idea of group II mediated autogenetic excitation of muscle has received further support from recent anatomical and physiological data showing that some secondary spindle endings can have monosynaptic

Figure 5.7: Demonstration of a Possible Role for Group II Spindle Secondary Endings in Production of the Stretch Reflex in the Soleus Muscle of a Decerebrate Cat. On the left, the muscle was stretched by 4mm to produce a background stretch reflex and then vibration was applied to the tendon in order to evoke further excitation of the primary group Ia afferents and a larger reflex tension. On the right, a larger stretch of 9mm was given so as to produce a reflex tension equivalent to that seen when the vibrator was applied on the left. If all spindle primary endings had been 'driven' by vibration, then it should be impossible to exceed this tension by applying vibration in the experiment on the right. However, this was not found to be the case. There was no occlusion between stretch and vibration. The response to vibration was similar despite the different levels of stretch reflex contraction (from Matthews, 1969; Figure 3, with permission)

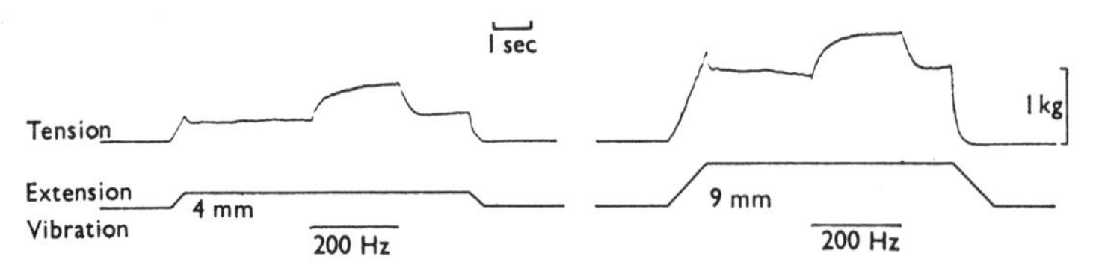

connections onto spinal alpha-motoneurones. The pattern of group II afferent termination in the spinal cord (see Figure 5.6) is consistent with this.

In great contrast to the detailed information available of the operation of the spindle secondary ending, it is still unclear at the present time as to what, if any, of the actions of group II spindle afferents predominate. In the intact animal, it seems likely that the actions will vary according to the control of spinal interneuronal machinery.

The Renshaw Cell

In the 1940s, Renshaw demonstrated that antidromic impulses in motor axons could reduce the excitability of alpha-motoneurones projecting to the same or synergistic muscles. This phenomenon was called *recurrrent inhibition* and has been shown to be due to activation of a group of interneurones by motor axon collaterals. These interneurones, named 'Renshaw cells' after their discoverer by J.C. Eccles, have been identified electrophysiologically and lie in the ventral horn of the spinal cord medial to the motor nuclei. They receive monosynaptic excitation from alpha-motoneurones and have a monosynaptic inhibitory connection back onto the homonymous and synergistic motoneurones (Figure 5.2). Each Renshaw cell may receive collaterals from many alpha-motoneurones. However, because such collaterals only distribute within a distance of less than 1mm from their parent cell bodies, the motoneurone input is relatively local.

Renshaw inhibition is distributed not only to motoneurones but to homonymous and synergistic gamma-motoneurones and Ia inhibitory interneurones, as well as to other Renshaw cells. The projection to the Ia inhibitory interneurone is unique since it is the only spinal interneurone interposed in a direct pathway to alpha-motoneurones which also receives monosynaptic Renshaw inhibition. The convergence of Renshaw and Ia afferents onto Ia inhibitory interneurones is used as a means of identifying these cells in electrophysiological experiments.

The function of the Renshaw cell is still a matter for debate. In the most general terms, its inhibitory feedback would tend to reduce the average resting potential of motoneurones. This would (1) reduce the frequency of alpha-motoneurone discharge below that expected in a system without recurrent inhibition, and (2) reduce the sensitivity of motoneurones to excitatory inputs. Some authors have suggested that with these actions, Renshaw cells might function to enhance the 'contrast' within a motoneurone pool. It would be a function analogous with the lateral inhibitory mechanisms of sensory systems.

To understand this idea, imagine what happens when a Ia discharge is produced by stretch of a single muscle. The Ia terminals are widely distributed to homonymous and heteronymous motoneurones, and the excitatory field is fairly large. The homonymous motoneurones would be activated most strongly and would exert strong recurrent inhibition on themselves and on the neighbouring heteronymous motoneurones. If the heteronymous motoneurones had been only weakly excited by the Ia input, then strong Renshaw inhibition might suppress their firing. In doing so, their own recurrent inhibition on the homonymous pool would be removed. The strongly activated motoneurones will therefore receive only their own recurrent inhibition, but because they are highly activated, their output (though diminished) will still be expressed. The nett effect would be that Renshaw cell activity would suppress activity in heteronymous motoneurones and restrict the spatial limits of excitatory inputs. Renshaw cells receive other inputs from supraspinal centres, and it is possible to imagine that the effectiveness of the 'motor contrast' mechanism could be controlled to suit different circumstances.

Another hypothesis of Renshaw cell action relies heavily on the possibility of supraspinal control of Renshaw excitability via the known anatomical projections from various brain centres (Figure 5.8). This views the Renshaw cell as a mechanism for controlling the sensitivity of the alpha-motoneurone (and Ia inhibitory interneurone) to other excitatory inputs. If the Renshaw cells were

Figure 5.8: Diagram Illustrating the Hypothesis that Renshaw Cells Might Serve to Regulate the 'Gain' of Motoneurones. A: input and output connections of motoneurone with supraspinal (–, +) control of Renshaw cells. B: graph showing how facilitation of Renshaw cells will decrease the slope of the input–output relationship of motoneurones, making them less sensitive to changes in size of the input signal (from Hultborn, Illert and Wigstrom, 1979; figure 1, with permission)

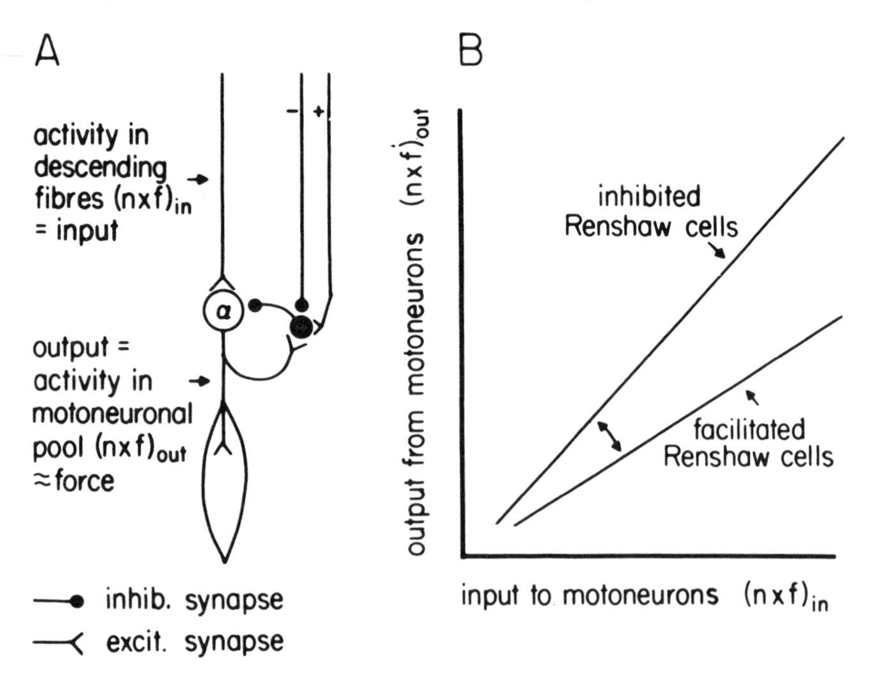

facilitated by descending inputs from the brain, the maximum output of the motoneurones would be reduced. Strong excitatory inputs to the motoneurone would produce strong Renshaw inhibition and prevent maximum excitation of the cell. To produce a given level of motoneurone output, there would have to be a much greater amount of excitatory input to the motoneurone than there would have been if the Renshaw cells were not facilitated. The slope of the relation between input and output could therefore be controlled by the effectiveness of Renshaw inhibition. With the Renshaw cells facilitated by descending commands, inputs to the motoneurone could play over a large range but cause only small changes in motoneurone output. At the level of the muscle this would mean that resolution of force changes would be high. If the Renshaw cells were inhibited, the motoneurone firing range would be large, and the resolution of muscle force changes by a given input would be reduced (see Baldissera *et al.*, 1981).

Integration in Spinal Reflex Pathways

For convenience, most of the reflexes above have been described in terms of private pathways to the spinal alpha-motoneurones, and it has been a feature of these reflexes that the more synapses that they traverse, the more difficult it has been to pin down a precise pattern of reflex action. The reason for this is that most pathways use interneurones in the spinal cord that are not, in fact, private to any particular reflex arc, but receive convergent inputs from many different sources. These may be from systems descending from the brain or from other afferent pathways. The nett result is that the excitability of the interneurones can be modulated and thus the strength of the reflex effects which they transmit can be changed. The study of 'integration' in these pathways has been one of the most important recent advances in spinal cord physiology and has been undertaken principally by Lundberg and his colleagues in Sweden (see Lundberg, 1975).

The method for investigating whether two different pathways share a common interneurone is shown in Figure 5.9. The upper diagram illustrates the experimental finding that low intensity stimulation of two different afferent inputs (which might separately be able to evoke a disynaptic response in a particular motoneurone if stimulated at high intensity), may have no effect on the motoneurone if stimulated alone. However, if both are stimulated together, the inputs summate and a response is recorded. The pathways are said to show *spatial facilitation*. This relies on the fact that addition of two subthreshold excitatory inputs at the membrane of the common interneurone will sometimes bring the interneurone above its threshold level and produce a nerve impulse. This activity will then evoke an EPSP in the motoneurone. Subthreshold addition of inhibitory and excitatory inputs may be demonstrated in the same way (Figure 5.9B) (see Lundberg, 1975).

Figure 5.9: Indirect Technique used to Investigate Convergence onto Interneurones of Reflex Pathways. A: Convergence between group I primary afferents (I test) and descending pathways (II.cond.). Stimulation of either input alone produces no EPSP in the motoneurone (M-n). Combined stimulation at the appropriate interval can evoke an EPSP. B: convergence between primary afferents (I) and inhibitory descending pathways (II). Stimulation of I at the appropriate intensity may evoke an EPSP, but stimulation of II may have no effect on the motoneurone potential. Combined stimulation at the correct interval may abolish the effect of primary afferent stimulation (from Lundberg, 1975; Figure 1, with permission)

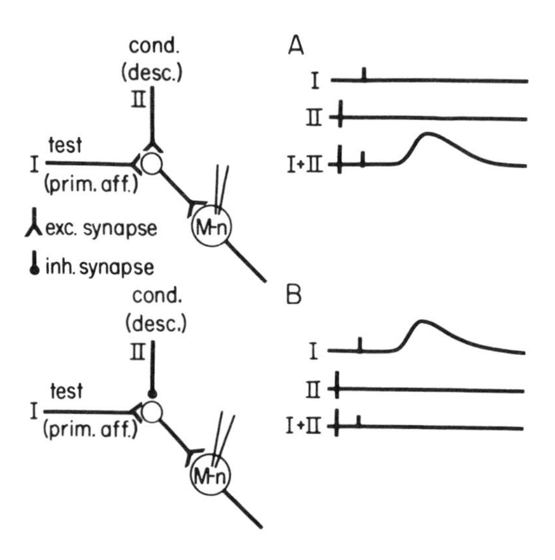

Convergence onto the Ia Inhibitory Interneurone

Because the Ia inhibitory interneurones lie in well defined disynaptic pathways within the spinal cord, and are activated by the largest diameter afferent fibres, they have proved relatively easy to study. They are one of the best characterised interneurones of the cord. They receive input from an enormous number of systems, including the corticospinal tract, flexor reflex afferents, cutaneous afferents, Ia afferents and Ia inhibitory interneurones of antagonist muscles. An example of Ia afferent convergence is shown in Figure 5.10. Muscle nerves from three different muscles of the cat hindlimb were stimulated at low intensity and found to send homonymous and heteronymous monsynaptic excitation to motoneurones of the semitendinosus muscle.

These same inputs also converged onto common Ia inhibitory interneurones. This was shown by recording from motoneurones innervating the antagonist muscle (vastocrureus), and applying pairs of subthreshold stimuli to the nerves of posterior biceps and semitendinosus, gracilis and semitendinosus, and posterior biceps and gracilis. All three pairs of conditioning stimuli gave IPSPs in vastocrureus motoneurones, but none of them gave an IPSP when stimulated alone.

Many inputs show common patterns of convergence onto groups of Ia

Figure 5.10: Convergence of Ia Afferents onto Common Motoneurones and Ia Inhibitory Interneurones. Scheme G illustrates the experimental arrangement, with stimulation of muscle nerves to semitendinosus (St), posterior biceps (PB) and gracilis (Grac) muscles of the cat hindlimb, and intracellular recording from motoneurones to semitendinosus and its antagonist vastocrureus (V-Cr). A to C show the EPSPs monitored in a semitendinosus motoneurone after low intensity stimulation of the three muscle afferent inputs. The EPSP is largest for St (homonymous) and smallest for Grac stimulation. The lower trace in each of A to C show records of dorsal root activity in entry zone of L7 segment. D, E and F show spatial facilitation of inputs producing a disynaptic IPSP in the V-Cr motoneurone. Traces start with a voltage calibration pulse of 0.5mV and 2ms. When each nerve is stimulated alone, there is little or no change in potential in the motoneurone. When any two inputs are stimulated together, an IPSP is evident (from Baldissera, Hultborn and Illert, 1981; Figure 9, with permission (after Eccles, Eccles and Lundberg, 1957 (A, B, C); and Hultborn and Udo, 1972 (D, E, F))

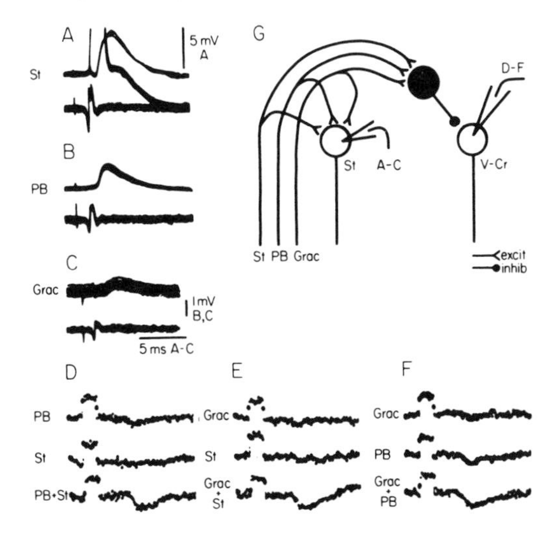

inhibitory interneurones and alpha-motoneurones (Figure 5.11). Examples are Ia afferents, descending inputs and connections from 'opposite' Ia inhibitory interneurones of the antagonist muscle. Because of this common pattern of connections, α-motoneurones and linked Ia inhibitory interneurones (and γ-motoneurones) are said to form a functional *unit* in the spinal cord. Inputs to this unit produce α-γ coactivation and reciprocal inhibition of antagonist muscles. This is termed *α-γ coactivation in reciprocal inhibition.*

The result is that inputs to motoneurones, such as homonymous Ia afferents, or corticospinal tract projections, not only activate the α-motoneurone: they also excite the Ia inhibitory interneurones, which secondarily produce disynaptic inhibition of antagonist motoneurones and the 'opposite' Ia inhibitory interneurones. In addition, γ-motoneurones are excited by some inputs (for instance, descending corticospinal tract inputs). With this combination of excitation and inhibition (which is relatively 'hard-wired' in the spinal cord), isotonic contraction of the agonist muscle will be accompanied by fusimotor activity to maintain Ia input from shortening spindles. Also, although the antagonist muscle

Figure 5.11: Schematic Drawing of Parallel Inputs (Dotted Lines) to α- and γ-motoneurones and Ia Inhibitory Interneurones (Larged Filled Circles) of One Muscle. Note mutual inhibition of 'opposite' Ia inhibitory interneurones (from Hultborn, Illert and Santini, 1976; Figure 3, with permission)

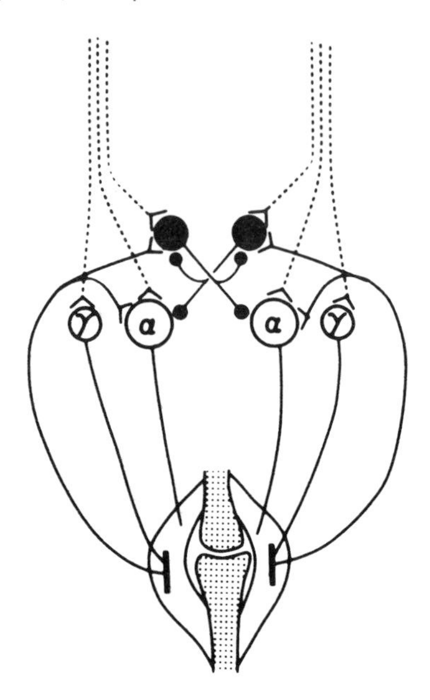

will be stretched, Ia firing will not evoke unwanted reciprocal inhibition of the agonist, since the antagonist Ia inhibitory interneurones are turned off. In addition, Ia input from the stretched antagonist is unlikely to evoke any monosynaptic excitation of antagonist motoneurones because they are actively inhibited by the agonist Ia inhibitory interneurones (see Hultborn, 1976; Hultborn, Illert and Santini, 1976; Hultborn and Pierrot-Deseilligny, 1979; plus review by Baldissera *et al.*, 1981).

Convergence onto Ib Interneurones

Like the Ia inhibitory interneurones, the Ib interneurones receive input from a wide range of sources. Of particular interest is the finding that Ib interneurones are facilitated at short latency by low threshold cutaneous and joint afferents. One possible consequence of this would be that if a limb movement were suddenly interrupted by a physical obstruction, the cutaneous input might facilitate autogenic inhibition of the agonist muscles and interrupt the contraction. This is not necessarily in contradiction to the servo-control of movement via the stretch reflex. It is simply part of the general hypothesis of this work on spinal cord mechanisms that under certain conditions, reflexes in different pathways might prevail. A relevant example might be that in reaching out to touch a delicate Meissen statuette, it might not be advisable for the CNS to employ the servo assistance mechanism to overcome the obstacle. Facilitation of the Ib pathway might be judged much more advisable in this instance.

Joint afferent input may also be of some functional importance. From the study of joint receptors, it is known that joint afferent discharge is particularly prominent at the extremes of joint movement. It may be that their action in facilitating the Ib interneurones could decrease agonist muscle force once the limit of a movement is approached. At the same time, the excitatory effects of Ib interneurones onto antagonist muscles might also contribute to this 'braking' process.

Many descending supraspinal systems converge onto the Ib interneurones, allowing the possibility of some kind of control of these pathways under different conditions. Recently, it has also been discovered that Ia afferents project onto the Ib interneurones, although the functional significance is not clearly understood.

Comments on Convergence

Only two examples of the convergent input to spinal interneurones have been destribed. There are very many more, still in the process of investigation. The main message of all this work is that a spinal reflex, involving one or more interneurones, cannot always be relied upon to be constant. Depending on the inputs from other afferents or other CNS centres, the efficacy of transmission in any pathway may change (see Baldissera *et al.*, 1981; Harrison, Jankowska and Johannisson, 1983).

Figure 5.12: Pattern of Presynaptic Inhibition of Ia, Ib and Cutaneous Afferent Fibres in the Cat Hindlimb. The width of the arrows indicates the strength of inhibition. A: Effects onto the terminals of Ia afferents to flexor (left) and extensor (right) motoneurones from ipsilateral Ia and Ib fibres. B: Effects onto terminals of Ib afferents (same for both extensor and flexor projections) from ipsilateral and contralateral afferents. C: Effects onto terminals of cutaneous afferents (same for both extensor and flexor projections) from ipsilateral and contralateral afferents (from Schmidt, 1971; Figures 24, 26 and 27, with permission)

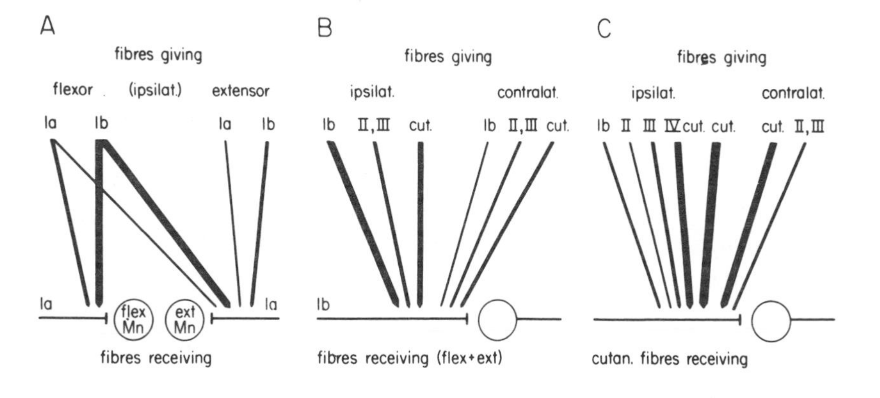

Presynaptic Inhibition

The section above has dealt with the possibility of modulating transmission in polysynaptic reflex pathways by affecting the excitability of the interposed interneurones. Presynaptic inhibition is another means of changing reflex transmission which can affect even monosynaptic pathways. The effects on the terminals of Ia, Ib and cutaneous afferent fibres have been investigated. All have their synaptic potency changed by presynaptic actions.

Presynaptic inhibition is caused by long-lasting depolarisation of the afferent terminal, which leads to a decrease in the number of quanta of transmitter released per nervous impulse. It can be detected by placing electrodes over the central end of a cut dorsal root and recording the depolarising potential which spreads electrotonically from the site of generation in the nerve terminals. This potential is known as the dorsal root potential. An alternative method is to stimulate the terminals within the cord with an electrical impulse. This sets up an antidromic volley which can be recorded in the dorsal roots. Paradoxically, because presynaptic inhibition involves depolarisation of nerve terminals, a given stimulus will produce a larger antidromic volley in the presence of presynaptic inhibition.

As a general rule, all three classes of afferent fibre presynaptically inhibit themselves (see Figure 5.12). These actions are long-lasting (up to 1s), widespread and produced via polysynaptic spinal pathways which have yet to be identified. As one might expect, the interneurones of these pathways show convergence from very many sources. A precise function for presynaptic inhibition in motor control has yet to be established (Brink, Jankowska and Skoog, 1984).

References and Further Reading

Review Articles

Baldissera, F., Hultborn, H. and Illert, M. (1981) 'Integration in Spinal Neuronal Systems' in V.B. Brooks (ed.), *Handbook of Physiology*, sect. 1, vol. II, part 1, Williams & Wilkins, Baltimore, pp. 509–97

Brown, A.G. (1981) *Organisation in the Spinal Cord*, Springer-Verlag, Berlin

Fyffe, R.E.W. (1984) 'Afferent Fibres' in R.A. Davidoff (ed.), *Handbook of the Spinal Cord*, vols. 2 and 3, Marcel Dekker Inc, New York

Lundberg, A. (1975) 'Control of Spinal Mechanisms from the Brain' in D.B. Tower (ed.), *The Nervous System, Vol. 1, The Basic Neurosciences*, Raven Press, New York

Patton, H. (1982) 'Spinal Reflexes and Synaptic Transmission' in T. Ruch and H.D. Patton (eds.), *Physiology and Biophysics*, vol. IV, W.B. Saunders, Philadelphia, pp. 261–302

Original Papers

Araki, T., Eccles, J.C. and Ito, M. (1960) 'Correlation of the Inhibitory Post-Synaptic Potential of Motoneurones with the Latency and Time Course of Inhibition of Monosynaptic Reflexes', *J. Physiol., 154*, pp. 354–77

Brink, E., Jankowska, E., McCrea, D.A., *et al.* (1983) 'Inhibitory Interactions Between Interneurones in Reflex Pathways from Group Ia and Group Ib Afferents in the Cat', *J. Physiol., 343*, pp. 361–73

——— Jankowska, E. and Skoog, B. (1984) 'Convergence onto Interneurones Subserving Primary Afferent Depolarisation of Group I Afferents', *J. Neurophysiol.*, *51*, pp. 432–49

Eccles, J.C., Eccles, R.M. and Lundberg, A. (1957a) 'The Convergence of Monosynaptic Excitatory Afferents on to Many Different Species of Alpha Motoneurones', *J. Physiol.*, *137*, pp. 22–50

——— (1957b) 'Synaptic Actions on Motoneurones Caused By Impulses in Golgi Tendon Organ Afferents', *J. Physiol.*, *138*, pp. 227–52

Eccles, R.M. and Lundberg, A. (1959) 'Synaptic Actions in Motoneurones by Afferents Which May Evoke the Flexion Reflex', *Arch. Ital. Biol.*, *97*, pp. 271–98

Harrison, P.J., Jankowska, E. and Johannisson, T. (1983) 'Shared Reflex Pathways of Group I Afferents of Different Cat Hind-limb Muscles', *J. Physiol.*, *338*, 113–27

Hultborn, H. (1976) 'Transmission in the Pathway of Reciprocal Ia Inhibition to Motoneurones and Its Control During the Tonic Stretch Reflex' in S. Homma (ed.), *Progress in Brain Research, Vol. 44. Understanding the Stretch Reflex*, Elsevier, Amsterdam, pp. 235–55

——— Illert, M. and Santini, M. (1976) 'Convergence on Interneurones Mediating the Reciprocal Ia Inhibition of Motoneurones', *Acta Physiol. Scand.*, *96*, pp. 193–201; 351–67; 368–91

Hultborn, H., Lindstrom, S. and Wigstrom, H. (1979) 'On the Function of Recurrent Inhibition in the Spinal Cord', *Exp. Brain Res.*, *37*, pp. 399–403

——— and Pierrot-Deseilligny, E. (1979) 'Changes in Recurrent Inhibition During Voluntary Soleus Contractions in Man Studied by an H-reflex Technique', *J. Physiol.*, *297*, pp. 229–51

Jankowska, E. and Lindstrom, S. (1971) 'Morphology of Interneurones Mediating Ia Reciprocal Inhibition of Motoneurones in the Spinal Cord of the Cat', *J. Physiol.*, *226*, pp.805–23

Laporte, Y. and Lloyd, D.P.C. (1952) 'Nature and Significance of the Reflex Connections Established by Large Afferent Fibres of Muscular Origin', *Am. J. Physiol.*, *169*, pp. 609–21

Lloyd, D.P.C. (1943) 'Neuron Patterns Controlling Transmission of Ipsilateral Hindlimb Reflexes in Cat', *J. Neurophysiol.*, *6*, pp. 293–315 (see also pp. 111–20 and 137–326)

Matthews, P.B.C. (1969) 'Evidence that the Secondary as well as the Primary Endings of the Muscle Spindles May Be Responsible for the Tonic Stretch Reflex of the Decerebrate Cat', *J. Physiol.*, *204*, pp. 365–93

Renshaw, B. (1940) 'Activity in the Simplest Spinal Reflex Pathways', *J. Neurophysiol.*, *3*, pp. 373–87

Schmidt, R.F. (1971) 'Presynaptic Inhibition in the Vertebrate Central Nervous System', *Ergeb. Physiol. Biol. Chem. Exp. Pharmakol.*, *63*, pp. 20–201

6 FUNCTIONAL CONSEQUENCES OF ACTIVITY IN SPINAL REFLEX PATHWAYS

The Stretch Reflex in Animals

Although the tendon jerk had been brought to the notice of scientists as early as 1875 by Erb and Westphal, much of the early work on the stretch reflex was performed on animals. This was principally due to the fact that the tendon jerk was for many years regarded as a direct response of the muscle to percussion rather than being a reflex event.

The study of the stretch reflex was begun in earnest by Sherrington and his colleagues who investigated the phenomenon in decerebrate cats (see Creed *et al.*, 1932). Intense rigidity develops if a complete transection is made through the brain at the level of the colliculi, particularly in the extensor or antigravity musculature. The limbs may become so rigid that animals may be made to stand unassisted, although they lack any postural reflexes and are therefore extremely unstable. Sherrington was the first to show that the rigidity of the muscles in the decerebrate animal is due to over-activity of stretch reflex mechanisms. When afferent input to the reflex arc is removed by sectioning the dorsal roots of the spinal cord, the rigidity disappears. He later went on to show that muscle afferent input was necessary for the rigidity to occur. The increased tone of the vastocrureus portion of quadriceps remained after all the other muscles and skin of the leg had been denervated. The only afferent input was then coming from receptors in the muscle, and these must have been responsible for the increased tone.

The phenomenon was investigated in more detail by Liddell and Sherrington in 1924. In a series of experiments (Creed, Denny-Brown, Eccles *et al.*, 1932), they showed that stretch, particularly of the antigravity extensor muscles, produced a reflex contraction of the same muscle which opposed the applied force. The response had a phasic component, proportional to the velocity of the stretch, and a tonic component, proportional to the final static degree of stretch (Figure 6.1). The muscular receptors which provided the input to the reflex later were shown to be the primary spindle endings. Their dynamic response to stretch probably accounts for the velocity–sensitivity of the reflex. However, it is not yet clear to what extent spindle secondary endings may be involved, since, as explained in Chapter 5, it has been discovered more recently that there are pathways available to the group II spindle secondary afferents which could contribute to the excitation of extensor muscles.

Why, then, does decerebrate rigidity occur? The usual explanation is that it is due to overactivity of the fusimotor system. This is supported by the presence of a high tonic spindle discharge in the decerebrate cat, and by a relatively high

Figure 6.1: Classical Myograph Record of the Stretch Reflex of the Quadriceps Muscle in a Decerebrate Cat. The tension exerted by the muscle through its tendon (M) was recorded by a rather cumbersome optical device. This device made it difficult to extend the muscle by pulling directly on the tendon. Instead, stretch was applied by moving the other end (the insertion) of the muscle. In fact, it proved easiest to leave the insertion of the muscle intact, and to move the whole cat by dropping the table a few millimetres (T). Curve M represents the reflex tension developed with the muscle nerve intact. Curve P shows the tension with the nerve cut — that is, tension due solely to the mechanical properties of the denervated muscle (from Creed, Denny-Brown, Eccles *et al.*, 1932; Figure 18, with permission)

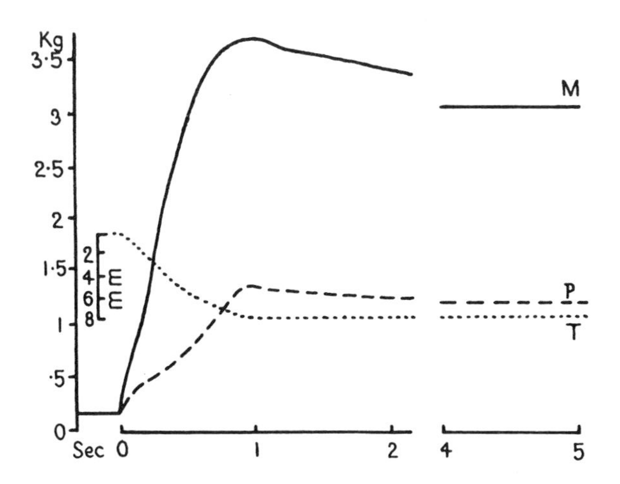

sensitivity of the receptors to stretch. In addition, Matthews and Rushworth (1958) showed that if procaine, a local anaesthetic, is applied to the muscle nerve, rigidity disappears before the large fibres are blocked. Small fibres are affected first by local anaesthetic and so it is supposed that blocking fusimotor efferents may reduce the sensitivity of spindles to stretch and decrease their background discharge.

Excess fusimotor activity certainly contributes towards the increased excitability of the stretch reflex arc in the decerebrate preparation. However, other factors also must be involved. One reason for this is the observation that vibration of a muscle, producing a large tonic Ia input to the cord, does not produce a tonic muscle contraction in spinal animals (nor in spinal man, see below). Thus, a large Ia input alone does not necessarily produce stretch reflex rigidity. It is thought that other spinal reflex pathways also have to be facilitated in order for decerebrate rigidity to occur. Details are not known; they may include facilitation of polysynaptic Ia connections to motoneurones, or decreased presynaptic inhibition of the terminals of Ia afferents.

The decerebrate preparation has particularly large stretch reflexes because of the unopposed action of centres in the brainstem which act on spinal reflex mechanisms after decerebration. The intercollicular section which produces decerebrate rigidity is thought to lead to an imbalance between descending facilitatory and inhibitory systems which project to flexor and extensor systems in the spinal cord. Section at the intercollicular level tips the balance towards facilitation of extensor systems, with a large portion of the excitability directed towards the fusimotor system. The effects are different if sectioned at other levels. No rigidity occurs if the section is made above the level of the colliculi, to include part of the red nucleus, or if the section is too caudal, below the level of the vestibular nuclei.

Finally, it should be noted that the cerebellum has an influence on rigidity in the decerebrate animal. Ablation of the spinal parts of the anterior cerebellum increases decerebrate rigidity and produces an animal with extreme extensor hypertonus (see Chapter 9). In these animals, the rigidity cannot be abolished by dorsal root section, and they are said to show *alpha-rigidity* in order to emphasise the preponderance of alpha over gamma effects. The anterior lobe of the cerebellum has an inhibitory projection to the neurones of the lateral vestibular nucleus of Deiters. Removal of the inhibition is responsible for a direct vestibulospinal excitation of extensor alpha-motoneurones.

Stretch Reflexes in Human Muscles

In normal human subjects, it is not possible in any muscle to evoke a prolonged stretch reflex which resembles that in the decerebrate cat. Several different types of response can be demonstrated.

The Tendon Jerk

In man the most common manifestation of the stretch reflex is the tendon jerk. In a relaxed muscle, sudden stretch produced by percussion of the muscle tendon with a tendon hammer gives rise to a brief contraction known as the tendon jerk. Sherrington referred to this as a 'fractional manifestation' of stretch reflex mechanism. The tendon hammer stimulus may stretch a muscle by only 50μm, but the rapid onset of stretch makes this a particularly powerful stimulus to excite the primary spindle endings.

Traditionally, the tendon jerk is regarded as monosynaptic; however, other pathways also may contribute to the response. Microneurographic recordings in the nerves of the leg following a tendon tap to the ankle show that in man the muscle spindle afferent volley may be dispersed by 20 to 30ms. This dispersion is due to (1) double firing of some receptors; (2) dispersion in the conduction velocity of the Ia afferents (from about 40–60ms^{-1}); and (3) differences in the times at which individual receptors discharge their first impulse. This dispersed afferent volley evokes a dispersed wave of depolarisation in spinal motoneurones, the duration of which can be estimated by analysing the firing pattern of single motor units. The time at which single units discharge after a tendon tap is not fixed; there is a jitter of 5–6ms from one trial to the next. Since motor units usually are made to fire during the rising phase of the composite EPSP, this means that the EPSP rise time also is of the order of 5–6 ms. Impulses arriving at any time within this 6ms period can contribute to excitation of the motoneurone. Thus it is possible that the first-arriving volley at the spinal cord could produce both monosynaptic and polysynaptic excitation within this period.

This is not proof that such polysynaptic inputs do contribute to the tendon jerk reflex. However, since the possibility exists, it is a warning not to attribute all pathologies of the tendon jerk to changes in the spinal monosynaptic reflex arc. It is possible that changes in polysynaptic pathways may be responsible for enhancement or depression of the tendon jerk in disease states in man.

A final point is, why is the EPSP rise time so much shorter than the temporal dispersion of the afferent volley? One reason is that later-arriving EPSPs may be superimposed upon the decay phase of EPSPs produced by earlier parts of the volley, and may therefore be missed. Another possibility is that the excitation is actively turned off by spinal inhibitory mechanisms involving, for example, Ib inhibition from the homonymous muscle.

A reflex analogous to that of the tendon jerk also may be evoked in some muscles by electrical stimulation of nerve trunks. In 1926 Paul Hoffman found that low intensity stimulation of the tibial nerve in the popliteal fossa could

Figure 6.2: Relationship Between the Strength of an Electrical Stimulus to the Median Nerve at the Elbow and the Size of the H-reflex and Direct M-wave in the Wrist and Finger Flexor Muscles of Man. In A, an H-reflex with a latency of some 20ms appears at low intensities. At 6.0mA, an earlier response appears. This is the M-wave, a muscle response caused by stimulation of efferent motor fibres in the mixed nerve. As the M-wave increases in size, the H-reflex gets smaller. This is shown graphically in B, where the peak-to-peak size of the H-reflex (●——●) and M-waves (▲----▲) are plotted on the same axes. C shows the effect on the H-reflex of vibrating the flexor tendons at the wrist. Vibration was applied 0.5s before the sweep and continued throughout. The H-reflex is abolished, although the M-wave remains unchanged (from Day, Marsden, Obeso *et al.*, 1984; Figure 1, with permission)

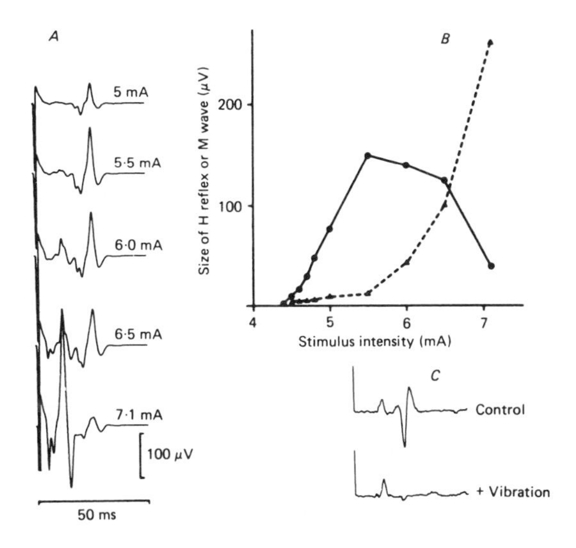

produce a reflex contraction of the triceps surae muscles without direct activation of the muscle via the alpha-motoneurones. The reflex had a latency of some 30ms and was believed to be an electrically elicited analogue of the tendon jerk. The reflex, now known as the H-reflex in this honour, occurs because the group Ia fibres are larger than, and have a lower threshold to electrical stimulation, than the alpha-motoneurones. Therefore, at very low stimulus intensities, Ia afferents may be the first fibres to be activated. The relatively pure and synchronous Ia afferent volley then produces reflex activation of triceps surae motoneurones. However, as the stimulus strength is turned up, the axons of alpha-motoneurones are stimulated and a direct muscle response (M-wave) is elicited. At high stimulus intensities, the H-reflex disappears. This is due to two factors: (1) antidromic firing of motor fibres renders the motoneurones refractory to the reflex input; and (2) the antidromic motor volley collides with the orthodromic reflex volley set up by the Ia input. H-reflexes can be obtained in many muscles. Figure 6.2 shows the behaviour of H-reflexes in the wrist and finger flexor muscles.

As with the tendon jerk, there are several caveats against assuming that the H-reflex is a purely monosynaptic reflex. Although there is less dispersion in the afferent volley set up by electrical stimulation of nerve (no repetitive firing and all fibres activated at the same time), the spread of conduction velocities in the Ia afferent fibres still ensures that the volley arriving at the spinal cord is not completely synchronous. Measurements of the EPSP rise time show it to be of the order of 2ms or so, giving time for the fastest conducting volleys to evoke di- or even trisynaptic excitation of the motoneurones in addition to the conventional monosynaptic response (Burke, Gandevia and McKeon, 1983).

Long-latency Stretch Reflexes

The tendon tap is a phasic stimulus and can evoke a response even in relaxed muscles. However, a more prolonged and slower muscle stretch, similar to that used to evoke a stretch reflex in the decerebrate cat, is without effect unless the muscle is activated. Stretch of such an actively contracting human muscle evokes a series of responses which can best be observed in the electromyogram (Figure 6.3). Three phases usually can be identified. The response begins with an EMG burst at short latency, similar to that of the tendon jerk in the same muscle. It ends with a prolonged burst beginning 100ms after onset of the stimulus, which is under conscious control of the subject, and which usually is considered to be a voluntary reaction. The origin of the middle part of the response (or the *long-latency* stretch reflex) has been the source of some controversy. It is particularly prominent in the muscles of the forearm, where it dominates the tendon jerk component. In these muscles it has a latency of approximately twice that of the tendon jerk and is sometimes known as the M2 response (with the tendon jerk component being the M1 response). Three main possibilities have been proposed to explain the extra latency of this response over and above that of the tendon jerk:

Figure 6.3: Reflex and Voluntary EMG Responses to Muscle Stretch in Flexor Pollicis Longus (FPL) and Biceps Brachii (biceps). Top traces represent position of the thumb or elbow; bottom traces are average (of 16) rectified EMG responses. Stretches were given by a torque motor every 5–6.5s and in separate sets of trials the subject was instructed either to do nothing (C), resist the stretch voluntarily as rapidly as possible (P) or to let go on perceiving the stretch (L). The two components (short-latency or tendon jerk component, S; long-latency LL) of the stretch reflex EMG response were little affected by these instructions. However, the voluntary response (VOL, beginning about 100ms after onset of stretch) is changed considerably. This affects the trajectory of the final portion of the position traces. Note how the long-latency reflex component is much larger than the short-latency reflex in the distal FPL muscle. Horizontal calibration 100ms; vertical calibration 30° or 150μV. In FPL, the short-latency component begins about 22ms after stretch, and the long-latency after some 40ms (from Marsden, Rothwell and Day, 1983; Figure 2, with permission)

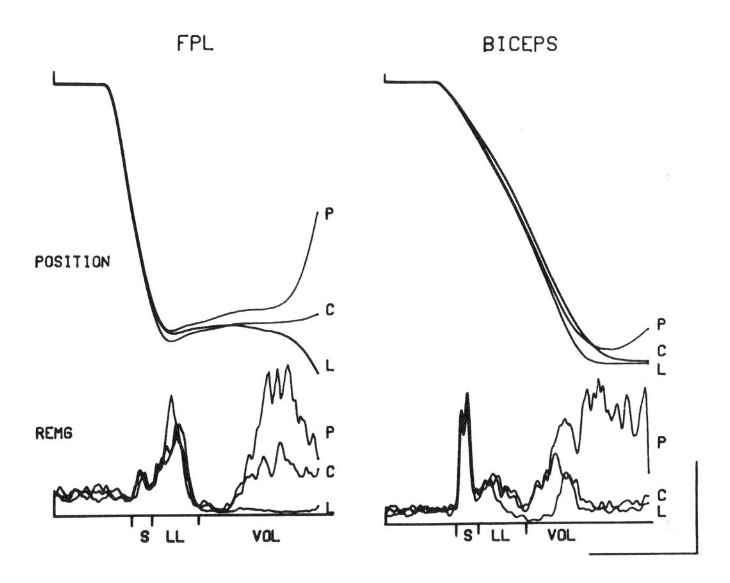

1. The segmented reflex EMG could be due to a segmented spindle afferent input;
2. The medium latency reflex could be mediated by more slowly conducting afferents (for example, group II spindle afferents) than those responsible for the tendon jerk;
3. The same afferents could be involved in both tendon jerk and the medium latency response, but the latter could be delayed because the impulses traverse a longer pathway (for example, polysynaptic spinal circuits or 'long loop' routes via supraspinal structures) than that of the monosynaptic reflex.

There is some evidence to support each of these possibilities and it is probably true to say that each may turn out to contribute to a greater or lesser extent to the long-latency, or M2 reflex.

It has been shown in recordings from normal human subjects that slow stretch of some muscles may evoke a segmented spindle discharge in primary spindle afferent fibres. The segmentation is believed to be due to underlying mechanical 'ripples' in the trajectory of muscle stretch which may be caused by the stretching device, or be a result of mechanical oscillations within the muscle-tendon compliance. The interval between the segmented bursts is of the order of 20 to 30ms which would be appropriate to produce long latency reflexes through repeated activation of the tendon jerk pathway. However, such repetitive activity, although contributory, is not sufficient on its own to produce segmented EMG responses. The reason for this is that very rapid muscle stretches, which are complete within 10 to 20ms, may under certain circumstances evoke a long latency reflex, presumably without giving rise to a segmented afferent volley. In addition, segmentation of the afferent volley does not explain the results of pathological studies described below.

Slowly conducting afferents also are unlikely to be wholly responsible for the medium latency response. If this were true, then those muscles nearest the spinal cord would be expected to have the shortest latency M2 responses. In an ingenious series of observations, Marsden, Merton and Morton (1973) showed that this was not the case. Stretch reflex responses were recorded in both the long thumb flexor and the infraspinatus muscle. The latency of the tendon jerk was about 22ms for the thumb and about 14ms for infraspinatus, yet the latency of the medium latency response was approximately the same at about 40ms in both muscles. If a slow-conducting afferent pathway were responsible, the M2 response in infraspinatus — a muscle very close to the spinal cord — would have been at a much shorter latency.

More recently, Matthews (1984) has argued against these conclusions and has suggested that secondary spindle afferents do indeed provide a strong input to the long latency stretch reflex. His observation is that stretch of a muscle may evoke a long latency stretch reflex, but vibration of the same muscle for 200ms or so only produces a short latency (spinal) response. Arguing that vibration stimulates primary endings far more than secondaries, he concludes that stretch, by stimulating both primaries and secondaries equally well, produces a long-latency stretch reflex because of additional input in group II fibres.

A strong contribution to the long-latency (or M2) stretch reflex responses probably is produced by activity in a long loop reflex pathway in addition to the spinal reflex pathways. It has been suggesteed by many authors that a trans-cortical route may operate: impulses in spindle primary afferent fibres may traverse a route through the dorsal column–medial lemniscal system to the somatomotor cortex and then back to the spinal motoneurones via the pyramidal tract. Such a route is known to exist anatomically and would explain the lack of difference in the latency of the long latency (M2) reflex in the long flexor of the thumb and infraspinatus. Strong evidence in favour of this explanation comes from clinical studies. Patients with lesions at any point within this hypothetical pathway, in the dorsal columns, sensory or motor cortex or in the pyramidal tract all have absent medium latency stretch responses. In contrast, the tendon jerk component is unaffected or may even be enhanced in some

Figure 6.4: Top: Tonic Vibration Reflex (TVR) in Quadriceps Muscle of Normal Man. A vibrator was applied to the ligamentum patellae and after a short delay produced a reflex activation of quadriceps. The effect is increased by performance of the Jendrassik manoeuvre or by applying a second vibrator to the muscle belly. Upper trace, muscle force; lower trace, quadriceps EMC. Time calibration is ls, vertical calibration 5kg or 1mV (from Burke, Andrews and Lance, 1972; Figure 2, with permission).
Bottom: Suppression of Quadriceps Tendon Jerks by Continuous Muscle Vibration in Normal Man. In A and B, the thick horizontal line indicates the period of vibration. Records are of the force of quadriceps contraction, and show the responses to repeated tendon taps given every 5s. In A, vibration evokes no tonic contraction, yet still suppresses the tendon jerks. In B, a TVR develops and again suppresses the tendon jerks. In C, the subject made a voluntary contraction similar to the reflex contraction of B. Voluntary contraction does not suppress tendon jerks. Time calibration is 10s, force calibration 0.4kg for A and 0.6kg for B and C (from DeGail, Lance and Nielson, 1966; Figure 1, with permission)

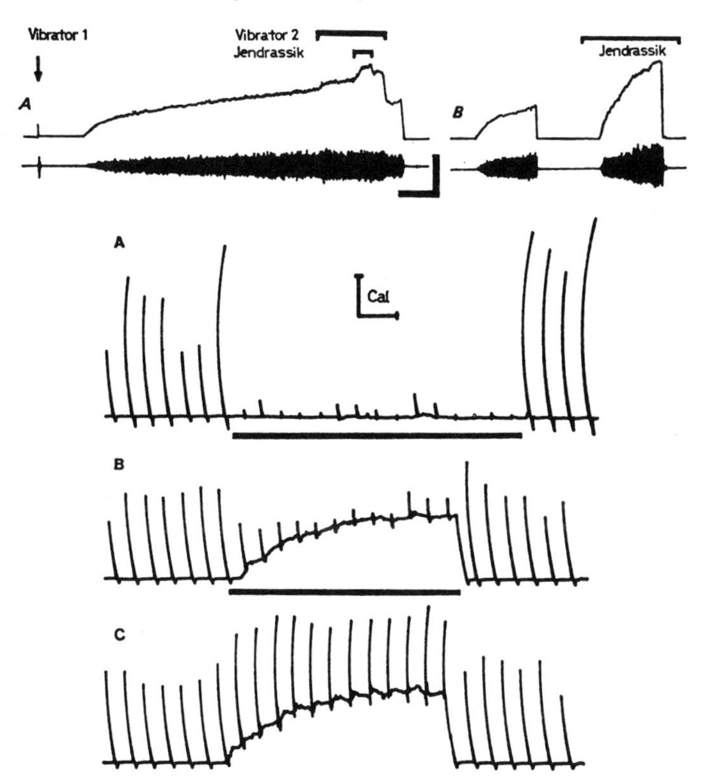

patients (for instance, in cases of hemiplegia).

Such evidence, although compelling, is not direct proof of the transcortical hypothesis and there are still arguments as to whether the long-latency pathway may involve polysynaptic spinal routes which are subject to descending influence from supraspinal structures. Removal, for example, of a hypothetical facilitatory influence on this pathway could explain the absence of latency reflex responses in the patients referred to above (see Eklund, Hagbarth, Hagglund *et al.*, 1982; Marsden, Merton and Morton, 1973; Marsden, Rothwell and Day, 1983; Matthews, 1984; for further details of work on long-latency stretch reflexes).

Vibration Reflexes

A final reflex from muscle spindles that may be mentioned here is the tonic vibration reflex (TVR). Vibration of a muscle belly or its tendon at 50–150Hz produces a slowly developing reflex contraction which is sustained throughout the period of vibration (Figure 6.4) and which subsides slowly when it is stopped. The reflex is seen best if the attention of the subject is distracted. Unlike the stretch reflex, it is possible voluntarily to prevent the reflex occurring if visual feedback is given. Muscle spindle Ia afferents are particularly well excited by such stimulation although in man, in contrast to the cat, many secondary spindle endings also are activated. The reflex pathway probably involves both the monosynaptic reflex arc as well as longer polysynaptic pathways, which would account for the slow onset and decay of the response. Some descending facilitation of these pathways must be necessary for the reflex to occur since no TVR may be recorded in the lower limbs of paraplegics with complete spinal cord transection. Recently, it has been demonstrated that vibration of purely cutaneous receptors may evoke a similar, although smaller, muscle contraction.

In addition to its excitatory effects on motoneurones, muscle vibration also has a concurrent inhibitory effect on the monosynaptic reflex (Figure 6.4). Thus, vibration of the triceps surae inhibits the ankle jerk or H-reflex while simultaneously producing a TVR in the muscle. This is not due to occlusion of the afferent volley in the H-reflex or tendon jerk by the constant high level of muscle afferent discharge produced by vibration (the 'busy-line' phenomenon). Recordings of gross neural discharge using near-nerve electrodes inserted through the skin show that the afferent volley from a tendon tap or H-reflex is preserved during vibration. Since the alpha-motoneurones are excited by vibration, the effect on the monosynaptic reflex is generally thought to be due to presynaptic inhibition of muscle spindle Ia afferent terminals. In experiments performed on anaesthetised cats, such presynaptic inhibition in the spinal cord can be recorded as a slow electrical potential from the dorsal roots (see dorsal root potential in Chapter 5).

Experiments with muscle vibration, therefore, illustrate that Ia fibres have polysynaptic as well as monosynaptic excitatory connections with spinal alpha-motoneurones. In addition, the fibres produce a presynaptic inhibition of their own terminals (see Lance, Burke and Andrews, 1973, for a review of vibration).

Figure 6.5: Diagram of Merton's 'Follow-up Servo' Theory of Muscular Contraction. Two routes were available from the brain to control the activation of spinal α-motoneurones: projections directly to the α-motoneurones (α-route) and projections to the γ-motoneurones (γ-route). The α-route was thought to be reserved for 'urgent' muscle activation. The γ-route could produce contraction more slowly by activating muscle spindles and exciting α-motoneurones reflexly via the spindle afferent fibres (from Hammond, Merton and Sutton, 1956; Figure 1, with permission)

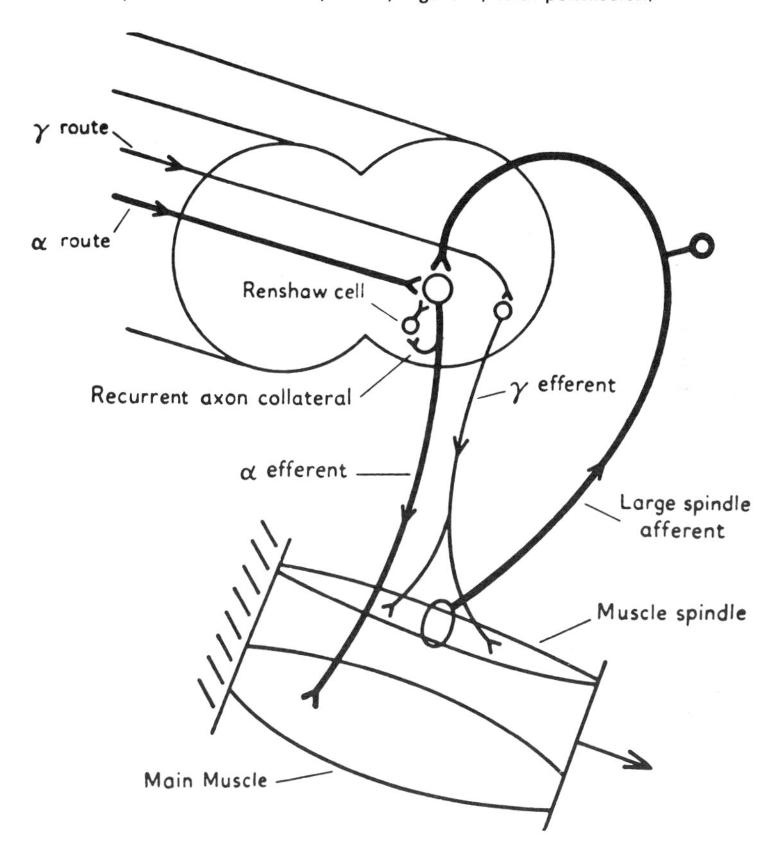

The Servo Hypothesis and Alpha-Gamma Coactivation

The Servo Theory

Ever since its discovery, there has been no generally agreed hypothesis to describe the role of the stretch reflex in the control of movement. The most influential has been the servo hypothesis put forward by Merton in 1953 (see also Eldred, Granit and Merton, 1953). In its simplest form, the theory envisaged that slow movements could be initiated and controlled by the activity of the gamma-efferents to the muscle spindles (Figure 6.5). Rather like power-assisted steering in a motor car, an increase in activity in the fusimotor (γ)

system would produce contraction of the spindle poles and stretch the sensory receptors. The increased afferent discharge would, via the stretch reflex, produce contraction of the main extrafusal muscle until the stretch on the spindle was removed. The extrafusal contraction would 'follow' that of the spindles.

This method of control would be rather slow because of the extra delay involved in traversing the servo loop (20ms in the arm, 30ms in the leg). However, the advantage of this system would be that the muscle servo would follow the spindle contraction even if the external load were to change. For example, if the load on the main muscle were increased during the contraction, so that the muscle was stretched, there would be an increased mismatch between extra- and intrafusal lengths. More power would be called up via the servo loop to overcome the increased mismatch and the obstacle would be overcome. This type of contraction was said to be via the gamma-route. Direct activation of muscles via the alpha-route, which would bypass the delay in the servo loop, was reserved for rapid, urgent muscle contraction.

In order for such a servo to work effectively, the 'gain' of the reflex must be fairly high. That is, the amount of extrafusal muscle contraction called up by a given mismatch of spindle and muscle length must be sufficient to produce movement and overcome any minor obstacle *en route*. In normal man, it seems that the gain of the stretch reflex generally is quite low. Thus, in Figure 6.3, the subject was trying to hold his thumb or elbow in a given position against a small load. According to the servo theory, the muscle power to maintain this position would be supplied by a very small mismatch between extra- and intrafusal muscle length driving the contraction via the reflex arc. When the force against the wrist was suddenly doubled, the reflex contraction in the flexor muscles was not sufficient to restore the wrist position. In fact, in this case, very little compensation was achieved by the reflex correction, which would be a poor performance to expect from a servo system. True compensation began with the onset of voluntary activity.

The most direct evidence against the simple version of the servo theory has come from recordings of spindle afferent discharge during movements in conscious man. The sequence of events preceding onset of movement in the servo theory would be: (1) gamma-efferent firing; (2) contraction of spindle poles; (3) stretch of sensory endings; (4) discharge in spindle afferent fibres; and (5) discharge of alpha-motoneurones. However, percutaneous recording of single spindle afferents during slow isometric finger movements showed that this sequence was not followed. Spindle discharge always began *after* the start of muscle activity, rather than *before* it (Figure 6.6). Slow voluntary movements are not initiated by the gamma-route.

Alpha-gamma Coactivation

Although spindle activation does not precede extrafusal contraction, Figure 6.6 suggests that the behaviour of spindles is not entirely passive during voluntary movements. If it was, then spindles lying in series with the extrafusal muscle

Figure 6.6: Discharge of a Single Muscle Spindle Afferent in the Finger Flexor Muscles (Top Trace, Direct Record; Second Trace, Instantaneous Frequencygram of Same Unit) During a Weak Voluntary Isometric Flexion of the Index Finger. The recording was made with a tungsten microelectrode, of about 10μm tip diameter, inserted into the median nerve above the elbow. Note how the afferent only begins to fire after the onset of EMG activity (bottom trace) in the muscle (from Vallbo, 1970; Figure 6, with permission)

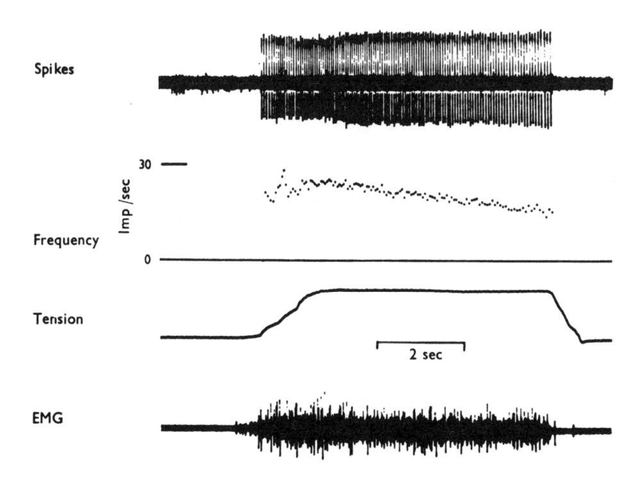

would be unloaded by the extrafusal contraction, causing a decrease, rather than an increase in spindle firing. (Even in isometric contractions, there is some spindle unloading due to stretch being taken up by the tendon compliance.)

The increase in spindle discharge seen in Figure 6.6 is evidence that there is an active contraction of the spindles which more than compensates for the extrafusal unloading. Both α and γ motoneurones must have been active in the movement. This concept usually is referred to as α-γ coactivation, meaning that the command to contract a muscle does not go solely to the α-motoneurones, or solely to the γ-motoneurones (as suggested by the follow-up servo theory), but to both sets of motoneurones at the same time.

One of the advantages of α-γ coactivation sometimes is said to be that gamma discharge during the contraction prevents spindle unloading, and enables spindles to respond to disturbances even during active muscle shortening. However, as shall be seen below, α-γ coactivation does not always achieve this ideal, particularly during rapid movements.

Although it is agreed that α-γ coactivation does occur, the main current area of controversy is whether α-γ coactivation is the rule or whether there are occasions in which α activity and γ activity can be controlled independently. In the cat, there are indeed sites within the brainstem, stimulation of which can produce pure γ activity, but the question remains as to whether the brain ever activates these centres independently of those producing α activity during normal movement. The very existence of a separate γ system, in addition to the shared skeletofusimotor β-system, which is all that is seen in amphibia, argues for the possibility of independent control.

Unfortunately, it is not easy to resolve the question of independence versus coactivation because it is very difficult to record directly from the small γ-efferent fibres during movement. Because of this, the main evidence for the existence of α-γ coactivation comes from observations like those above in which the activity of fusimotor neurones is deduced from the spindle afferent discharge. α-γ coactivation is presumed to occur if the spindle discharge during active movements is different from that seen in passive movements of the same kind.

α-γ Coactivation in Man

In human studies, using the technique of microneurography, there is very little evidence that the γ system can be activated independently of the α system. However, the converse situation of α activity without γ activity has been documented quite frequently. Pure α activity is responsible for the contraction of the tonic vibration reflex and for reflex contractions produced by cutaneous stimuli. Thus muscle vibration or cutaneous stimulation can produce a contraction in which spindles are unloaded, as they would be during pure passive shortening of muscle.

Independent activation of the γ system is more controversial. An initial candidate for supposed γ independence was the Jendrassik manoeuvre. The action of clenching teeth and fists is well known to increase tendon jerks in remote muscles and it was thought that this could have been caused by a general increase in γ drive to muscles not involved in the contraction, thereby increasing the sensitivity of their spindles to stretch. (In particular, it should be noted that this would have been a specific γ_d effect, since γ_s stimulation decreases the sensitivity of primary endings to rapid stretch.)

Microneurographic findings in man have been divided. Some groups claim to have recorded an increase in spindle discharge to standard tendon taps given during a Jendrassik manoeuvre. Others, however, insist that this is never seen unless there is a small amount of EMG activity, and hence α-γ coactivation, in the muscles under test. They suggest that in truly relaxed muscles, the spindle discharge is not affected by a distant Jendrassik reinforcement. This viewpoint is gaining ground, and it appears that the Jendrassik manoeuvre increases the tendon jerk by facilitating transmission at central synapses rather than by increasing spindle sensitivity.

There has never been any evidence from human studies for independent activation of the γ system during the course of normal voluntary contractions. Because of the difficulty in maintaining a stable recording, most of the movements that have been examined have been relatively slow. Spindle discharge usually increases in such movements, indicating activity in both α and γ systems. However, in faster contractions, the spindles may silence, which suggests that they have been unloaded by the extrafusal shortening. The probable explanation for this is not that the γ fibres are inactive during rapid contractions, but that the fibres of muscle spindles cannot contract as rapidly as the extrafusal muscle. Indeed, Prochazka and Hulliger (1983) have shown in experiments on cats that even when static γ axons are stimulated at 100Hz, with the intention of maintaining a strong intrafusal contraction, primary endings are silenced during rapid contractions. In order to include muscles of different lengths, they normalised the speed of shortening in terms of the resting length (RL) of the muscle. As a general rule, they found that when a muscle contracts at a velocity greater than 0.2RL/s, then its spindles are usually silenced, even if they are maximally activated by the fusimotor system.

A final addendum to the story of α-γ coactivation in man is that although it has so far not been possible to reveal independent activation of the γ system, it is possible to change the balance between α and γ activity. In stereotyped isometric slow voluntary contractions, each spindle ending begins to accelerate its discharge at a particular level of contractile force. This force threshold is reproducible for that ending, but can differ markedly between endings. Threshold differences may be caused (1) by the particular position of the spindle in relation to the active extrafusal muscle fibres, or (2) by a recruitment threshold for γ motoneurones innervating that ending.

Whatever the reason, given that a contraction is always the same, changes in the threshold at which a spindle is activated give a measure of the level of α-γ balance. Two procedures have been shown to affect this balance. Caloric stimulation of the vestibular organs (by passing cold water into the ear) and vibratory stimulation of cutaneous receptors. Both manoeuvres increase the threshold for spindle excitation, and therefore probably decrease the γ drive to the spindles relative to the α activity. Thus different systems (that is, cutaneous afferents and vestibulospinal connections) can activate α and γ systems in different proportions.

α-γ Coactivation in the Cat

There is more evidence for independent control of both α and γ systems during normal movements in the cat than there is in man. Two types of movement have been examined in conscious animals, walking and chewing. In the former, deductions about γ activity have been made by examining spindle discharge during movement, whereas in the latter it has proved possible to record directly from both α and γ fibres.

In the studies of normal walking, the pattern of spindle firing has been compared to that seen in spindles from an anaesthetised cat subjected to exactly the same passive joint movements. This type of experiment is remarkably difficult to perform. Stable recording electrodes have to be implanted into the dorsal roots of intact cats, together with miniature length gauges to monitor the length of the muscles under study. When this has been done, single spindle afferents, muscle length and muscle EMG can be recorded during normal walking, jumping and so on. These records are then compared with the afferent discharge recorded from spindles in the same muscle of an anaesthetised cat.

Under anaesthetic, the experimenter can control the level of fusimotor stimulation to the spindles by stimulating appropriate ventral root filaments. Thus he can find the precise level of fusimotor discharge required to produce a given spindle afferent response when a particular length change is applied to the muscle. This fusimotor discharge is then compared with the EMG discharge recorded from extrafusal muscle in the intact animal. Mismatch between the EMG and fusimotor input indicates that for the movement studied, there must have been separate control of α and γ motor fibres.

The general finding was that most unobstructed and familiar movements were accompanied by a relatively low and steady level of either dynamic or static fusimotor activity, which was independent of the phasic activation of extrafusal muscle. This meant that the spindles behaved like passive stretch receptors throughout the movement. For example, during active shortening of the extrafusal muscle the spindle discharge would silence (unlike the records of Figure 6.6); if the shortening was obstructed, there was no increase in spindle discharge (as seen during α-γ coactivation). Although fusimotor activity usually stayed at a steady level, there were circumstances in which it could change to become phasically active. This was seen when the cat performed unusual or novel tasks. It may be that when new tasks are learned, the cat makes use of the servo assistance which a phasically active fusimotor system can provide.

α-γ independence also has been documented during rythmic jaw movements in the cat. Direct recording from fusimotor neurones revealed two classes of discharge. One type of neurone showed a sustained increase in firing rate during chewing, whereas the other class discharged phasically together with the extrafusal EMG. The two types probably were γ_d and γ_s, respectively. Thus, unlike the limb movements described above, the phasically modulated units (γ_s) provided an α-γ linkage during normal movements. This was not strong enough to prevent unloading of the spindles during rapid extrafusal shortening, but was sufficient to increase spindle discharge if the shortening movements were obstructed.

Such experiments suggest that α and γ motoneurones can be independently controlled in the cat, and the type of control varies with the type of movement that is being made (see articles in Barnes and Gladden, 1985; Taylor and Prochazka, 1980; and articles by Burke, Gandevia and McKeon, 1983; Prochazka and Hulliger, 1983; Vallbo, 1970; for further details of the work covered in this section).

The Regulation of Stiffness Hypothesis and Ib Effects

Although the servo theory and its variations have been the mainstay of speculation of stretch reflex function, other theories have been proposed, the most important of which is the regulation of stiffness hypothesis put forward by Houk and his colleagues (see Houk and Rymer, 1981). Their main proposal is that it is not muscle *length* which is regulated by reflex actions, but muscle *stiffness*. Furthermore, they propose that because stiffness is the regulated variable, not only is the stretch reflex involved, but also the Ib reflex arc.

These investigators studied the mechanical properties of muscles in the decerebrate cat after elimination of the stretch reflex by dorsal root section. They stimulated the motor axons to a hindlimb muscle so that it generated a small active tension, and then either stretched or released the muscle by varying amounts. When the muscle was stretched there was a sudden increase in tension (due to the short-range stiffness) followed by a slow yielding. After release by the same amount, there was no equivalent 'yielding' and the tension drop was much greater than the tension increase during stretch. In the decerebrate cat with intact stretch reflexes, this asymmetry was abolished because of an opposite asymmetry in stretch reflex action, which was much greater to stretch than to release (Figure 6.7). A more linear relationship between length and tension was obtained, that is, the stiffness of the muscle remained constant.

The central idea of this theory is that to achieve a constant stiffness, the nervous system must be provided with some information on the force generated by the muscle. This was supposed to be provided by Golgi tendon organ input, which, together with stretch reflex mechanisms, acted to maintain a constant stiffness of muscle. Thus the attraction of this theory is that it proposes a useful role for the Ib pathway, that is, to co-operate with the stretch reflex to regulate a combined variable of length and tension (stiffness).

The control of stiffness theory was proposed before the discovery of widespread actions of the Ib fibres. Since that time, more subtle roles for the Ib system have been proposed. Convergence of inputs onto the Ib interneurone plays an important part in these theories since it allows for the possibility of controlling transmission in the Ib pathway according to the demands required of the system. Some actions of the Ib system have been revealed in man by Pierrot-Deseilligny and Maziers (1984) using H-reflex testing in the muscles of the leg.

H-reflexes are easily elicited in the triceps surae muscles and can serve as an indicator of spinal alpha-motoneurone excitability if presynaptic effects on the Ia terminals are excluded. It can then be shown that stimulation of Ib afferents in muscle nerves to the same muscles can depress the H-reflex with a latency compatible with disynaptic inhibition in the spinal cord. In man, such Ib inhibition from gastrocnemius also crosses the knee joint to depress H-reflexes in quadriceps and biceps femoris.

The Ib inhibition from gastrocnemius to quadriceps can be affected by cutaneous inputs. Stimulation of the anterior part of the foot sole depresses Ib inhibition to quadriceps when the subject is at rest, although it has no effect on the inhibition from gastrocnemius to gastrocnemius and soleus (triceps surae). This effect may be brought about by convergence of cutaneous inputs onto specific groups of Ib interneurones in the spinal cord: those projecting to quadriceps being excited by cutaneous input, while those to triceps surae are

Figure 6.7: Asymmetry of Reflex Action to Stretch and Release of a Soleus Muslce in a Decerebrate Cat. Two experiments were performed on the same cat. In the first, symmetrical stretches and releases were given to the muscle (bottom traces) with the stretch reflex arc intact. This produced the changes in muscle force shown in the upper graphs (net reflex response). Then, the dorsal roots were cut and the same stretches given to the muscle. Stretch and release produced much smaller changes in muscle tension (muscle mechanical response). The important feature is that under these areflexic conditions, there was an asymmetry in the force changes experienced by the muscle: the force drop on release was much greater than the force increase on stretch. In the presence of afferent feedback, reflex action (the shaded areas between the curves) compensated for this asymmetry in mechanical response so that stretch and release both produced the same force changes in the muscle (from Nichols and Houk, 1976; Figure 6, with permission)

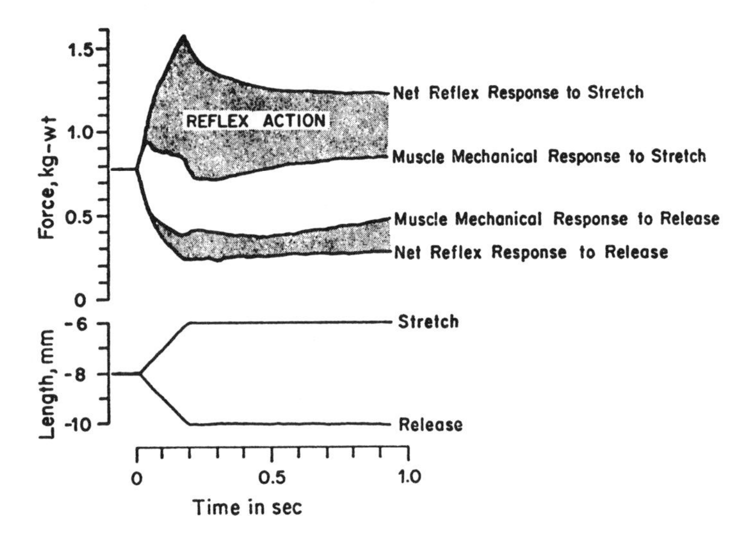

unaffected.

When tested during contraction of the triceps surae, the cutaneous facilitation of Ib interneurones to quadriceps is *reversed*: cutaneous stimulation increases Ib inhibition. Reversal of the cutaneous effect is thought to be caused by descending control of interneurones in the cutaneous reflex pathway. By this means, the weighting of cutaneous effects onto the Ib interneurone may be reversed by voluntary contraction. Exploratory movements of the foot could provide a possible role for this circuitry. These movements are produced by activity in quadriceps and triceps surae. If the big toe or anterior part of the foot comes into contact with an object, the cutaneous input would facilitate Ib inhibition of quadriceps and thereby depress its contraction. The effect would be to reduce the effect of stubbing one's toe during a voluntary exploratory movement.

The explanation is not so far-fetched as it may seem. For example, if one is walking around the house at night, trying to avoid children's toys scattered about on the floor, there are powerful reflexes which inhibit one's exploratory movements when an unexpected obstacle is encountered. In daylight, however, one is free to point and press one's toes against any desired object or to kick a ball with full force. The difference may be that the excitability of spinal circuits has been reset in a totally different way (see reviews by Pierrot-Deseilligny and Maziers, 1984; Pierrot-Deseilligny *et al.*, 1981 for further details on possible Ib actions in man).

Other Spinal Reflex Pathways

Pierrot-Deseilligny, Bergego, Katz *et al.* (1981) have used the H-reflex technique to examine two further pathways in man: the disynaptic reciprocal Ia inhibition between agonist and antagonist muscles, and Renshaw inhibition. The advantage of working in man rather than in legally dead (that is, decerebrate) animals, is that in man it is possible to investigate the effect of voluntary activity on transmission in defined reflex pathways. This type of work has shown that spinal reflex circuits are not static, hard-wired and unadaptable. On the contrary, they can be modulated subtly by activity in other pathways so that they function in a way appropriate to the task in hand.

Reciprocal Ia Inhibition

Tanaka (1974) was the first to document reciprocal Ia inhibition in man. The technique is relatively simple: H-reflexes are elicited in one muscle, and low intensity electrical stimuli, designed to stimulate only the largest afferent fibres (the muscle spindle Ia afferents), are applied to the nerve supplying the antagonist muscle. If the antagonist stimulation is given at the appropriate time, it is possible to inhibit the test H-reflex. Timing considerations show that the spinal latency for this effect is appropriate for disynaptic inhibition in the spinal cord.

In normal subjects it is possible to demonstrate convergence onto this pathway from supraspinal centres. For example, in the leg it is not possible to reveal any Ia reciprocal inhibition from the tibialis anterior onto the triceps surae when subjects are at rest. A single stimulus to the common peroneal nerve, which supplies the anterior tibial muscles, cannot inhibit the H-reflex in triceps surae. However, this is not true if the stimulus is given to the common peroneal nerve during or even slightly before a small voluntary contraction of the tibialis anterior muscles. Under these conditions, a single stimulus can inhibit triceps surae reflexes. The 'opening' of the Ia reciprocal inhibitory pathway during voluntary contraction is considered to be due to a facilitation of the Ia inhibitory interneurones by descending input. This facilitation makes them accessible to the conditioning volley in the peroneal nerve. Similar effects can be demonstrated on the reciprocal inhibition from extensor to flexor muscles of the human forearm (Figure 6.8). This type of experiment is good evidence for the theory that groups of α, γ and Ia inhibitory interneurones receive common inputs, linking them together in a functional unit.

Figure 6.8: Disynaptic Ia Inhibition Between the Extensor and Flexor Muscles of the Human Forearm. The traces on the right show H-reflexes recorded in the wrist and finger flexor muscles after stimulation of the median nerve at the elbow. In the right column are superimposed averages of 10 control H-reflexes and 10 H-reflexes which had been conditioned by a low intensity stimulus applied to the antagonist, radial, nerve in the spiral groove. The H-reflex is inhibited if the radial nerve stimulus is given at the same time as the median nerve test stimulus. Because the spiral groove is nearer the cord than the cubital fossa, the radial nerve volley reaches the cord first and has enough time to inhibit the flexor H-reflex via the disynaptic Ia inhibitory pathway. The time course of this effect is shown on the left (solid lines) where the amount of inhibition is shown as a percentage of the control H-reflex size. The dotted lines on the same graph show the reciprocal inhibition recorded during a willed voluntary extension of the wrist. The radial nerve was blocked with local anaesthetic distal to the spinal groove so that no movement actually occurred. Despite this, the amount of inhibition was greatly increased, indicating facilitation of the Ia inhibitory interneurones from extensor to flexor muscles. This can also be seen in the raw data on the right (from Day, B.L., Obeso, J.A. and Rothwell, J.C., unpublished observations)

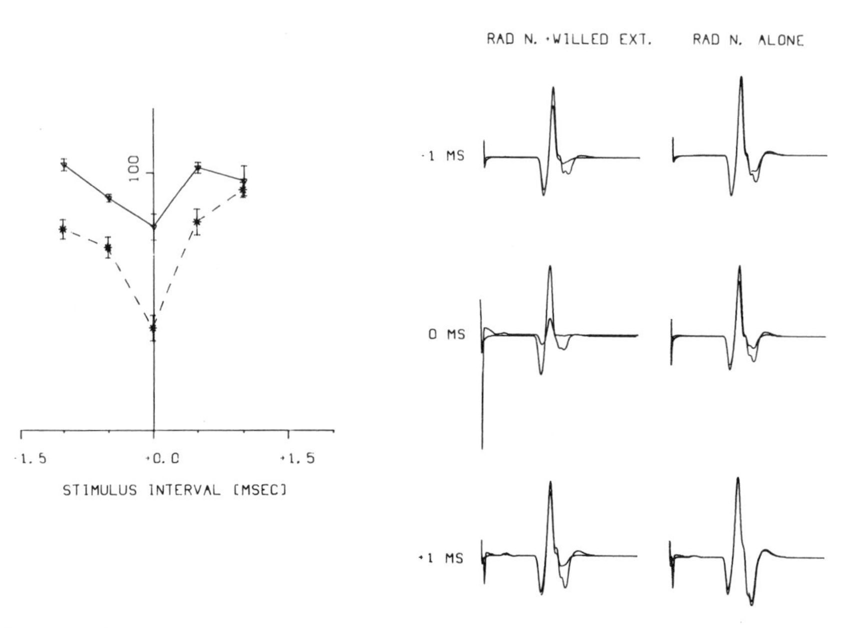

Renshaw Inhibition

The technique used to demonstrate Renshaw inhibition in man is too complex to detail here. It is discussed fully in the paper by Katz and Pierrot-Deseilligny (1982) and by Hultborn and Pierrot-Deseilligny (1979) (see end of Chapter 5). Like Ia reciprocal inhibition, it is possible to describe Renshaw inhibition, and to show that voluntary contraction of the muscles reduces the amount of Renshaw inhibition which can be recorded. The greater the force of contraction, the greater the depression of Renshaw inhibition. This is interpreted as an inhibition of the Renshaw cells by the descending voluntary command (see Chapter 5).

Cutaneous Reflexes

In contrast to the enormous literature on the reflex effects of muscle afferents, cutaneous reflexes have been very little studied. Their actions depend on the type or modality of the stimulus, its location on the body surface and, because most cutaneous reflexes are polysynaptic, they depend very much on descending supraspinal and other types of afferent inputs.

In the decerebrate cat, stimuli which would be regarded as painful usually result in flexion of the limb to withdraw it from the stimulus, accompanied by extension of the contralateral limb. This is known as ipsilateral flexion with crossed extension, which would presumably be of use in the intact animal, allowing the contralateral leg to take the weight of the animal after ipsilateral withdrawal. These flexion reflexes are considerably enhanced in the spinal preparation and may be triggered off by less noxious stimuli. In a higher decerebrate animal, with an intact red nucleus, more complex cutaneous reflexes may be elicited, involving non-noxious input from mechanoreceptors. An example of this is the placing reaction which can be evoked by stroking the dorsum of a paw against the edge of a table while suspending the cat above it. The leg will flex and then extend to lift its paw onto the surface. This may then be followed by the positive supporting reaction induced by pressure sensors in the ventral surface of the paw, leading to co-contraction of flexor and extensor muscles of the leg and the formation of a rigid supportive pillar.

One of the most complex cutaneous reflexes is the scratch reflex of the spinal dog. Very light, moving stimulation of the hairs, mimicking the passage of an insect, can give rise to a vigorous scratching by the ipsilateral foot. The scratching is rhythmic, continues until the stimulus is removed and is directed quite specifically towards the site of stimulation. Despite their range, very little is known about the pathways which mediate these cutaneous effects.

Nociceptive Cutaneous Reflexes in Man

In man, similarly, cutaneous reflexes may be divided into those evoked by noxious stimuli, and those produced by relatively innocuous tactile stimuli. The former are easily demonstrated even in relaxed muscles, whereas the effect of the latter is best seen in the reflex modulation of muscle activity during tonic voluntary contractions. As noted above, cutaneous inputs also modulate transmission in reflex pathways from other afferent fibres.

Noxious mechanical or electrical stimulation sufficient to evoke a distinct sensation of pain in the skin gives rise to withdrawal responses. These are related to the flexion reflexes described in animals, in that they produce withdrawal of the limb from the stimulus. However, careful study in man has revealed that these reflexes are rather less stereotyped than the conventional flexion reflex. The rule of thumb is that the reflex elicited by noxious stimulation of a single point on the skin will produce reflex muscle activity which is appropriate to *withdraw that area* from the stimulus (Figure 6.9). If the stimulus

Figure 6.9: Top: Schematic Drawing of the Reflex Pattern of Movement Obtained by Painful Cutaneous Stimulation of the Ball of the Great Toe (left) and Buttock (right). Filled areas denote the trunk muscles involved in the reflexes.
Bottom: EMG Recordings of Reflex Responses from Flexor Hallucis Brevis (Upper Trace of Each Pair in a and b) and Extensor Hallucis Brevis (Lower Trace of Each Pair). In column A, the stimulus was given to the ball of the foot and in B it was given to the ball of the great toe. Row a shows responses recorded at rest: they begin with a stimulus artefact produced by the train of electrical stimulating pulses. These are followed by a reflex in the flexor muscle in A, whereas in B, the response is in the extensor muscle. When standing, such activity would be appropriate to withdraw the ball of the foot from the stimulus in A (the normal plantar response of clinical neurology) and withdraw the ball of the toe from the stimulus in B. Records in row b were made during contraction of the extensor (A) and flexor (B) muscles. They show reciprocal inhibition of the extensor muscles in A and of the flexor muscles in B. Note the different time scales in A and B (50Hz marker on bottom trace) (from Kugelberg, Eklund and Grimby, 1960; Figures 1 and 10, with permission)

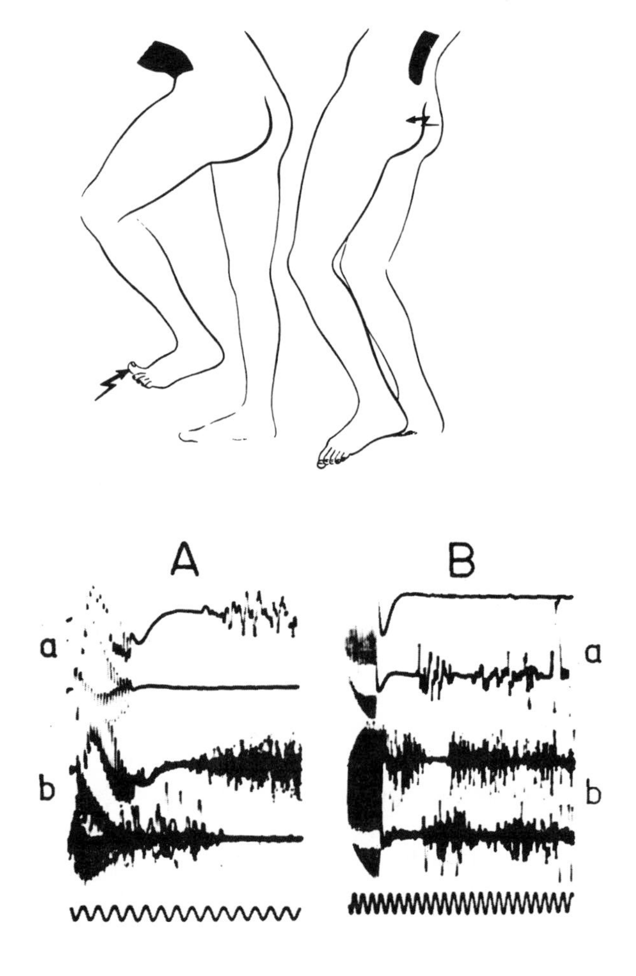

is, for example, over the front of the thigh, the legs will be extended at the hip whereas if it is on the back of the thigh, the legs will be flexed.

In man, particularly interesting variations on this theme occur after stimulation of the sole of the foot. Normally, this results in flexion at the knee, plantar flexion at the ankle and flexion of the toes. In clinical neurology, this *plantar response* is usually produced by firm stroking of the lateral aspect of the sole with a blunt object. It serves to withdraw the sole of the foot from the site of stimulation. However, the response changes after damage to the pyramidal tract. Physiological flexion is produced at all joints: the knee flexes and both the ankle and toes dorsiflex. This is known as Babinski's sign ('the upgoing big toe'). It is a pure flexion reflex quite distinct from the subtle response in normal man. The normal response is believed to be adapted from the true flexion reflex to allow withdrawal of the foot sole, while maintaining contact of the toes with the ground. With man's bipedal gait, this adaptation is of obvious advantage in postural stability. In fact, it is only when noxious stimuli are applied directly to the ball of the toe that a full flexion reflex is elicited in normal man. These nociceptive reflexes, sometimes known by electromyographers as RA III reflexes, have a latency of 100ms or so, and probably involve the Aδ category of afferent fibres. Unmyelinated C fibres, conducting at $1ms^{-1}$, are too slow to be involved in the initial part of the response (see Kugelberg, Eklund and Grimby, 1960, for original experiments).

Low Threshold Cutaneous Reflexes in Man

Tactile reflexes are difficult to elicit in relaxed muscles. When detected in the electromyogram, they usually have a shorter latency and a lower threshold than the nociceptive responses, and adapt very rapidly to repeated presentation of the stimulus. They are labelled RA II responses by some authors. Because they adapt so rapidly, these responses are more easily seen as fluctuations in the level of activity in tonically contracting muscles. The muscles of the hand and forearm have been extensively studied in this way. The procedure is to ask the subject to produce a sustained contraction of one or more muscles and then to give small electrical stimuli to the cutaneous nerves of the fingers. An average response of one of the hand muscles is shown in Figure 6.10. There are three main phases: a short latency excitation (E1) followed by inhibition (I1) and later excitation (E2).

The E1 and I1 components are believed to be mediated by spinal circuits, whereas the E2 response is thought to traverse a transcortical reflex pathway. Thus the E2 response is small or absent or delayed in patients with lesions of the motor cortex, dorsal columns or pyramidal tract. Motor cortical damage also reduces the size of the I1 component. Because of this it has been suggested that the I1 response depends on tonic facilitation of spinal interneurones from the brain. In confirmation of this, the I1 response is present at normal latency in patients with intact, but slowly conducting descending pathways, although their E2 response is delayed.

Figure 6.10: Cutaneous Reflex Responses from the Left (A) and Right (B) First Dorsal Interosseous Muscle Following Electrical Stimulation of the Digital Nerves of the Index Finger. The records are from a patient with a cerebral angioma in the region of the right motor cortex. The patient was required to make a steady isotonic contraction of the muscle while 512 stimuli were given at 3Hz to the index finger. The traces represent the average modulation of EMG activity that the stimulus produced. On the right, intact, side (B), the response consists of an initial excitation, followed by a depression and later excitation. These are the E1, I1 and E2 phases of the reflex. The onset of E1 (35ms) occurs at a latency compatible with operation of a spinal reflex arc. This can be shown by measuring the F + M latency in the same muscle (arrow). E2, which begins at about 60ms and I1 are not present on the left side. E2 is thought to be due to activity in a transcortical reflex loop. I1 is believed to be a spinal reflex, dependent upon the presence of tonic descending input from the brain. Damage to the motor cortex prevents the appearance of both components (from Jenner and Stephens, 1982; Figure 5, with permission)

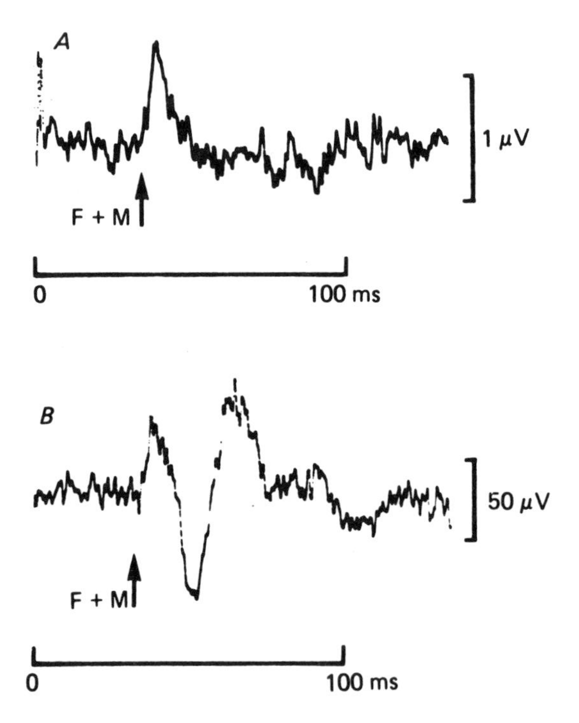

The E2 component also shows an interesting development with age. It is completely absent at birth, and only begins to appear in the second year of life. This parallels the development and myelination of the corticospinal tract during the same period, so that maturation of the reflex may depend on maturation of its anatomical pathway.

A second feature of these cutaneous effects is their functional distribution to different fractions of the motoneurone pool. At the level of the single motor unit, it has been shown that electrical stimulation of cutaneous nerves can produce overall inhibition of first-recruited units and excitation of later-recruited units.

The implications of this result on the firing order of motor units have been discussed in Chapter 4 (see Jenner and Stephens, 1982, for further details of cutaneous reflexes).

Pathophysiology of the Stretch Reflex: Disorders of Muscle Tone

If a normal joint is moved passively while the subject is at rest, it will be felt that it is not completely flaccid, but has some degree of resistance to movement. Clinicians refer to this as the *tone* of the muscles around the joint. In pathological conditions, this may be increased (hypertonus) or decreased (hypotonus).

Two mechanisms contribute to the resistance felt during passive manipulation of limb segments: the inherent viscoelastic properties of the muscle itself and the tension set up in the muscle by reflex contraction caused by muscle stretch. In a normal muscle at complete rest, the stretch reflex component of tone is not present. This is partly because of the very low level of background fusimotor discharge in relaxed human muscles. Because of this, the afferent response to such relatively slow movements is small. In addition, the central level of α-motoneurone excitability is very low and together these factors explain the lack of reflex muscle activity during manipulation. In fact, muscle tone is the same in relaxed normal individuals as it is in totally anaesthetised patients, in whom fusimotor discharge is absent.

Hypotonus

From this, it might be assumed that for hypotonus to occur, there must be some fundamental change in the intrinsic mechanical properties of muscle. However, this is probably a relatively rare occurrence. Under most conditions, it is extremely difficult to ensure that muscles are completely relaxed during manipulation, and a small degree of contraction probably leads to some stretch reflex contribution to muscle tone. This is the normal state of patients during clinical examination, particularly if they are at all anxious. Patients with detectable hypotonus are either perfectly relaxed (and hence normal) or, more commonly, they have a diminished stretch reflex contribution to muscle tone.

This can be seen in patients with lesions of the cerebellar hemispheres (see Chapter 9). In addition to a lack of tone on passive manipulation, patients with cerebellar hypotonia often have what is known as a 'pendular' knee jerk. In normal subjects, the leg movement dies away fairly rapidly after the initial jerk. However, in cerebellar patients, oscillations continue for some time. This is due to a decreased stretch reflex activation of the muscles during the jerk, which normally damp out the oscillations of the limb. In monkeys with surgical removal of the cerebellum, the decrease in tone has been shown to be due to a reduction in the tonic sensitivity of spindle afferents to muscle stretch. This has been attributed to a decrease in the γ_s discharge to spindles.

Hypotonia also is seen in the acute stage (two or three weeks) of spinal transection. In contrast to the cerebellar patient, the tendon jerk is absent during this period. This cannot be because of lack of fusimotor drive, since this is at such a low level in normal individuals, that it does not affect the size of spindle afferent volley to a rapid tendon jerk. In these patients, lack of the tendon jerk is probably due to the absence of any background excitation at central synapses in the tendon jerk pathway.

Hypertonus

Spasticity v. Rigidity. Hypertonus may be caused either by increase in the sensitivity of the stretch reflex or by the inability of patients to relax their muscle completely during examination. As described in Chapter 2, activation of muscle increases its stiffness, and this can be felt quite clearly in some rare conditions which result in continuous involuntary muscle activity such as Isaac's syndrome or 'Stiff Man' syndrome.

Most commonly, however, hypertonus is caused by increased activity in stretch reflex pathways. Two forms of hypertonus are recognised: spasticity and rigidity. There are several differences between them, although they both have in common an increased resistance to passive limb movement. Spasticity is seen after lesions of the 'upper motoneurone' or in the chronic stage of spinal transection. It is characterised by increased tendon jerks, the clasp-knife phenomenon and a particular *distribution* of increased tone to the flexors of the arm and to the extensors of the leg. Rigidity is seen in patients with Parkinson's disease. Tendon jerks are normal, there is no clasp-knife phenomenon and the increase in tone is distributed evenly between flexors and extensors.

(Unfortunately, the experimentally produced condition in animals which corresponds most closely to spasticity in man is known as decerebrate *rigidity*. It is a confusion of terms which it is now too late to change. Decerebrate rigidity does not correspond to clinically defined rigidity in man.)

Several theories have been put forward to explain the presence of exaggerated tendon jerks in spasticity but not rigidity. For example, both forms of hypertonus could be caused by an increase in the resting discharge to the spindles. In rigidity, a specific increase in γ_s discharge would mean that the spindle sensitivity to rapid stretches such as a tendon tap would, if anything, be decreased, whereas the sensitivity to more slowly applied static stretch would be increased. The increased tendon jerk of spasticity could be due to additional γ_d action. Neither of these explanations is correct. Experimental models of spasticity in monkeys have been produced by lesions of the motor cortical areas, and in these animals tone is increased while the sensitivity of the spindles is unchanged. Similarly, the microneurographic studies so far performed in man show no evidence of increased spindle sensitivity in patients with spasticity. Recording from rigid patients with Parkinson's disease may show an increased spindle sensitivity to stretch, but this is only because of the difficulty which these patients have in relaxing completely. The spindle discharge is compatible with the low resting levels of EMG activity, rather than an increased γ-drive to the spindles. The hypertonus of both spasticity and rigidity appears to be due to changes in the excitability of central synapses.

This does not clear up the problem of the differences between spasticity and rigidity. If the increased tone is due to hyper-excitability of central mechanisms, then why are tendon reflexes enhanced in spasticity but not in rigidity? The most likely explanation is as follows. The tendon jerk depends on a phasic afferent input which excites motoneurones predominantly via the monosynaptic stretch reflex pathway. (Careful measurements actually show that there is time for other di- or trisynaptic linkages to participate as well as the monosynaptic route.) However, clinical examination of tone depends on giving relatively slow stretches to the muscle. This produces a maintained afferent input to the cord which may activate motoneurones via polysynaptic pathways quite separate from the monosynaptic route. Such pathways would include, for example, the transcortical stretch reflex (see, for example, Tatton, Bedingham, Verrier *et al.*, 1984) or activity in circuits receiving muscle spindle secondary input from group II afferents.

In spasticity, there is an increased gain in the monosynaptic reflex arc. This certainly leads to an increased gain of the spinal stretch reflex, and may contribute to the increase in clinically evaluated muscle tone. Increased excitability in other pathways probably also contributes, although of course, the transcortical route cannot be active in patients with lesions of the upper motoneurone. In rigidity, the monosynaptic reflex arc is normal (hence the normal tendon jerks). Increased tone is due solely to heightened excitability in the polysynaptic and/or slowly conducting stretch reflex pathways, at both spinal and cortical level.

The other differences between spasticity and rigidity are the distribution of increased tone and the presence of the clasp-knife phenomenon. The reason for the former is not known in detail, but an analogy can be drawn with the distribution of tone in the decerebrate cat. In both cat and man, the spasticity is predominant in anti-gravity muscles. In the cat these are the extensors of all four limbs, whereas in man they include the flexors of the upper limb and the physiological extensors of the lower limb. (Among the latter are the anatomical *flexors* of the toes, which are anti-gravity muscles, and which for physiological purposes are always considered together with the other anatomical extensors of the calf.) (See Burke, 1983, for a critical discussion of the mechanisms of spasticity.)

Reasons for the Increased Tendon Jerks in Spasticity. The increased tendon jerk of the spastic patient is due to an increased excitability of central synapses involved in the reflex arc. The question remains as to which synapses are changed and how are they changed. At present, it is thought that both pre- and postsynaptic factors are involved.

Spasticity is caused by interruption of descending input to the cord from the brain. The physical destruction of axons causes degeneration of their terminals on the spinal motoneurones and interneurones. When this occurs, two things happen in the spinal cord. Over a short time interval, a proportion of the remaining intraspinal synapses, which had until the lesion been inactive, become active. These so-called 'silent synapses' appear to have no function in the intact animal, but following the lesion are called into action by some unknown mechanism. Second, over a longer time period, the remaining spinal synapses sprout new terminals which take over the spaces left vacant by degenerating terminals of descending axons. The nett effect of both these phenomena is to increase the effectiveness of spinal cord connections. In the case of the pathway involved in the tendon jerk, this includes the Ia-motoneuronal synapses and perhaps the di- and trisynaptic relays from afferents to motoneurones that could be involved in the tendon jerk.

Synaptic efficiency may also be increased by presynaptic mechanisms, although less is known about these. Vibration inhibition of the H-reflex provides one piece of evidence in favour of this hypothesis. In normal man, vibration of a muscle produces inhibition of homonymous H-reflexes by a premotoneuronal mechanism (see above). In spasticity, this effect is reduced, as if the amount of presynaptic inhibition of Ia afferent terminals is decreased. If so, then it may be that removal of *tonic* levels of presynaptic inhibition can increase the effectiveness of the Ia-motoneuronal synapse and increase the size of the tendon jerk.

Clasp-knife Phenomenon. The clasp-knife phenomenon is best seen in the knee extensor muscles of spastic patients. As the knee is flexed, resistance builds up gradually, and then as a certain point is reached, the resistance suddenly melts away and the remainder of the stretching movement can be completed with little force. The mechanism of this effect has been debated for some time.

It was originally suggested that the inhibitory actions of Ib afferents might mediate the clasp-knife reaction. It was thought that these fibres had a high threshold to muscle tension and hence could account for the abrupt onset of stretch reflex inhibition, which is characteristic of the clasp-knife effect. The phenomenon was believed to be a protective reaction to prevent production of dangerously high tensions in muscle. However, with the discovery that tendon organs are very sensitive to actively produced muscle tension, this theory could no longer hold: tendon organs do not fire *only* when dangerously high muscle tensions are produced. Also, it was found that the underlying inhibition responsible for the clasp-knife reaction far outlasts the actual reduction of muscle tension, whereas Ib activity declines in parallel with muscle tension. Ib effects are not now believed to be the sole factor in the phenomenon, although they may contribute to its initiation.

Closer examination of the clasp-knife phenomenon has suggested other mechanisms. Figure 6.11 is one of the few published records of a clasp-knife phenomenon in spastic man. What appears to happen is that on flexing the knee

Figure 6.11: Spastic Stretch Reflex and the Clasp-knife Phenomenon. Quadriceps EMG (bottom trace) was elicited by stretching the muscle from the extended position of the knee joint to 90° flexion. Velocity of stretch is shown (second trace), with joint position (third trace) and muscle tension (fourth trace) (estimated by force exerted by the examiner on a transducer at the ankle). On flexing the knee, tension and EMG build up rapidly, and then subside almost completely before flexion is completed. This sudden decrease in resistance to movement is known as the clasp-knife phenomenon (from Burke, Gillies and Lance, 1970; Figure 1, with permission)

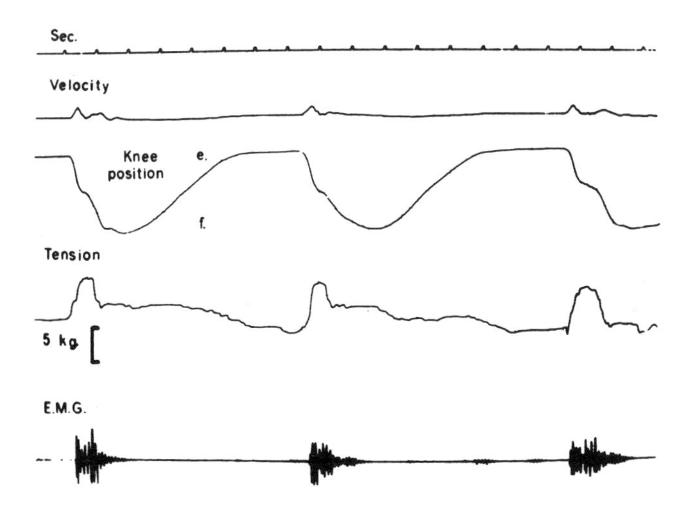

from the extended position there is a slow, velocity-sensitive build-up of tension in the quadriceps muscle which opposes movement. As the reflex tension increases, it begins to oppose the movement imposed by the examiner, and sometimes is so powerful as to stop the movement entirely. The reduction of stretch velocity reduces the velocity-sensitive input to the reflex, and reflex tension declines. The examiner, noticing the decrease in speed of movement, then resumes a rapid knee flexion. However, at this angle of the knee, the movement is no longer opposed by reflex muscle contraction. It is the absence of reflex contraction at this stage that constitutes the clinical impression of the clasp-knife phenomenon. The melting away of resistance in the clinical situation therefore is due to two factors: (1) reduction in stretch velocity by the opposing muscle contraction; and (2) a decrease in sensitivity of the quadriceps stretch reflex.

The latter effect has been described in some detail by Burke and colleagues who showed that the velocity-sensitive stretch reflex in quadriceps is smaller if stretch is imposed when the knee joint is flexed, than it is when the knee is extended. They refer to this as a position-dependent inhibition of quadriceps stretch reflex, and argue that this is all that is necessary to produce a clinical clasp-knife phenomenon. The reasoning is that quadriceps stretch reflex activity decreases throughout the range of joint movement, but that this is only noticed by the examiner when he tries to increase the rate of knee flexion following a reflex-induced slowing. The novelty of the explanation is that the clasp-knife phenomenon is no longer regarded as a *sudden* onset of active inhibition. Instead, it is possible to get the same clinical impression from a gradual build-up of reflex inhibition.

Receptors responsible for this inhibitory activity have been suggested to be group II afferents from muscle spindle secondary endings. All group II afferents once were thought to have inhibitory effects on extensor muscles (that is, they belonged to the flexor reflex category of afferent fibres), and it was postulated that with increasing muscle stretch, there would be a length-dependent inhibition of the extensor muscles. However, this explanation in its simplest form fell from favour when it was discovered that group II spindle afferents from extensor muscles could have, like the group Ia fibres, autogenetic excitatory actions (see Chapter 4).

More likely candidates are the group II and III non-spindle afferents. Indeed, some afferents of the group III class have recently been discovered to respond to muscle stretch only when the tension reaches a high level. Such afferents are ideal fibres to sharpen the point at which inhibition becomes noticeable as the clasp-knife phenomenon. Thus the clasp-knife phenomenon is no longer regarded as a pure Ib effect, but is thought to be a complex phenomenon resulting from the actions of many different classes of afferent fibre. The reason that the reaction is not present in rigidity is usually ascribed to a lack of facilitation or active inhibition of the requisite spinal interneuronal pathways (Rymer, Houk and Crago, 1979).

Changes in Muscle Tone Produced by Changes in the Mechanical Properties of Muscle. In the past, the discussion of muscle tone has focused very closely on the contribution of reflex mechanisms to the increased stiffness of muscle. However, it is becoming clear that other, purely muscle, factors may also contribute to the increased tone of spastic muscles. Disuse of weak muscles because of lack of voluntary power leads to changes within the muscle, including fibrosis and probably even changes in the properties of the muscle fibres themselves (see Chapter 3). Such changes make the muscle stiffer to move, even in the absence of any muscle activity. The contribution of these factors to the clinical impression of increased tone in spasticity is, however, very much a matter of controversy (see Berger, Horstmann and Dietz, 1984; Dietz, Quintern and Berger, 1981).

Pathophysiology of Other Reflex Pathways

Effects of Spinal Cord Transection

In all mammals, acute transection of the spinal cord always produces an initial period of areflexia, during which all reflexes have extremely high thresholds. The duration of this areflexia varies, being only an hour or so in the cat, to days or weeks in man. It is caused by the sudden removal of facilitatory descending inputs to the cord from supraspinal structures, and is followed by the gradual reappearance and eventual hypersensitivity of some reflexes.

In man, acute transection is immediately followed by flaccid paralysis and areflexia. The first signs of recovery are the automatic reflex emptying of the bladder and rectum, followed by the appearance of cutaneous reflexes in some muscles. These invariably result in flexion withdrawal of the limb and may be accompanied by emptying of the bladder and rectum (the *mass* reflex). Later on, tendon jerks appear and, after some months, the enhanced flexor reflexes may lead to a permanent posture of flexion of the hip, knee and ankle, resulting in paraplegia in flexion. The stretch reflexes are enhanced in this state, but the predominant final effect of spinal transection is to enhance transmission in flexor reflex pathways.

Disynaptic Reciprocal Inhibitory Pathway

This pathway appears to be affected in cases of capsular hemiplegia. These patients have a marked spasticity of lower limb extensor muscles, with relatively preserved voluntary power. The flexor muscles are weak and although they are not usually considered clinically to show signs of spasticity, there is evidence that the monosynaptic arc is hyperexcitable in these muscles, just as it is in the extensors. The reason for saying this is that H-reflexes, which in normal man cannot be found in the anterior tibial muscles, are readily demonstrated in spastic patients. However, in clinical examination, the increased flexor stretch reflexes are concealed by the overwhelming spasticity of the extensor muscles. The imbalance further favours the extensors by the distribution of reciprocal inhibition. There is no inhibition between the anterior tibial muscles and triceps surae, while the inhibition from triceps surae onto the tibialis anterior is very strong. Thus, the flexor muscle group is said to be doubly weakened in spasticity due to capsular lesions, by a smaller direct facilitation of the flexor α-motoneurones and also by a powerful disynaptic inhibition from spindles of the extensor muscles.

Patients with spasticity of spinal origin do not show this imbalance of reciprocal inhibition. Inhibition is distributed fairly evenly from extensors to flexors and vice versa. There is thus no tendency to cancel out the release of flexor reflexes which occurs with lesions below the brainstem (see Yanagisawa, Tanaka and Ito, 1975).

Renshaw Cell Inhibition

Renshaw inhibition also has been explored in spastic patients with lesions of the upper motoneurone. Unlike reciprocal Ia inhibition, no abnormalities have been seen at rest. However, as might be expected because of the lesion in descending pathways, the control of Renshaw cell activity during voluntary contraction is severely diminished. Renshaw cell activity remains high even during voluntary activation of muscles (see Katz and Pierrot-Deseilligny, 1982).

References and Further Reading

Review Articles and Books (see also references to Chapter 5)

Barnes, W.J.P. and Gladden, M.H. (eds.) (1985) *Feedback and Motor Control in Invertebrates and Vertebrates,* Croom Helm, London

Burke, D. (1983) 'Critical Examination of the Case For or Against Fusimotor Involvement in Disorders of Muscle Tone' in J.E. Desmedt (ed.), *Advances in Neurology, Vol. 39,* pp. 133–50

Lance, J.W., Burke, D. and Andrews, C.J. (1973) 'The Reflex Effects of Muscle Vibration' in J.E. Desmedt (ed.), *New Developments in Electromyography and Clinical Neurophysiology, Vol. 3,* Karger, Basel, pp. 444–62

Marsden, C.D., Rothwell, J.C. and Day, B.L. (1983) 'Long-latency Automatic Responses to Muscle Stretch in Man: Origin and Function' in J.E. Desmedt (ed.), *Advances in Neurology, Vol. 39,* pp. 509–39

Merton, P.A. (1953) 'Speculations on the Servo-control of Movement' in J.L. Malcolm, J.A.B. Gray and G.E.W. Wolstenholme (eds.), *The Spinal Cord,* Little Brown, Boston, pp. 183–98

Pierrot-Deseilligny, E. and Maziers, L. (1984) 'Circuits Reflexes de la Moelle Epiniere Chez l'Homme. Controle au Cours du Mouvement et Role Fonctionnel', *Rev. Neurologique, 140,* pp. 605–14 and 681–94

Prochazka, A. and Hulliger, M. (1983) 'Muscle Afferent Function and Its Significance for Motor Control Mechanisms During Voluntary Movements in Cat, Monkey and Man in J.E. Desmedt (ed.) *Advances in Neurology, Vol. 39,* Raven Press, New York, pp. 93–132

Taylor, A. and Prochazka, A. (1980) *Muscle Receptors and Movement,* MacMillan, London

Original Papers

Berger, W., Horstmann, G. and Dietz, V. (1984) 'Tension Development and Muscle Activation in the Leg During Gait in Spastic Hemiparesis: Independence of Muscle Hypertonia and Exaggerated Stretch Reflexes', *J. Neurol. Neurosurg. Psychiat., 47,* pp. 1029–33

Burke, D., Andrews, C.J. and Lance, J.W. (1972) 'The Tonic Vibration Reflex in Spasticity, Parkinson's Disease and Normal Man', *J. Neurol. Neurosurg. Psychiat., 35,* pp. 477–86

——— Gandevia, S.C. and McKeon, B. (1983) 'The Afferent Volleys Responsible for Spinal Proprioceptive Reflexes in Man', *J. Physiol., 339,* pp. 535–52

——— Gillies, J.D. and Lance, J.W. (1970) 'The Quadriceps Stretch Reflex in Human Spasticity', *J. Neurol. Neurosurg. Psychiat., 33,* pp. 216–23

Creed, R.S., Denny-Brown, D., Eccles, J.C. *et al.* (1932) *Reflex Activity of the Spinal Cord* (2nd edn), copyright Oxford University Press (1972), Oxford

Day, B.L., Marsden, C.D., Obeso, J.A. *et al.* (1984) 'Reciprocal Inhibition Between the Muscles of the Human Forearm', *J. Physiol., 349,* pp. 519–34

DeGail, P., Lance, J.W. and Nielson, P.D. (1966) 'Differential Effects on Tonic and Phasic Reflex Mechanisms Produced by Vibration of Muscles in Man', *J. Neurol. Neurosurg. Psychiat., 29,* pp. 1–11

Dietz, V., Quintern, J. and Berger, W. (1981) 'Electrophysiological Studies of Gait in Spasticity', *Brain, 104,* pp. 431–49

Eklund, G., Hagbarth, K.-E., Hagglund, J.V., *et al.* (1982) 'Mechanical Oscillations Contributing to the Segmentation of the Reflex Electromyogram Response to Stretching Human Muscles', *J. Physiol., 32,* pp. 65–77 (see also pp. 79–90)

Eldred, E., Granit, R. and Merton, P.A. (1953) 'Supraspinal Control of the Muscle Spindles and its Significance', *J. Physiol.,* pp. 498–523

Hammond, P.H., Merton, P.A. and Sutton, G.C. (1956) 'Nervous Gradation of Muscular Contraction', *Br. Med. Bull., 12,* pp. 214–18

Houk, J.C. and Rymer, W.Z. (1981) 'Neural Control of Muscle Length and Tension' in V.B. Brooks (ed.), *Handbook of Physiology, Sect. 1, Vol. II, Part 1,* Williams and Wilkins, Baltimore, pp. 257–323

Jenner, J.R. and Stephens, J.A. (1982) 'Cutaneous Reflex Responses and their Central Nervous Pathways Studied in Man', *J. Physiol., 333,* pp. 405–19

Katz, R., and Pierrot-Deseilligny, E. (1982) 'Recurrent Inhibition of γ-Motoneurones in Patients with Upper Motor Neuron Lesions', *Brain, 105,* pp. 103–24

Kugelberg, E., Eklund, K. and Grimby, L. (1960) 'An Electromyographic Study of the Nociceptive Reflexes of the Lower Limb. Mechanism of the Plantar Responses', *Brain, 83,* pp. 394–410

Marsden, C.D., Merton, P.A. and Morton, H.B. (1973) 'Is the Human Stretch Reflex Cortical Rather than Spinal?' *Lancet, i,* pp. 759–61

Matthews, P.B.C., and Rushworth, G. (1958) 'The Discharge from Muscle Spindles as an Indicator of γ Efferent Paralysis by Procaine', *J. Physiol., 140,* pp. 421–6

Matthews, P.B.C. (1984) 'Evidence From the Use of Vibration that the Human Long-latency Stretch Reflex Depends Upon Spindle Secondary Afferents', *J. Physiol., 348,* pp. 383–416

Nichols, T.R. and Houk, J.C. (1976) 'The Improvement in Linearity and the Regulation of Stiffness that Results from the Action of the Stretch Reflex', *J. Neurophysiol., 39,* pp. 119–42

Pierrot-Deseilligny, E., Bergego, C., Katz, R., *et al.* (1981) 'Cutaneous Depression of Ib Reflex Pathways to Motoneurones in Man', *Exp. Brain Res., 42,* pp. 351–61

Rymer, W.Z., Houk, J.C. and Crago, P.E. (1979) 'Mechanisms of the Clasp-Knife Reflex Studied in an Animal Model, *Exp. Brain Res., 37,* pp. 93–113

Tanaka, R. (1974) 'Reciprocal Ia Inhibition During Voluntary Movements in Man', *Exp. Brain Res., 21,* pp. 529–40

Tatton, W.G., Bedingham, W., Verrier, M.C., *et al.* (1984) 'Characteristic Alterations in Responses to Imposed Wrist Displacements in Parkinsonian Rigidity and Dystonia Musculorum Deformans', *Can. J. Neurol. Sci., 11,* pp. 281–7 (see also pp. 228–96)

Vallbo, A.B. (1970) 'Slowly Adapting Muscle Receptors in Man', *Acta Physiol. Scand., 78,* pp. 315–33 (see also *80* pp. 552–66)

Yanagisawa, N., Tanaka, R. and Ito, Z. (1975) 'Reciprocal Ia Inhibition in Spastic Hemiplegia of Man', *Brain, 99,* pp. 555–74

7 ASCENDING AND DESCENDING PATHWAYS OF THE SPINAL CORD

Ascending Pathways

There are a large number of ascending pathways, all of which have important direct projections to areas of the brain concerned with movement. In this chapter, a short summary of the types of sensory fibre which contribute to each tract, and the cells of origin of the tracts, will be given. The data is mostly from anatomical studies in the cat (and monkey); comparable human studies have not yet been performed.

Afferent Input to the Spinal Cord (Figure 7.1)

The site at which sensory fibres terminate within the dorsal horn of the spinal cord is related to the diameter of the afferent fibre. Large diameter fibres enter more medially than small fibres within the dorsal root entry zone, and descend deeper into the grey matter before forming synapses. Many fibres branch on entry to the cord and give off an ascending and descending axon which terminate in nearby segments. Rexed's lamina I and the dorsal part of lamina II receive input from the smallest myelinated (Aδ) fibres innervating mechanosensitive nociceptors in the skin. Lamina II receives input from unmyelinated nociceptive (C) fibres, and the deeper laminae III-VI receive input from larger diameter (Aα-β) myelinated fibres. Group I muscle afferent axons send most of their input to lamina V and deeper. Apart from the majority of fibres of the dorsal column pathway, the main ascending tracts arise from cells in specific laminae of the spinal cord, and in some instances from particular subgroups of cells within a lamina.

Pathways to Thalamus and Cerebral Cortex

Dorsal Column Pathway (Figure 7.2). The axons which travel within the dorsal columns of the spinal cord are a more heterogenous group than was previously supposed. The 'classical' dorsal column system is made up of large myelinated dorsal root fibres which ascend immediately on entering the cord and terminate in the dorsal column nuclei of the medulla. Their cell bodies lie in the dorsal root ganglion and their axons, the longest in the whole body, are known as first-order (or primary) afferent fibres. At the segment of entry to the dorsal columns, new fibres occupy the most lateral position, but as they ascend, they are displaced medially by fibres entering at successively higher levels. Afferents from the leg travel in the gracile fasciculus, and afferents from the arm in the cuneate fasciculus. This orderly behaviour results in a somatotopic organisation of the dorsal column fibres within the gracile and cuneate fasciculi.

Figure 7.1: Diagrammatic Representation of the Distribution of Cutaneous and Muscle Afferent Fibres to the Spinal Grey Matter (From Brown, 1981; Figure 10.1, with permission)

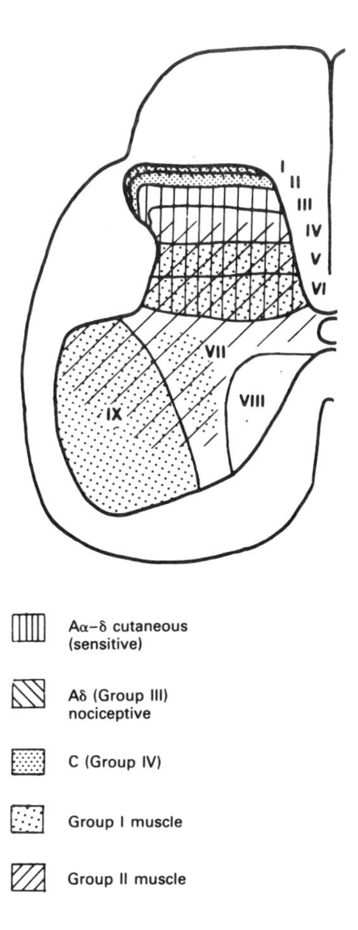

In addition to the first-order afferent fibres, the dorsal columns also contain many axons of second-order neurones which also terminate in the dorsal column nuclei. These fibres arise from neurones which have their cell bodies in lamina IV of the dorsal horn and receive their afferent input trans-synaptically from first-order dorsal root fibres. They are known as second-order fibres. Some of them ascend in the dorso-lateral funiculus, rather than in the dorsal columns and then terminate in the dorsal column nuclei.

Propriopsinal neurones also travel in the dorsal columns. These are fibres which connect together segments of the cord without projecting as far as the brain. Those within the dorsal columns are usually branches of first-order afferent fibres. These fibres may be of any diameter, not necessarily being limited to large Aα or Aβ range. The smaller the diameter, the less distance they

Figure 7.2: Main Anatomical Features of the Dorsal Column-Medial Lemniscal Pathway. A: Main routes taken by fibres carrying impulses from low threshold cutaneous and rapidly adapting deep receptors. Note that many dorsal column fibres are axons of second-order neurones with cell bodies in the dorsal horn. Other second-order axons ascend in the dorso-lateral columns to terminate in the dorsal column nuclei.
B: Main Routes Taken by Fibres Carrying Impulses from Slowly Adapting Receptors in Joints and Muscles. Note that those from the lower limb only travel in the dorsal columns for a short distance before synapsing in Clarke's column and ascending the dorsolateral columns with the dorsospinocerebellar tract. The axons relay in nucleus z and not in the dorsal column nuclei proper (from Brodal, 1981; Figure 2.10, with permission)

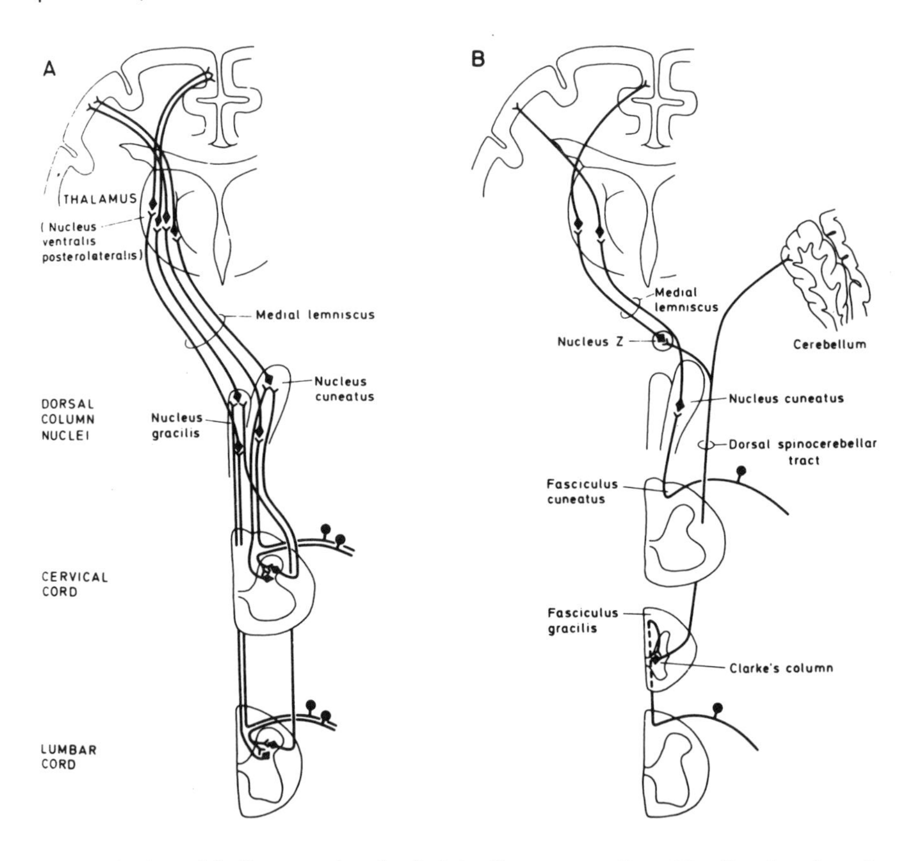

appear to travel before passing back into the grey matter. Finally, the dorsal columns also appear to contain some descending fibres, arising from the dorsal column nuclei.

The ascending dorsal column fibres terminate in the cuneate and gracile nuclei in the medulla. Anatomical and physiological studies suggest that there are two patterns of termination in the dorsal column nuclei. A 'cluster' region receives input mainly from the distal parts of the body, via first-order afferent fibres, whereas a 'reticular' zone receives more proximal inputs with a large

Figure 7.3: Main Pathways Taken by A, the Spinocervical Tract and B, the Spinothalamic Tract. Note that the spinothalamic tract may also terminate in parts of the thalamus other than the VPL nucleus (from Brodal, 1981; Figure 2,12, with permission)

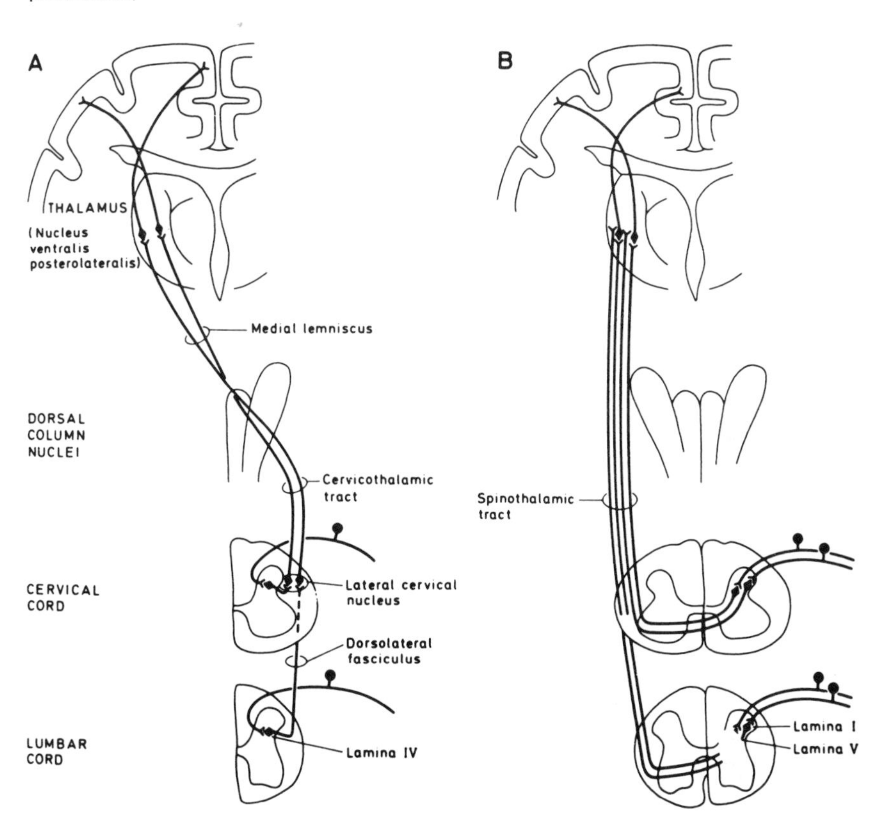

contribution from second-order afferents.

Cells in the 'cluster' region are modality specific and tend to have smaller receptive fields than those in the 'reticular' zone. The latter have larger receptive fields and often receive inputs from various types of receptor, including painful ones. Input to the 'cluster' region comes mainly from large diameter peripheral afferent fibres: units innervating low threshold mechanoreceptors in skin (hair follicles, Meissner corpuscles, Merkel discs, Ruffini endings and Pacinian corpuscles), joint receptors and muscle receptors.

Exceptions to these rules of termination are afferents from the slowly adapting joint and group I muscle receptors from the lower limbs. Afferents from these receptors only travel in the dorsal columns for a short distance. Above lumbar levels, the fibres travel in the dorso-lateral funiculus. Many of them are collaterals of the dorsal spino-cerebellar tract, and as well as sending fibres to the spinal parts of the cerebellar cortex, also terminate in nucleus z, just rostral to the gracile nucleus. Nucleus z, then, is an important relay site transmission of proprioceptive input from the legs to the cerebral cortex.

Afferent fibres from the 'reticular' and 'cluster' zones of gracile and cuneate nuclei and from nucleus z, ascend in the medial lemniscus in the brainstem, cross the midline in the medulla and terminate in the VPL nucleus of the thalamus (see below). The fibres do not give off any collateral branches to other structures en route.

Spinocervical Pathway (Figure 7.3A). The spinocervical tract has been recognised only for the past 30 years. It appears to be particularly prominent in the cat, and rather smaller in monkey and in man. The cervical nucleus consists of a column of cells in the first and second cervical segments of spinal cord, just ventrolateral to the dorsal horn. In the cat it consists of a well-defined mass of cells (the lateral cervical nucleus), but in man there is no such distinct structure. It is assumed that the cells which form the lateral cervical nucleus in the cat are more diffusely scattered in man. It receives input ipsilaterally from all levels of spinal cord, from second order afferents whose cell bodies lie mainly in lamina IV of the dorsal horn, and whose axons ascend in the dorso-lateral funiculus. In this position, they are intermixed with axons of many other ascending (dorsal spino-cerebellar) and descending (corticospinal, rubrospinal and raphe-spinal) tracts, making it impossible to determine their role using lesioning techniques.

The main source of input to the spinocervical tract in the cat comes from hair follicle receptors, although there are cells in the cervical nucleus which also respond to noxious stimuli. Such noxious input is under the control of descending systems, which probably act presynaptically on the input to the cells of origin in the dorsal horn. It is not known which descending tracts mediate this effect.

The efferent axons of the cervical nucleus cross in the high cervical cord and travel with lemniscal fibres to the VPL nucleus of the thalamus without synapsing in the dorsal column nuclei.

Spinothalamic Tract (Figure 7.3B). The spinothalamic tract arises from dorsal and intermediate laminae neurones of the spinal grey matter whose axons cross the cord at segmental level and ascend in the contralateral ventrolateral funiculus. Fibres join the tract medially, but are pushed lateral as they ascend the cord by fibres entering the tract at higher levels. In man, the spinothalamic tract is large: larger than in the monkey and much larger than in the cat. In the monkey, the cells of origin lie in layers I and V (and adjacent laminae IV and VI) in the cervical cord. At lumbar levels, a substantial number are found in layers VII and VIII.

The afferent axons from cells in lamina I tend to be of smaller diameter than those of deeper layers. They respond to intense mechanical or painful heat stimuli, as expected from the As terminations in layer I. Afferents from layer V

Figure 7.4: Three-dimensional Reconstruction of the Right Human Thalamus. The posterior part is separated from the rest by a cut, to display some of the internal structure. Only the rostral tip of the reticular nucleus (R) is shown. Input from trigeminal, medial leminiscal and spinothalamic afferents is shown entering the VPM (ventroposterior medialis) and VPL (ventroposterior lateralis) nuclei. Cerebellar input ends in the VL (ventrolateral) and pallidal input in the VA (ventroanterior) nucleus. More details of the precise regions of termination are described in the text and in Chapter 8 (from Brodal, 1981; with permission)

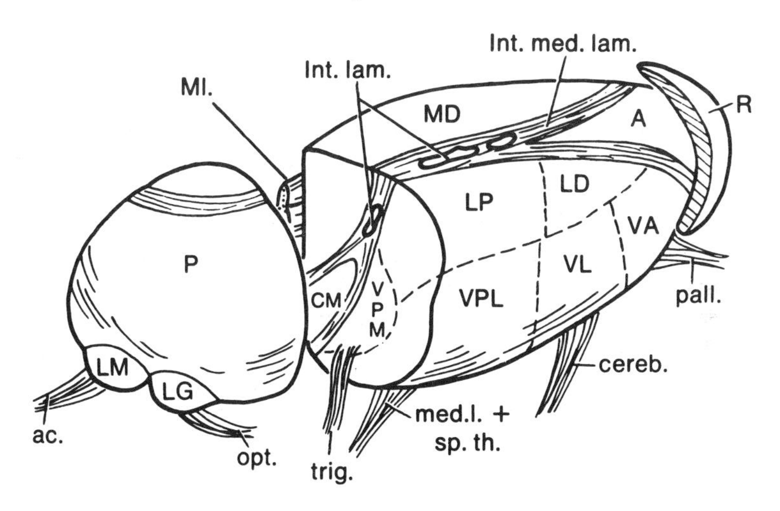

are larger and are activated by both light tactile and also by intense noxious stimuli, and are said to have a 'wide dynamic range'. Spinothalamic fibres from muscular and joint receptors arise from cells in laminae V, VII and VIII. The tract ascends through the medulla and pons occupying a space more dorsal than the leminiscal fibres, and terminates mainly in the VPL nucleus of the thalamus. Its fibres give off collaterals to the reticular formation en route.

The spinothalamic tract is sometimes subdivided into two components: the neospinothalamic tract and the paleospinothalamic tract. The neospinothalamic tract is said to convey sensations of touch and deep pressure, and is located in the ventromedial portion of the tract. In contrast, the paleospinothalamic tract is said to convey sensations of pain and temperature, and occupies the dorsolateral portion of the tract. However, there is some debate over this, with some authors claiming that the two types of fibre are completely intermixed within the anatomical spinothalamic tract. Even if this is true, it may be that the two types of fibre terminate in different portions of the thalamus (see below).

Thalamic Terminations of Somatic Afferents. Laminae of white matter subdivide the thalamus into its three major grey masses: the anterior, lateral and medial nuclear groups. More detailed subdivision has been attempted by many authors on anatomical grounds, the most commonly accepted being that of Walker. Walker's classification was developed from a study of monkey and chimpanzee and will be used here (Figure 7.4). The nomenclature for human thalamus was developed from this by Hassler (1959), and is considerably more detailed.

The ventral portion of the lateral cell mass of the thalamus is the main region of termination of somatosensory afferents. It is subdivided into an anterior (VA), lateral (VL) and posterior part, the posterior part itself being further subdivided into the ventralis posterior lateralis (VPL), ventralis posterior medialis (VPM) and the ventralis posterior inferior (VPI). All of these nuclei send efferent fibres to the cerebral cortex, with the exception of parts of VA and VPI. There are two other smaller areas of the thalamus where somatosensory afferents have been found to terminate: the nucleus centralis lateralis (CL), a part of the intralaminar nuclear group, and the medial part of the posterior thalamic nucleus (POm). This latter nucleus is a rather ill-defined cell mass at the caudal thalamus lying outside the more classifically defined ventroposterior and geniculate nuclei.

The vast majority of fibres in the medial leminiscus (which include afferents from the gracile and cuneate nuclei as well as from nucleus z and the lateral cervical nucleus) terminate in the VPL nucleus. Fibres from the gracile nucleus end most laterally and those from the cuneate nucleus more medially. Trigeminal afferents end in the VPM nucleus. Some small diameter leminiscal fibres also are found in the POm region, but their terminals are sparse. Spinothalamic afferents were once believed to terminate in VPL and to overlap with the leminiscal afferents. However, more recent studies have shown that the situation probably is not quite so simple as this. Spinothalamic fibres in the monkey have quite a different pattern of termination in VPL than those of the medial leminiscus. Endings are sparse in the central region and more dense along the borders of VPL. In particular, there is a region of dense innervation in the rostral part of VPL in the border region with VL. Smaller regions of spinothalamic termination are in the POm and the CL nuclei. The POm nucleus has been extensively studied as a possible 'pain' relay for the small diameter fibre component of the spinothalamic tract (the paleospinothalamic tract).

There have been many studies of the physiological properties of the cells in the VPL nucleus. As a rule, cells respond only to one modality of sensory input and are grouped together according to modality in continuous layers of tissue. Cells in the rostral and caudal parts of VPL are responsive to input from deep receptors (muscle spindles, tendon organs, joint receptors), whereas those in the central part of the nucleus receive cutaneous inputs. Because of the superimposed (mediolateral) topographical arrangement of the afferent fibres, such an organisation means that a small group of cells in any one part of the VPL nucleus responds to stimulation of one particular kind coming from one particular part of the body.

In deeply anaesthetised animals, the receptive fields of VPL neurones are very small but more recent investigations, using conscious animals, have tended to suggest that receptive field sizes in freely moving animals are larger. In particular, cells responding to noxious stimuli have been found in VPL. Such cells had previously been seen only in recordings from the posterior nuclear group (POm) of anaesthetised animals, and the finding of similar cells in VPL of conscious preparations has tended to focus attention away from POm as a possible candidate for specific transmission of painful sensations.

Pathways to the Cerebellum

Dorsal Spinocerebellar and Cuneocerebellar Tracts (see Figure 9.8 in Chapter 9). The thick myelinated axons of the dorsal spinocerebellar tract arise from the cells of Clarke's column (which is also known as the nucleus dorsalis). Clarke's column lies at the base of the dorsal horn in spinal segments T1 to L2 in man (to L3/L4 in cat) and receives monosynaptic input from group Ia and Ib muscle afferents and joint receptors. There is also an input from cutaneous, touch and pressure receptors and a smaller input from group II spindle afferents. As Clarke's column finishes at the second lumbar segment, afferent input from lower levels reaches it via the dorsal columns of the lower lumbar cord. The axons of cells in Clarke's column cross into the ipsilateral lateral column where they ascend dorsolaterally to the lateral corticospinal tract. Physiological studies have shown that their receptive fields are small and that many of the fibres carry modality-specific information.

There are two subgroups of neurones in the dorsal spinocerebellar tract. One group receives proprioceptive input from group Ia, Ib or II muscle afferents, the other receives input from cutaneous and high threshold muscle afferents. Within each group cells may receive convergent input from the various submodalities, but there are very few cells which receive convergence from both groups of inputs. Proprioceptive inputs sometimes are particularly strong. A single primary afferent fibre may give off 50 synapses onto the proximal dendrites of a dorsal spinocerebellar tract cell. Corresponding EPSPs are large (5mV), and the coupling between input and output strong. The fibres course through the inferior cerebellar peduncles and end as mossy fibres chiefly in the hindlimb region (see Chapter 9) of the intermediate zone of the ipsilateral cerebellar cortex.

The dorsal spinocerebellar tract does not convey any information from the forelimbs. Clarke's column only begins at the level of the first thoracic segment: its cervical cord analogue is the external (or accessory) cuneate nucleus, which lies in the medulla, lateral and rostral to the cuneate nucleus. This forelimb analogue of the dorsal spinocerebellar tract is known as the cuneocerebellar tract. It receives input from branches of dorsal root fibres entering the cord from C1-T4/T5, as well as a small component from cells in the cuneate and gracile nuclei. The bulk of the efferent fibres end in the forelimb regions of the ipsilateral cerebellar cortex, although a smaller number of fibres recently have been found to project to the rostral part of the VPL nucleus of the contralateral thalamus. This unexpected projection therefore is a further pathway in which proprioceptive input can reach the main somatosensory relay nucleus of the thalamus.

Ventral and Rostral Spinocerebellar Tracts. Like the dorsal and cuneocerebellar tracts, the ventral and rostral spinocerebellar tracts convey impulses from hind and forelimbs, respectively. The cells of origin of the ventral spinocerebellar tract are believed to be the 'spinal border cells' situated in the dorsolateral part of the ventral horn, and some cells situated laterally in Rexed's laminae V, VI and VII. Their axons are smaller in diameter than those of the dorsal spinocerebellar tract, and the majority cross the midline to ascend in the contralateral lateral fasciculus. Physiological studies indicate that the input to the ventral spinocerebellar tract is distinct from that of the dorsal pathway. Although there is a small contribution from group Ia and Ib muscle afferents, the main input comes, polysynaptically, from flexor reflex afferents. The cells have wide receptive fields and in contrast to the cells of origin of the DSCT, there is convergence between proprioceptive and cutaneous afferents. In addition, there are inputs from descending systems, including the corticospinal, rubrospinal and vestibulospinal tract. In fact, many of the inputs to the spinal α-motoneurones also project in parallel to the ventral spinocerebellar tract cells. The activity in these cells is, for example, modulated during locomotion in deafferented cats, while that of DSCT cells is abolished. Thus, the dorsal spinocerebellar fibres are much more dependent upon peripheral input for their discharge than those of the VSCT.

The complex convergence of inputs from many different sources onto the VSCT makes it difficult to envisage precisely what it is signalling back to the cerebellum. To the extent that the motoneurones receive the same input, the VSCT input might be regarded as a form of efference copy signal. Another possibility, which emphasises the fact that these neurones receive input both from spinal inhibitory interneurones as well as the afferent fibres which excite them, is that they monitor the excitability of spinal interneuronal circuits. The ventral spinocerebellar tract axons traverse the medulla and most of the pons, entering the cerebellum via the superior peduncle. Their terminations are mainly in the contralateral hindlimb areas of cerebellar cortex, with a sparse termination also in ipsilateral cortex.

The forelimb equivalent is the rostral spinocerebellar tract. Its cells of origin are probably located in lamina VII of the cervical cord, and receive flexor reflex afferent (FRA) input from the forelimb. The axons ascend ipsilaterally, unlike those of the ventral spinocerebellar pathway, and pass partly in the brachium conjunctivum and partly in the restiform body to end as mossy fibres in cerebellar cortex. FRA afferents produce mainly excitatory actions on the cells of the rostral spinocerebellar tract, whereas their action is predominantly inhibitory onto the ventral pathway.

Spino-olivary-cerebellar Tracts. The axons of all the cells of the inferior olive go to the contralateral cerebellum, to end in the cerebellar nuclei and as climbing fibres in the cerebellar cortex. The inferior olive is subdivided into a principal nucleus and a dorsal and medial accessory nucleus. Spinal inputs pass only to the two accessory olivary nuclei. There are two main pathways by which spinal afferents reach the olive: a direct route via the spino-olivary tract, which ascends contralaterally in the ventral funiculus of the cord, and an indirect route via a relay in the contralateral dorsal column nuclei. These two routes correspond to the ventral funiculus spino-olivary-cerebellar tract (VF-SOCT) and the dorsal funiculus spino-olivary-cerebellar tract (DF-SOCT) as defined physiologically by Oscarsson (see Oscarsson and Sjkolund, 1977). The input to both these pathways is mainly from flexor reflex afferents.

Ascending Pathways to Other Structures

Spinoreticular Tract. Contrary to earlier beliefs, there is a substantial afferent pathway to the reticular formation from the spinal cord which is not formed from collaterals of other pathways. The medial leminiscus does not give off any collaterals to the reticular formation, nor are the spinothalamic collaterals very dense. The main input comes from fibres which ascend in the ventrolateral fasciculus, intermingled with spinothalamic afferents. In the brainstem the axons deviate from the spinothalamic tract and run dorsally and medially, giving off branches which drop perpendicularly to the axis of the brainstem, to contact the vertically oriented dendrites of the reticular formation cells. The projection is mainly bilateral and many fibres have terminations throughout the length of the reticular formation. Two particularly dense regions of spinal afferent terminals are in the medulla (the nuclei reticularis pontis caudalis and oralis) and the pons (nucleus reticularis gigantocellularis).

Of particular interest in motor control is the somatotopically organised spinal input to the lateral reticular nucleus, which lies lateral to the inferior olive. Physiological studies show that the input is polysynaptic and mainly from flexor reflex afferents with extremely large receptive fields, sometimes covering all four limbs. The efferents from the nucleus travel to the cerebellar nuclei and also end as mossy fibres in the cerebellar cortex.

Spinovestibular and Spinotectal Pathways. The number of spinovestibular fibres is small in comparison to the number of vestibulospinal tract fibres. They are thought to be collaterals of spinocerebellar tract fibres, and end in parts of the vestibular nuclei which receive no direct input from vestibular fibres. Some of the areas of termination project to the cerebellum, or VPL thalamus. Spinal afferents also project to the deep layers of the superior colliculus, although their function is unknown.

Descending Motor Pathways

Several areas of the brain can directly influence the activity of the spinal cord via their descending fibre connections. The main fibre tracts are shown in Figure 7.5, and comprise the corticospinal, rubrospinal, interstitiospinal, tectospinal,

Figure 7.5: Summary of Fibre Systems Descending to the Spinal Cord from the Brain. Shown here are the corticospinal, rubrospinal, interstitiospinal, tectospinal, pontine and medullary reticulospinal and vestibulospinal pathways (from Brodal, 1981; Figure 4.1, with permission)

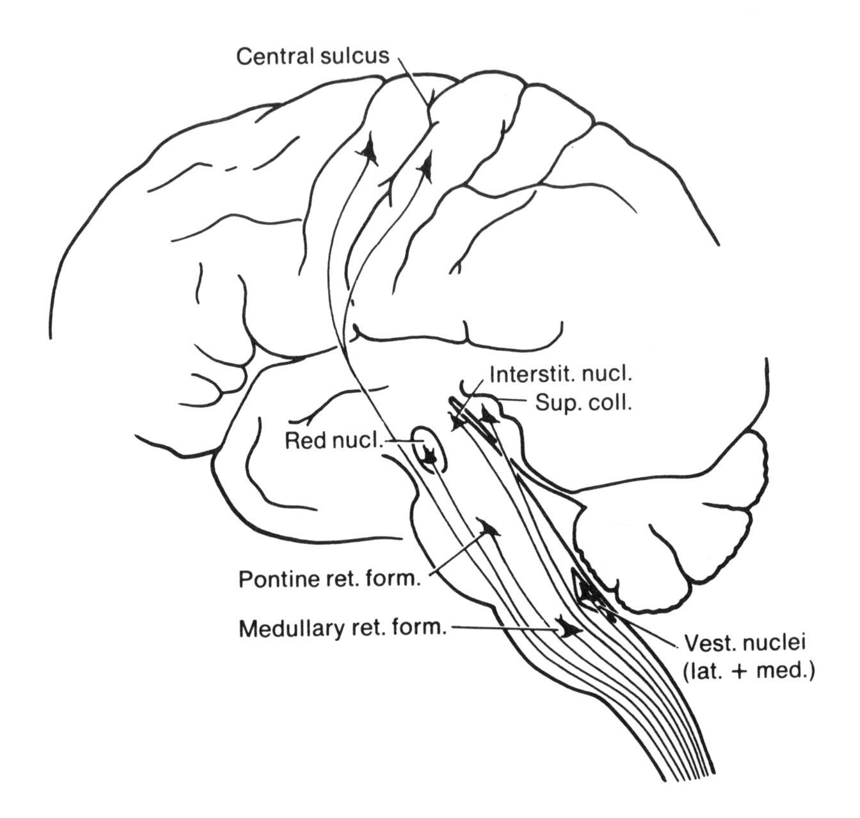

pontine and medullary reticulospinal and vestibulospinal tracts. Although each fibre tract will be described separately, it must be remembered that, under normal physiological conditions, they will never act alone. This is because each region of origin receives input from many other areas of brain (as well as afferent input from sense organs). The cerebral cortex, for example, does not influence the spinal cord solely by way of the corticospinal tract, but also via its interconnections with the red nucleus, reticular formation, tectum and interstitial nucleus (see Nathan and Smith, 1955, 1982, for review of descending tracts in man).

At one time it was believed that all the descending tracts were involved only with efferent, motor commands. However, over the past 30 years it has been shown that these tracts also have important actions on afferent systems, and on reflex pathways within the spinal cord. They may regulate transmission of sensory information from spinal cord to brain and influence the excitability of spinal interneurones, as well as providing a motor output. It is as if the engine of a motor car, besides powering the driving wheels, also adjusted the damping on the suspension or the sensitivity of the steering mechanism at the same time.

Pyramidal Tract

The pyramidal tract was one of the first large fibre bundles to be recognised as a particular tract in the brain. By definition, it consists of those fibres with axons running in the medullary pyramids. The vast majority of these fibres come from the cerebral cortex and continue to the spinal cord. These fibres are known as the *corticospinal* tract. Other fibres leave the tract in the pyramids or even rostral to the pyramids, and innervate the cranial nerve motor nuclei. This portion of the fibres is known as the *corticobulbar* tract. Although it is not strictly true, because some corticobulbar fibres do not travel as far as the pyramids, the term *pyramidal* tract usually is taken to be a common name for the corticospinal and corticobulbar fibres.

Pyramidal tract fibres originate in both motor and sensory regions of cerebral cortex. Degeneration studies and anatomical methods using retrograde axonal transport suggest that in the monkey, 60 per cent of fibres in the pyramidal come from motor and supplementary motor (areas 4 and 6) cortex. The other 40 per cent of fibres come from primary somatosensory cortex (areas 3, 1, 2) and parietal cortex (areas 5 and 7). The figures for man are likely to be similar, although the results of surgical excision indicate that as many as 60 per cent of the fibres may arise in area 4. Unlike the pyramidal tract of the cat, no fibres have been found to arise from cells in prefrontal or secondary somatosensory cortex. Pyramidal tract fibres from precentral cortex have a motor function in the spinal cord, whereas those from the postcentral cortex appear to be involved in the regulation of afferent inputs into the cord.

About one million fibres are found in the pyramidal tract in man. This is more than in any other animal. There are about 800,000 in the chimpanzee, 400,000 in the macaque monkey and 186,000 fibres in the cat. In relation to weight, man has no special superiority in numbers, but it is probably more appropriate to relate the figures to the number of muscles to be controlled, which is about the same in all these species. Six per cent of the pyramidal tract fibres in man are unmyelinated: their origins and functions are unknown. The remainder of the fibres are myelinated and have a range of diameters. Some 2 per cent are large diameter (11–20μm) and conduct impulses at velocities of 50ms^{-1} or so. They are known to electrophysiologists as *fast* corticospinal fibres. Ninety per cent of axons have diameters in the range of 1–4μm. These are the *slow* corticospinal fibres, and conduct at speeds of about 14ms^{-1}.

All pyramidal tract fibres have cell bodies in layer V of cerebral cortex. By chance, the corticofugal cells of layer V in any area of cortex also are known to microscopists, on the basis of their shape, as pyramidal cells. Thus, the pyramidal-shaped cells in layer V of pre- and postcentral cortex give rise coincidentally to the pyramidal tract. There is an ill-defined subclass of giant pyramidal-shaped cells in the motor cortex known as Betz cells. In man, there are some 34,000 of these Betz cells, and it is the axons of such cells which probably account for the large diameter fibres in the pyramidal tract. The majority of small diameter fibres are made up of the axons of small pyramidal shaped cells of cortical layer V.

On leaving the cerebral cortex, the pyramidal tract axons descend in a small region of the posterior limb of the internal capsule. They then course through the middle portion of the cerebral peduncle, and into the pons. The pyramidal tract fibres split into many parts to pass between the pontine nuclei before rejoining as a distinct bundle in the medullary pyramids. Most of the other fibres which run together with the pyramidal tract in the cerebral peduncles terminate in the pons. The majority of the pyramidal tract fibres decussate in the brainstem and travel down to all levels of the spinal cord in the dorsolateral columns (the lateral corticospinal tract). A small number of uncrossed fibres descend within the ventromedial part of the spinal cord (the ventral corticospinal tract), but terminate in man at thoracic levels. Finally, there is a proportion of uncrossed fibres which join the crossed fibres of the lateral corticospinal tract. Throughout most of their course (except within the pons, and perhaps even within the medullary pyramid), the corticospinal fibres are thought to preserve an approximate somatotopic arrangement. Fibres to the leg are found lateral to those to the arm and trunk in the cerebral peduncles and dorsolateral spinal cord. In the internal capsule, the fibres to the leg are found most posteriorly.

In the cat, lateral corticospinal fibres terminate mainly in the intermediate zone of the spinal grey matter (Figure 7.6). There are very few direct contacts with spinal motoneurones of lamina IX and activation of motoneurones is achieved via spinal interneurones. Precentral fibres terminate ventral (laminae VI and VII) to those from postcentral cortex (laminae IV and V), consistent with a role of postcentral fibres in regulating afferent input within the dorsal horn of the spinal cord. In contrast to the cat, in the monkey some fibres of the lateral corticospinal tract end directly on spinal motoneurones. Such monosynaptic connections from cerebral cortex to motoneurones are thought to be more important in man, particularly to the muscles of the hand and forearm.

Within the cervical and lumbar enlargements of the spinal cord, the motoneurones are arranged somatotopically (see Figure 7.10). Those innervating the axial muscles lie most medially and their cell bodies form a ventromedial column in the grey matter, which runs the whole length of the cord. Motoneurones which innervate the extremities are situated in the lateral part of the ventral horn. In the caudal segments of the cervical and lumbar enlargements, the most dorsally located cells in the ventral horn supply the most distal muscles of the limbs. They are nearest the lateral corticospinal tract and receive the densest corticospinal projections. The uncrossed fibres of the ventral corticospinal tract terminate in the ventromedial cell group and innervate the axial muscles. Unlike the lateral tract, the medial corticospinal fibres commonly have a bilateral projection to motoneurones on both sides of the cord.

Corticospinal fibres do not only give excitatory projections to motoneurones;

Figure 7.6: Distribution of Corticospinal Fibres from the Left Hemisphere to Low Cervical Segments in Four Different Species. Note how the pattern of termination changes from innervation of the dorsal horn, to intermediate zone and, finally, in the rhesus monkey and chimpanzee, to the motor nuclei (lamina IX) of the ventral horn. The locations of the axons in the lateral and ventromedial tracts also are shown. Note the predominantly crossed component of the lateral tract and the uncrossed projection in the ventromedial tract (from Kuypers, 1981; Figure 12, with permission)

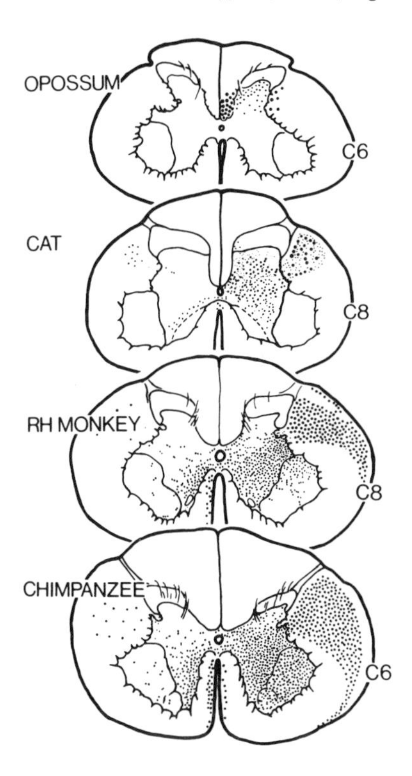

they also project strongly to spinal inhibitory interneurones such as the Ia interneurone and Renshaw cells. In addition, they also innervate gamma motoneurones as well as the large alpha motoneurones.

Rubrospinal Tract (Figure 7.7)

Opinions differ considerably over the importance and even the presence of a rubrospinal tract in man (Nathan and Smith, 1982). For convenience, it is best to describe first the tract and its origins in the cat, where it is undoubtedly an important descending projection to spinal cord.

The red nucleus contains cells of various size. Small cells are located in the rostral (parvocellular) portion and large cells in the caudal (magnocellular) part. In the cat, descending spinal projections arise from cells of all sizes in both rostral and caudal parts. As expected, the rubrospinal tract consists of axons with a wide range of diameters, with a maximum conduction velocity

Figure 7.7: Diagram of the Principal Features of the Rubrospinal and Corticorubral Projections in the Cat. A: Somatotopic projection from motor cortex (anterior sigmoid gyrus in the cat) to the dorsal and ventral regions of the red nucleus. In the cat, it is thought that rubrospinal tract fibres arise from both small and large cells of the rostral and caudal portions of the nucleus. Accordingly, the rubrospinal projection is shown to consist of axons with large and small diameters. In the monkey, this is not the case: only large cells contribute axons to the rubrospinal tract. B: Topographical arrangement of projection neurones in the red nucleus. C: Main region of termination of the rubrospinal tract within the intermediate zone of the spinal grey matter (from Brodal, 1981: Figure 4.6, with permission)

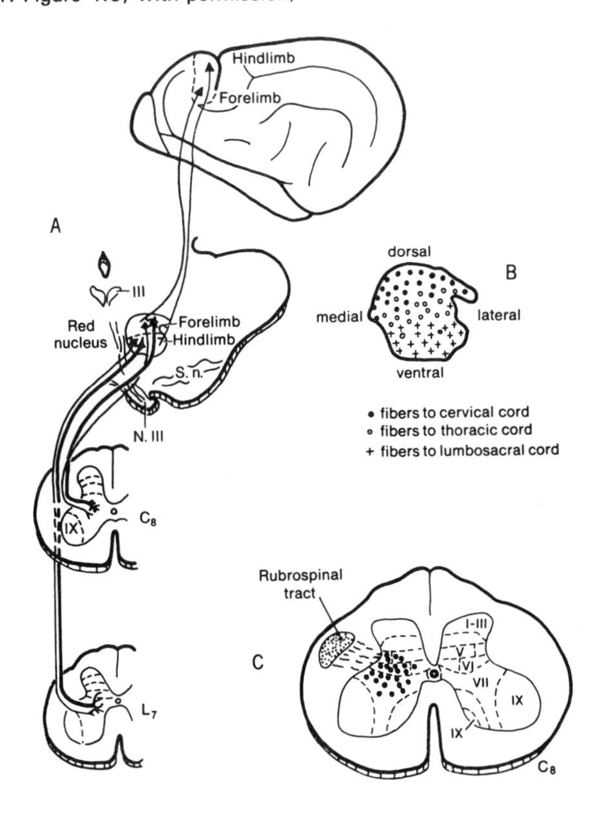

(120ms^{-1}) exceeding that of the pyramidal tract. It decussates near its origin and travels the contralateral brainstem in a ventromedial position (von Monakow's bundle) where it gives off collaterals to many areas, including the interpositus nucleus of the cerebellum and the vestibular nuclei. In the spinal cord, it lies ventral and slightly lateral to the lateral corticospinal tract. The two tracts intermingle and in the cat terminate together on interneurones in laminae V, VI and VIII of the spinal grey. Some of these interneurones are shared between the two tracts so that impulses in one descending system can facilitate transmission in the other. Monosynaptic transmission to spinal alpha-motoneurones does not occur in the cat, although such connections are seen in the monkey. Electrophysiologically, stimulation of the cat red nucleus tends to produce di- or trisynaptic excitation of contralateral flexor motoneurones and inhibition of extensor motoneurones.

The tract is arranged somatotopically. Cells from the dorsomedial part of the nucleus give rise to axons which travel medially in the cord and innervate cervical segments. Cells from the ventrolateral parts of the nucleus have axons which terminate in lumbar regions. Two of the main inputs to the red nucleus preserve this somatotopy. They come from the ipsilateral cerebral (particularly motor) cortex and from the contralateral interpositus nucleus of the cerebellum. For example, inputs from the arm area of motor cortex and from cerebellar regions receiving afferent input from the forelimb both project to cells in the dorsomedial red nucleus and thence to the cervical cord. Such projections from cortical motor areas obviously give rise to the possibility of a cortico-rubro-spinal pathway operating in parallel with the direct corticospinal route. Synapses from cortex are, however, located more distally on the dendrites than those from the interpositus, and probably provide the weaker input. The projection from interpositus nucleus is particularly strong. In the cat, the great majority of its output travels via the superior cerebellar peduncle (brachium conjunctivum) to the red nucleus. Electrical stimulation of the peduncle is a better way of exciting rubral cells than stimulation with the nucleus itself.

In contrast to the situation in man, the crossed rubrospinal tract of the cat is the major output of the red nucleus. However, there are other outputs and one of these, an ipsilateral connection to the olive, takes particular predominance in man. The human red nucleus, to a much greater extent than the cat, can be clearly divided into a rostral parvocellular and a caudal magnocellular part. In man and monkey, the most recent work with retrograde labelling techniques shows that the rubrospinal tract takes origin only from the large (and possibly medium-sized) cells of the caudal part. Since there are only 150 to 200 large cells in this region in man, the rubrospinal tract is probably very small indeed. In contrast, the parvocellular region of the red nucleus is much larger than in lower animals. Its afferent and efferent connections are quite separate from those of the caudal nucleus. Descending axons travel ipsilaterally in the central tegmental tract, giving off projections to the reticular formation and terminate in the olivary nuclei. This tract is much larger in man than in the cat, where it has usually received little attention.

The major input to the rostral part of the red nucleus comes from the ipsilateral dentate nucleus. Thus, since there is a strong projection from the olive (via the inferior cerebellar peduncle) to cerebellar cortex, the red nucleus forms part of a cerebello-rubro-olivary-cerebellar loop. Lesions to the central tegmental tract in man, which interrupt this loop, give rise to palatal myoclonus. This consists of rapid twitch-like contractions of the soft palate in the pharynx at about $120 min^{-1}$. A possible explanation is that they arise from instability in the loop circuit outlined above.

Vestibulospinal Tracts

As with the rubrospinal tract, there are little data available on the human vestibulospinal tract. However, what results there are, from pathological investigations, are in agreement with results from experimental animals. Almost all the work has been performed on the cat.

There are two vestibulospinal tracts: medial and lateral. The names derive from the nuclei of origin within the vestibular complex (Figure 7.8). The lateral tract is larger and arises from both large and small cells in the lateral vestibular (or Deiter's) nucleus. The cells are somatotopically arranged within the nucleus. Rostroventral parts project to the neck and forelimb, dorsolateral parts of the hindlimb. The fibres descend ipsilaterally in the ventrolateral columns of the spinal cord, occupying a progressively more medial position from cervical to lumbar segments. Terminations are distributed mainly to laminae VIII and VII. The medial vestibulospinal tract is said to arise only from the medial vestibular nucleus, but large contributions from both the lateral and descending vestibular nuclei have been noted electrophysiologically. Axons travel in the medial longitudinal fasciculus together with some reticulospinal fibres and descend only to upper thoracic levels of the spinal cord in the most medial part of the ventral white matter. The area of termination is the same as that for the lateral vestibulospinal tract. Both tracts produce disynaptic excitation of extensor motoneurones and inhibition of flexor motoneurones.

The input to the vestibular nuclei comes from two main sources: cerebellum and labyrinth. Some fibres also arise in the reticular formation. Input from the labyrinth travels via the vestibular branch of the VIIIth nerve. Fibres from the semicircular canals terminate mainly in the superior vestibular nucleus and the rostral part of the medial and descending nuclei. Fibres from the utricle terminate in the same regions and also in the rostroventral part of Deiter's nucleus. The descending spinal projections from these regions travel in both medial and lateral vestibulospinal tracts and end in the cervical cord. They are the basis for strong reflex connections between labyrinthine receptors and the neck musculature. Indeed, monosynaptic connections (both excitatory and inhibitory) from the vestibular nuclei to neck motoneurones have been described. Ascending fibres also arise from the rostral part of the medial nucleus and from the superior nucleus and travel rostrally in the medial longitudinal fasciculus to the oculomotor nuclei. They are involved in the labyrinthine control of eye movements.

The cerebellar projections to the vestibular nuclei are extensive and very complex. There are direct ipsilateral pathways from spinal and vestibular regions of cerebellar cortex, as well as ipsi- and bilateral pathways from the fastigial nucleus, to all four vestibular nuclei. In particular, these are the main inputs to the dorsolateral part of Deiter's nucleus, which provides the only vestibulospinal fibres to innervate motoneurones of the lumbar region.

Figure 7.8: Vestibulospinal Pathways in the Cat. The major tract is the lateral vestibulospinal tract, which arises somatotopically (B) in Deiter's nucleus (the lateral vestibular nucleus, L). Fibres terminate in the medial portion of the ventral horn of the spinal cord (C) at both cervical and lumbar levels. The medial vestibulospinal tract is much smaller and only travels as far as the cervical segments. Its cell bodies lie in the medial vestibular nucleus (M) (from Brodal, 1981); Figure 4.7, with permission)

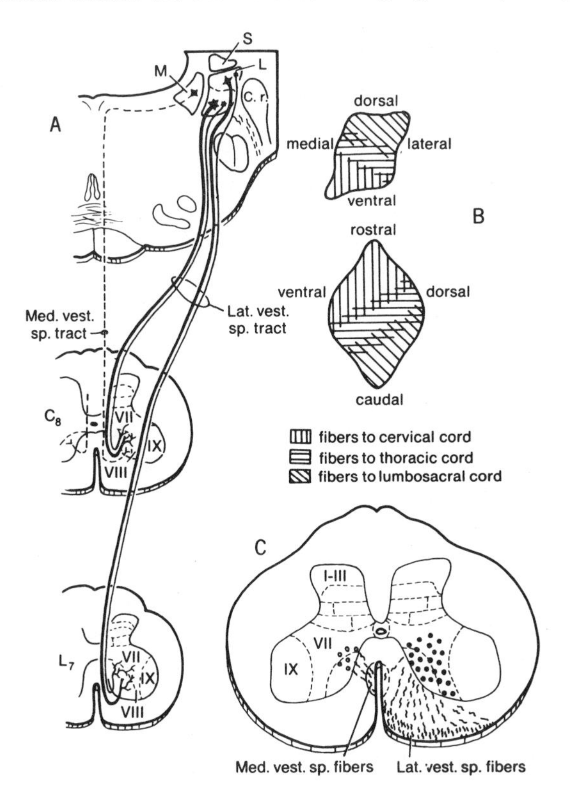

Reticulospinal Tracts (Figure 7.9)

The reticulospinal tracts arise from an area in the bulbar reticular formation known as the medial tegmental field. This is the rostral continuation of spinal grey matter in the medial part of the intermediate zone. Several nuclear groups can be distinguished within it and include the paramedian and interfascicular nuclei in the medulla and the gigantocellular, para-gigantocellular and nucleus reticularis pontis in the pons.

The fibres which arise from cells in the pontine medial tegmental field descend bilaterally at first through the medullary tegmental field. They continue, mainly ipsilaterally, into the ventral funiculus of the spinal cord. This tract, which runs the length of the cord is known as the medial reticulospinal tract. The fibres of cells in the medullary medial tegmental field descend bilaterally (but mainly ipsilaterally) to the spinal cord. In the brainstem these medullary

Figure 7.9: Transverse Section of a Cat Spinal Cord at the C8 Level Showing the Pattern of Termination of Reticulospinal Tracts Within the Grey Matter and the Location of Their Axons in the White Matter. Projections from the left reticular formation are shown (from Nyberg-Hansen, 1965; Figure 9, with permission)

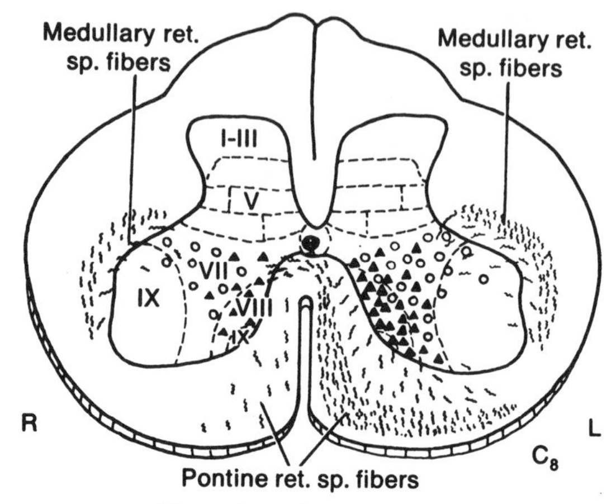

reticulospinal fibres constitute a vertically orientated band running just lateral to the medial longitudinal fasciculus. The fibres in the dorsal portion of this band enter the spinal ventral fasciculus, running together with the pontine fibres in the medial reticulospinal tract. The fibres from the ventral portion of the band enter the ventrolateral funiculus of the spinal cord, forming the lateral reticulospinal tract.

Unlike other descending tracts, there is no somatotopic organisation in either of the reticulospinal projections. Electrophysiological studies reveal that two thirds of fibres terminating at cervical levels also project to the lumbar enlargement. The main areas of termination are in the ventral grey matter (laminae VIII and medial lamina VII), with the fibres of the ventrolateral columns tending to be located dorsal to those of the ventral columns. Some studies also indicate that there may even be direct monosynaptic connections with motoneurones of neck and axial muscles.

Some of the main inputs to the reticular regions which give rise to the descending tracts are the sensorimotor areas of cerebral cortex. This input is bilateral. Two other important inputs come from ascending spinal cord fibres and from the fastigial nucleus of the cerebellum (particularly to the medullary region). There is also another area of the reticular formation which gives rise to long descending pathways. It is located in the ventrolateral (rather than medial) pontine tegmentum close to the rubrospinal tract. It gives rise to a crossed reticulospinal tract which travels in the spinal dorsolateral funiculus with the corticospinal and rubrospinal tracts. It is known sometimes as the pontine

Figure 7.10: Diagram of the Approximate Patterns of Termination of the Three Major Categories of Descending Pathways: Corticospinal and Group A (Ventromedial Brainstem Pathways) and B (Lateral Brainstem Pathways) Brainstem Pathways. The sections are from the lower cervical cord of a monkey. The bottom section shows the arrangement of motoneurones and interneurones of the intermediate zone. Lateral brainstem pathways terminate in regions of the intermediate zone that connect with motoneurones supplying distal muscles. Ventromedial brainstem pathways terminate on regions which project to proximal and axial motoneurones. The corticospinal tract has terminals directly onto both types of motoneurones as well as interneurones (from Lawrence and Kuypers, 1968; Figure 3, with permission)

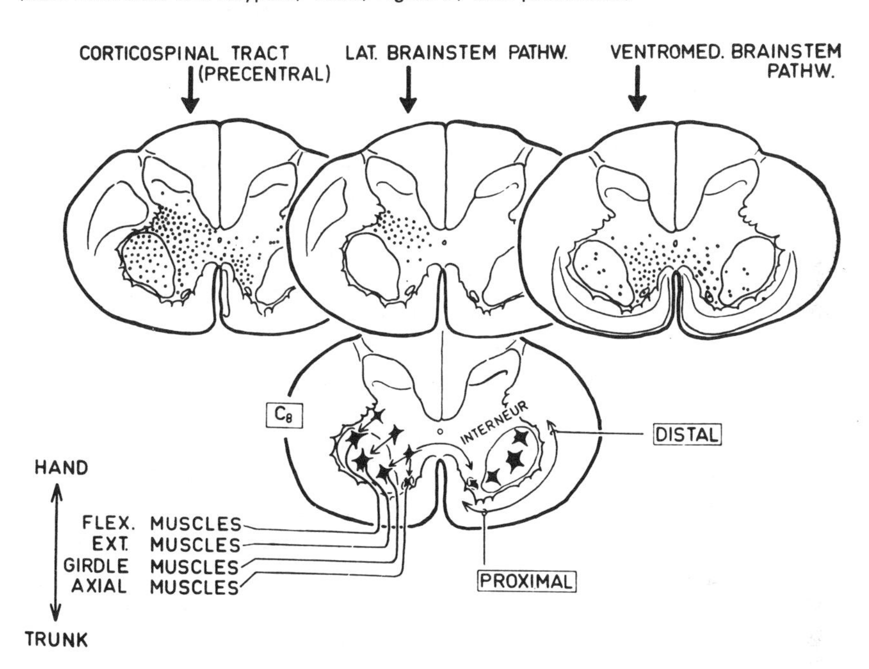

component of the rubrospinal tract.

Tectospinal and Interstitiospinal Tracts

There are two other main descending tracts from brain to spinal cord. Little work has been performed on either, in comparison with the other descending systems, and almost nothing is known about them in man.

The tectospinal tract comes from cells in the deep layers of the superior colliculus. Its fibres cross at the dorsal tegmental decussation of the midbrain and travel just ventral to the medial longitudinal fasciculus. In the cord they are found in the medial part of the ventral funiculus and travel only as far as cervical segments. Most of the terminations are in laminae VI, VII and VIII of the high cervical segments, and the effects are probably directed via interneurones to motoneurones of the neck muscles. The main input to the superior colliculus is from visual cortical areas, and thus the tract may take part in head and neck orienting reactions to visual stimuli.

The interstitiospinal tract arises from cells in the interstitial nucleus of Cajal in the rostral midbrain. The nucleus lies ventral to the medial longitudinal fasciculus and the efferent axons descend ipsilaterally in this bundle to the ventromedial funiculus of the spinal cord. Terminations are given off to all levels of the cord, mainly in laminae VII and VIII, overlapping with the area of termination of vestibulospinal fibres. The input to the interstitial nucleus comes mainly from the vestibular nuclei and cerebral cortex.

Summary of Descending Pathways (Figure 7.10)

The mass of descending pathways described above may be simplified by dividing them into three groups. This classification was devised by Kuypers and is important because the anatomical distinction between the pathways also reflects functional differences between their actions on spinal motoneurones (see Chapter 8 for effects of lesions of these pathways). The three groups are the corticospinal tract and the two groups (A and B) of descending brainstem pathways. The group A brainstem pathways comprise the interstitiospinal, tectospinal, lateral and medial vestibulospinal and the reticulospinal tracts from the pontine and medullary medial tegmental areas. All these tracts travel in the ventral and ventromedial funiculi of the spinal cord and terminate to a greater or lesser degree, bilaterally in laminae VII and VIII. This is a region of long propriospinal neurones. The fibres have many collaterals, and seem especially well suited for the synergistic activation of large numbers of muscles. Many elements of these pathways also have direct connections with neck and axial musculature. The group B brainstem pathways consist of the rubrospinal and the crossed reticulospinal tract from the pontine lateral tegmental region. They travel in the contralateral dorsolateral funiculi of the spinal cord, terminating in the intermediate zone in laminae V-VIII. These fibres have relatively few collaterals, ending either directly or motoneurones or on short propriospinal fibres of the spinal cord. Their connections are directed mainly to distal muscles.

The corticospinal (CST) tract in monkey and man can be regarded as having two components which parallel the two descending groups of brainstem pathways. They terminate on both distal extremity muscles (lateral CST) and on axial and proximal muscles (ventral CST). In contrast to the brainstem pathways (especially group A), the corticospinal pathway is exceptionally well organised, with a high degree of topographical representation. It would seem to provide a highly differentiated system superimposed on those from the brainstem and probably contributes the ability to produce fractionated movements in both these groups of muscles.

References and Further Reading

Brodal, A. (1981) *Neurological Anatomy in Relation to Clinical Medicine,* Oxford University Press, Oxford

Brown, A.G. (1981) *Organisation in the Spinal Cord,* Springer-Verlag, Berlin

Hassler, R. (1959) 'Anatomy of the Thalamus' in G. Shaltenbrand and P. Bailey (eds.), *Introduction to Stereotaxis with an Atlas of the Human Brain,* Georg Thieme, Stuttgart

Kuypers, H.G.J.M. (1981) 'Anatomy of the Descending Pathways' in V.B. Brooks (ed.), *Handbook of Physiology, Sect. 1, Vol. II, Part 1,* Williams and Wilkins, Baltimore, pp. 597–666

Lawrence, D.G. and Kuypers, H.G.J.M. (1968) 'The Functional Organisation of the Motor System in the Monkey. Parts I and II, *Brain, 91,* pp. 1–14 and 15–36

Nathan, P.W. and Smith, M.C. (1955) 'Long Descending Tracts in Man. I. Review of Present Knowledge, *Brain, 78,* pp. 248–303

——— (1982) 'The Rubrospinal and Central Tegmental Tracts in Man', *Brain, 105,* pp. 223–69

Nyberg-Hansen, R. (1965) 'Sites and Modes of Termination of Reticulospinal Fibres in the Cat. An Experimental Study with Silver Impregnation Methods', *J. Comp. Neurol., 124,* pp. 71–100

Oscarsson, O. and Sjkolund, B. (1977) 'The Ventral Spino-olivocerebellar System in the Cat. I. Identification of Five Paths and Their Termination in the Cerebellar Anterior Lobe', *Exp. Brain, Res., 28,* pp. 469–86

8 CEREBRAL CORTEX

Structure of Cerebral Cortex

The cerebral cortex is the most distinctive feature of the human brain. It is composed almost entirely of neocortex, the most recent part of the cortex to develop in the evolutionary time scale. Archicortex (hippocampus) and paleocortex (olfactory cortex) form only a small fraction of cerebral cortex in man.

The internal structure of cortex is covered in detail in many textbooks. Only a brief summary is given here. There are two main types of cell in the cerebral cortex: pyramidal and stellate (or granule) cells. Most of the pyramidal cells have axons which leave the cortex, whereas the stellate cells (of which there are several different types) are the true cerebral interneurones. The relative position and preponderance of the cell types gives the cortex a laminar appearance when seen in vertical sections (Figure 8.1). Six layers are defined:

1. The molecular or superficial plexiform layer. This lies immediately below the pia and contains very few cell bodies. Its main components are the axons and apical dendrites of neurones in deeper layers;
2. The external granular layer. This is a layer of densely packed small cells, including small pyramidal and stellate cells;
3. The external pyramidal layer. A layer of medium sized pyramidal cells;
4. The internal granule layer, composed of densely packed stellate and pyramidal cells;
5. The ganglionic layer. A layer of large- to medium-sized pyramidal cells;
6. The multiform layer. Irregularly shaped cells, many with axons leaving the cortex.

The pyramidal cells in all layers tend to have apically directed dendrites which may extend into layer I. Their axons leave the cortex through a layer of basal dendrites which may extend horizontally for several millimetres. These dendrites form a conspicuous band in layers II, III and IV. Afferents to cortex come from either the thalamus or from other areas of cortex, via cortico-cortical connections. Afferents from the specific sensory nuclei of the thalamus terminate in discrete zones in layer IV, whereas those from the nonspecific nuclei branch in the subcortical white matter and end widely in layer I with horizontally directed axons. Cortico-cortical afferents are found in several layers.

In addition to the obvious horizontal layering of cortex, the vertically oriented dendrites of the pyramidal cells give the cortex strong perpendicular connections. Indeed, electrophysiological studies have shown that groups of cells work together in vertical units called cortical columns. Within the columns, the cells share common input and output connections, forming the basic unit of cortical

Figure 8.1: Diagrams Showing Structure of Cerebral Cortex in Cross-section.
This page: Horizontal layering as seen with three different staining techniques: Golgi (left), Nissl (middle) and myelin sheath (right) (from Brodal, 1981; Figure 12.1, with permission).
Opposite: Golgi preparation from an adult cat showing vertical orientation of apical dendrites of pyramidal cells and the horizontal layering of their basal dendrites in layer 5 (from Schiebel, Davies, Lindsay *et al.*, 1974; Figure 1, with permission)

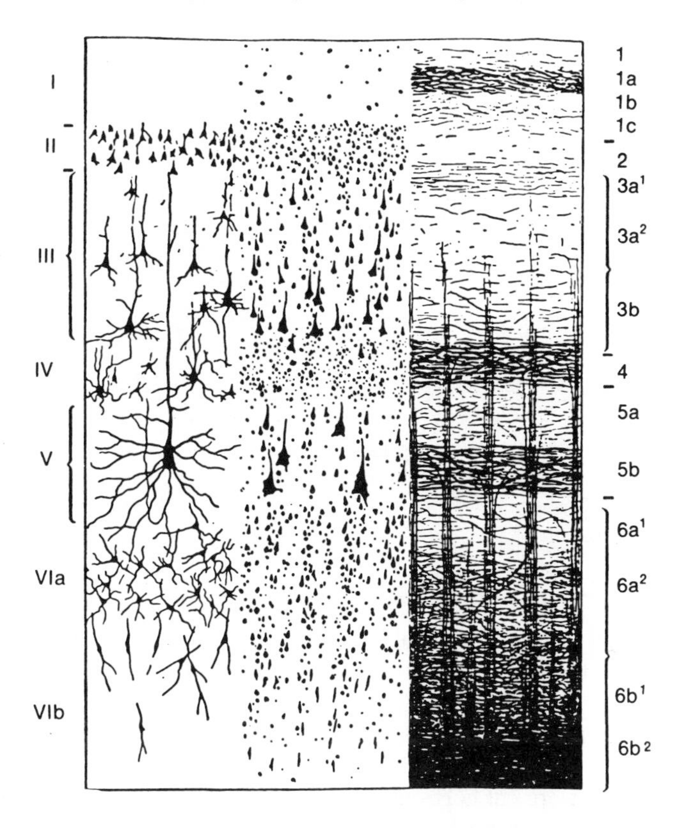

processing. Specific thalamic afferents terminate on the dendrites of stellate cells in layer IV so that there are few direct connections between specific inputs and the pyramidal output cells of the cortex. Instead, many of the stellate cells of layer IV have vertically oriented axons which make contacts with the apical dendrites of pyramidal (and other stellate) cells. These vertical connections may form the basis of a strong input–output coupling within a cortical column.

There are also some strong horizontal connections between columns. Some stellate cells have processes which extend for many millimetres, as do the basal dendrites of pyramidal cells. These connections might allow for interaction, such as lateral inhibition, between nearby vertical columns (see Porter, 1981, for more detail on the fine structures of motor cortex).

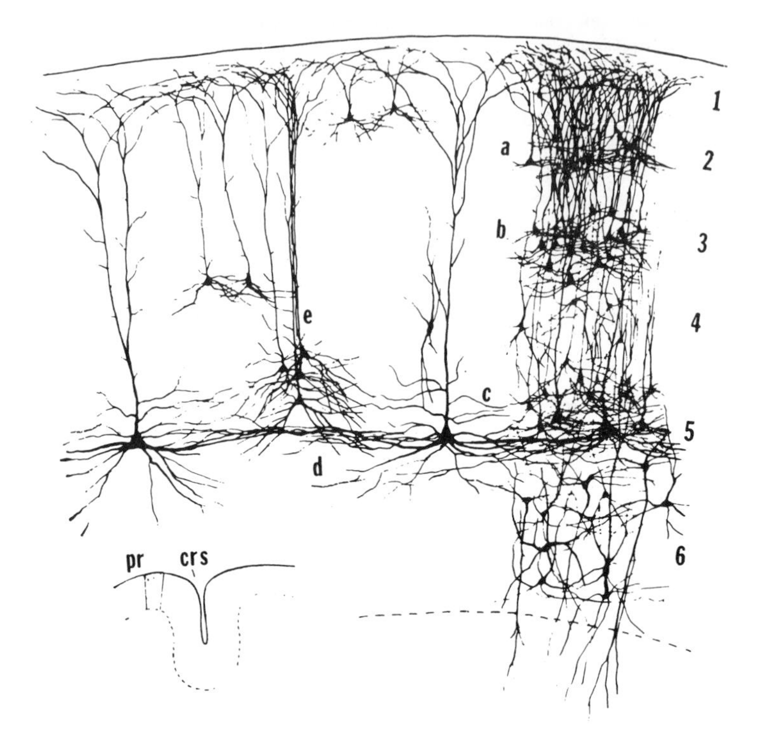

Sub-areas of Cerebral Cortex

The description above of cortical layering only gives a 'typical' picture of cortical organisation. There are considerable variations in the number and thickness of the layers in various regions of cortex. It was on this basis that Brodmann distinguished more than 50 separate regions. The numbering of Brodmann's areas seems confusing at first sight. This is because the sections of brain that he examined were cut horizontally, beginning at the vertex. Precentral and postcentral regions appeared in the first few sections and were given low numbers, jumping from front to back as new cytoarchitectonic features were encountered. This began on the crown of the postcentral gyrus with areas 1 and 2, then area 3 in the depths of the sulcus, followed by area 4 in the precentral region, and so on.

There have been other attempts at parcellation of the neocortex on similar principles. Figure 8.2 compares the maps of monkey brains made by Brodmann, Vogt and Vogt, and von Bonin and Bailey. In modern literature different authors may use any of these three nomenclatures. The important point to note is that the lines which separate one cortical region from another are not the same in each system. As shall be seen later, in the motor system the differences between area 6 of Brodmann and area 6 of Vogt and Vogt pose a particularly difficult problem (see Wiesendanger, 1981, for further discussion).

Figure 8.2: Motor (Shaded) Areas of the Brain Shown in Relation to Classic Cytoarchitectonic Maps of Monkey Cortex. Brodmann's maps (left) show a lateral and medial view of the brain. Note the subdivision of area 6 into 6Aα and 6Aβ by Vogt and Vogt, and into FC and FB by von Bonin and Bailey (from Wiesendanger, 1981; Figure 1, with permission)

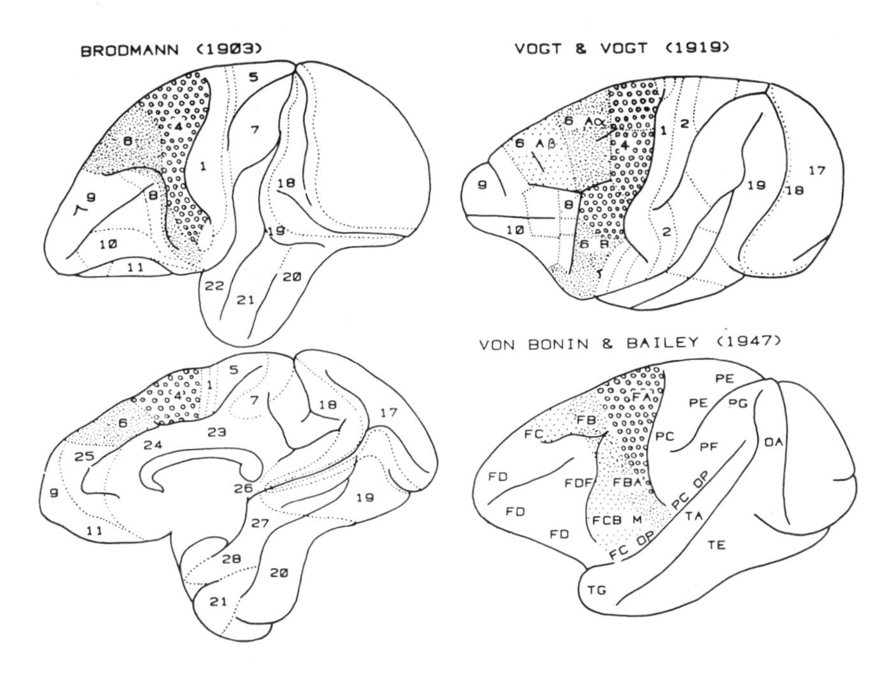

Which are the Motor Areas of Cortex?

The conventional definition of a cortical motor area is that electrical stimulation of that region produces movement of a part of the body. It is not a particularly useful definition, since stimulation of almost any area of cortex with sufficient intensity and under the appropriate conditions can evoke movement. Such effects depend critically on the depth and type of anaesthesia used and, if strong stimuli are used, there is always the possiblity that the stimulus will spread electrically or via cortico-cortical connections to other areas distant from the stimulating electrode. Despite these limitations, the commonly accepted electrically excitable areas of cortex do correspond with what are nowadays regarded as those areas most closely involved in the control of movement.

The areas with the lowest threshold for electrical stimulation are Brodmann's areas 4, 6 and 8. At higher intensities, movements also can be elicited from areas 1, 2 and 3. These latter are conventional sensory areas of cortex and it is believed that much of the movement provoked by electrical stimulation is mediated by cortico-cortical connections to the precentral cortex. There is large pyramidal projection from these sensory areas to spinal cord, but this is probably used in modulating transmission of sensory input to brain rather than in production of movement.

The electrical excitability and localisation of motor cortex was discovered independently in the 1870s by Ferrier and by Fritsch and Hitzig in experiments on the brains of monkeys and dogs. Detailed maps of the monkey motor cortex were made later by Woolsey, Settlage, Meyer *et al.* (1952) and in man by the neurosurgeon Penfield and his colleagues (Penfield and Jasper, 1954). In man, the investigations were performed during surgery for epilepsy and were necessarily limited in time by ethical considerations. Final maps of human cortex were made by combining the results obtained from many different patients.

In these pioneering investigations, electrical stimulation of the exposed cortex with 60Hz alternating current was used to provoke visible movements of the body. The movement that was produced was mapped as each point on the brain was stimulated. *Three* separate representations of the body were found in both human and monkey, in the region of cortex outlined by Brodmann's areas 4 and 6 (Figure 8.3). Movements produced by stimulation in each of these areas had different characteristics and electrical thresholds. The largest representation, with the lowest electrical threshold, was the primary motor area, lying along the precentral gyrus. This was termed MI by Woolsey *et al.* (1952). They found that the body was mapped topographically along this strip: movements of the legs were elicited by stimulation of the most medial parts, whereas movements of the arms and face by the most lateral parts. As well as this medio-lateral projection, there also was a rostral-caudal gradient. Movements of the most distal body parts were represented nearest the central sulcus, whereas movements of proximal parts were represented more rostrally. In man, distal movements of the fingers are buried deep within the central sulcus and therefore are difficult to investigate in detail.

An important feature of the motor map in MI is the relative sizes of areas devoted to movement of particular body parts. The finger and hand representations are much larger than those for the shoulder and arm. This probably reflects the large number of finely graded and independent movements that the hand and fingers can perform in comparison with the shoulder. The large area contains many cortical output cells to the distal muscles as well as many local interneurones.

Stimulation of MI produces simple 'flick-like' or twitch movements; purposeful movements are never obtained. Although most movements are contralateral, there are exceptions: bilateral activation of the soft palate, larynx and masticatory muscles, and frequent bilateral activation of axial muscles of the trunk, for example. The degree to which ipsilateral as well as contralateral musculature is stimulated is thought to depend on the proportion of crossed and uncrossed corticofugal connections to spinal cord (see Chapter 7).

A second motor area (MII) is found in the most lateral part of the precentral gyrus, extending into the lip of the Sylvian fissure. The two body representations in MI and MII lie lip to lip, with that of MII being reversed compared to MI.

Figure 8.3: Somatotopic Organisation of Motor Areas in the Monkey. The central and longitudinal fissures have been opened out on this diagram to show the representation more fully. Dotted lines represent the floor of a fissure and the solid lines the lip of a fissure. There are three representations. The most lateral is a second face area, MII. The middle, precentral representation is MI. The medial representation is the supplementary motor area (SMA) (from Woolsey, Settlage, Meyer *et al.,* 1952; Figure 131, with permission)

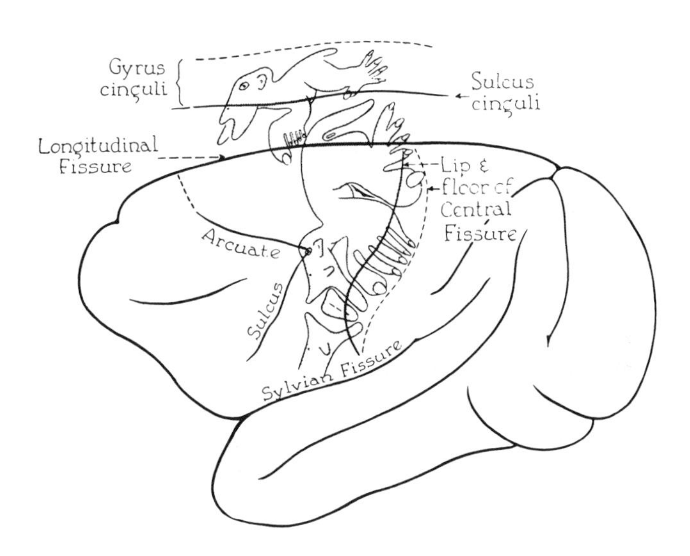

MII has been little investigated in recent times. There is a strong bilateral representation of the face, which may be a factor in the relative immunity of the facial muscles to hemiplegia.

The third motor area is found on the medial surface of the hemisphere, and is termed the supplementary motor area (SMA) rather than MIII. It is quite different from MI. The threshold for stimulation is higher than for MI, and somatotopic representation of the body parts is poor. The movements which are elicited are complex, involving activation of the muscles at many joints, consisting of the assumption and maintenance of limb postures. Complex acts like yawning or vocalisation occur, and bilateral movements are common. The SMA will be discussed in more detail later in this chapter.

There are two other motor areas of the brain which are not illustrated on the map of Woolsey above. One is the frontal eye field, which corresponds roughly to area 8 of Brodmann. In fact, in Woolsey's simunculus, the representation of the eyes spreads over the arcuate sulcus into area 8. Nowadays, the whole of area 8 is regarded separately from MI. Stimulation here produces conjugate deviation of the eyes to the contralateral side. Other less predominant movements include vertical and oblique eye movement, pupillary constriction and lachrymation.

The final motor area is the premotor area, which lies between MI and the frontal eye fields of area 8. The premotor area is not defined at all on Woolsey's map, principally because it is now believed that Woolsey's forward boundary for MI is placed a little too far rostral. Part of the axial representation of the body on these maps actually lies within what is presently understood to be the premotor cortex. This consists anatomically of the lateral part of area 6 of Brodmann, or area 6Aβ of Vogt and Vogt. Stimulation of this area produces movements like those of the SMA or of the frontal eye fields. The premotor cortex is discussed in more detail later in this chapter (see Phillips and Porter, 1977; Wiesendanger, 1981).

Cytoarchitecture and Connections of Cortical Motor Areas

Cytoarchitecture. Areas 4 and 6 of Brodmann are examples of agranular cortical regions, with a very small or non-existent internal granular layer IV. They are bounded anteriorly by area 8 (the frontal eye field) and posteriorly by area 3 (part of the sensory cortex). Both these regions are granular cortex. The rostral junction between area 6 and area 8 is relatively clear, but the caudal boundary between areas 3 and 4 rather more difficult. The problem is that area 3 is itself subdivided into a caudal 3b and a rostral 3a. Area 3b is typical koniocortex with a prominent granular layer, and a grainy appearance under the microscope. Area 3a, which abuts onto area 4 in the depths of the central sulcus is a 'transitional' type of cortex, with approximately equal development of all layers, which makes it difficult to distinguish from area 4. This difficulty is responsible for some of the controversies surrounding the roles of areas 3a and 4 in movement that have been reported in the literature.

Agranular areas 4 and 6 contain four separate motor cortical regions as distinguished on physiological grounds (MI, MII, SMA, and premotor area), and it has been a persistent problem for the anatomist to detect whether these physiological differences are reflected in cytoarchitecture. Area 6 is said to be distinguished from area 4 by its lack of giant Betz cells. This is relatively easy in medial (leg) representation of area 4 since the largest Betz cells are found here. However, the Betz cells in the arm and face area are smaller than in the leg area so that the division of area 4 from area 6 is more difficult in lateral parts of the cortex. Area 6 itself was subdivided by Vogt and Vogt into a lateral 6B and medial 6Aα and 6Aβ on similar difficult criteria. As a working definition, it is usually said that MI (excluding the most rostral portions of Woolsey's map) corresponds to area 4 of Brodmann. In the monkey, this largely coincides with areas 4 and 6Aα of Vogt and Vogt. MII may be the most lateral part of area 6 (=6B). SMA is that portion of area 6 on the mesial surface of the hemisphere and the premotor cortex is area 6Aβ of Vogt and Vogt or the remainder of the lateral part of Brodmann's area 6 (see Wiesendanger, 1981).

Thalamic Inputs of Cortical Motor and Sensory Areas (Figure 8.4). The thalamic inputs to motor and sensory cortex come principally from the ventrobasal nuclei. Until recently, the connections were summarised by saying that somatic afferents projected via VPL to postcentral areas, whereas cerebellar and basal ganglia efferents projected via VL to precentral motor cortex. However, over the past ten years, the use of new tracer techniques has refined this picture so that there is a more detailed picture of the input to various subregions of sensory and motor cortex.

In the monkey, it is not, unfortunately, true to say that the VPL nucleus projects to sensory cortex, and the VL nucleus to motor cortex, even though this may be true for many other species. In the monkey, there has been a shift in the position of thalamic groups which project to sensorimotor cortex so that only the caudal part of VPL (VPL_c) sends efferents to postcentral cortex. The remainder of VPL (known as VPL pars oralis, VPL_o) projects to precentral cortex, together with the caudal part of VL (VL_c). The rest of VL, the pars medialis (VL_m) and pars oralis (VL_o) projects to mesial area 6, the SMA. Finally, a neighbouring thalamic nucleus X sends fibres to the lateral portion of area 6, the premotor cortex. More anterior thalamic nuclei (for instance, VA) send fibres to progressively more rostral areas of cortex.

The inputs to thalamic nuclei are well defined with little or no overlap (Figure 8.4). Neither are there any interconnections between the cell groups in nearby nuclei. Somatic afferents from the body and face terminate in VPL_c and VPM, respectively. Within their fields of termination, fibres of different sensory modalities remain separate. A 'core' of cutaneous afferents in VPL_c projects to sensory areas 3b and 1; the surrounding cells transmit input from deep structures in muscle and joint to areas 3a and 2. At the cortical level, this means that there are four parallel representations of the body surface. Muscle afferents end in 3a, slowly adapting cutaneous afferents end in 3b, rapidly adapting cutaneous afferents end in area 1 and joint afferents end in area 2. Thalamic inputs to the motor areas of cortex contain relatively few direct sensory afferents. Lemniscal (medial and trigeminal) fires only terminate in VPL_c and VPM hence have no direct route to precentral cortex. Only spinothalamic afferents, some of which terminate in VPL_o are in a position to send direct sensory information to motor areas of cortex.

The two nuclei which send fibres to precentral cortex, VPL_o and VL_c, receive most of their input from cerebellar efferents (mainly from the rostral portions of the cerebellar nuclei). VPL_o innervates caudal area 4 (that is, the part in which distal muscles are represented), and VL_c innervates rostral area 4. Inputs from the internal segment of globus pallidus and the substantia nigra pars reticulata, which were once thought to overlap with cerebellar terminations, project quite separately and end in VL_m, VL_o and VA. Inputs from caudal parts of the internal segment of globus pallidus (which represents the output of regions of basal ganglia processing input from sensorimotor cortex) end in VL_m and VL_o and project to the SMA. Inputs from rostral globus pallidus (processed information from frontal regions of cortex) end in VA and project to prefrontal cortex (see Schell and Strick, 1984; Wiesendanger, 1981).

Figure 8.4: Thalamic Inputs to Pre- and Postcentral Areas of Cerebral Cortex in the Monkey. This page: Inputs to areas 1 to 4, with the thalamic nuclei shown in saggital section. In the VPL_c, leminiscal inputs separate into a core of afferents from cutaneous receptors, surrounded by a shell of afferents from deep receptors (including muscle afferents). All of these fibres then project to postcentral areas of cortex. Spinothalamic afferents also terminate in VPL_c but also project to VPL_o and VL_c, whence they may travel to precentral cortical areas. These nuclei also are the relay station for cerebellar afferents to motor cortex (from Friedman and Jones, 1981; Figure 20, with permission).
Opposite: Summary of thalamic projections to precentral motor areas, including MI (MC), SMA and the premotor (here called the arcuate premotor area, APA) cortex. As in the diagram on this page, cerebellar input is shown travelling via the VPL_o nucleus to MI. In this case, extra detail is included to make the point that this input comes mainly from the rostral portions of the deep cerebellar nuclei (DNr). Caudal portions of the deep cerebellar nuclei (DN_c) project via nucleus X to the premotor (APA) area. Input from basal ganglia travels via VL_m and VL_o to the SMA (from Schell and Strick, 1984; Figure 16, with permission)

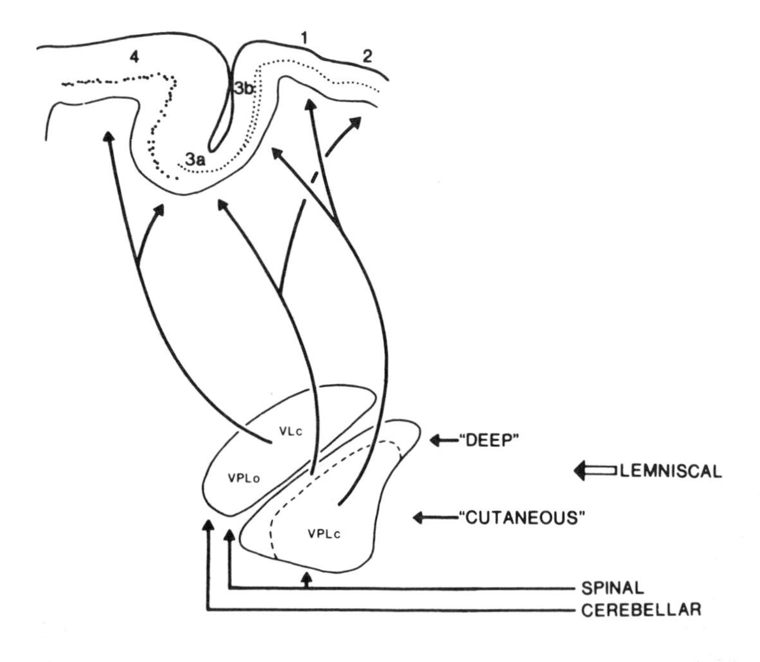

Cortico-cortical Inputs of the Motor Areas. Although there are very few direct sensory inputs to the motor areas, there are substantial cortico-cortical connections within the white matter from postcentral sensory areas. Area 3b receives the most dense thalamic input and sends cortico-cortical fibres to less densely innervated areas 1 and 2. These connections are not reciprocal, and some authors regard areas 1 and 2 as having a 'higher level' sensory function than area 3b. Area 3a also projects backwards to area 1. None of the sensory strips 3a, 3b or 1 send any fibres to precentral areas in the monkey (although the situation is different in the cat where 3a projects forwards to area 4). In the monkey, only area 2, the most caudal sensory strip, sends fibres forwards to area 4, and

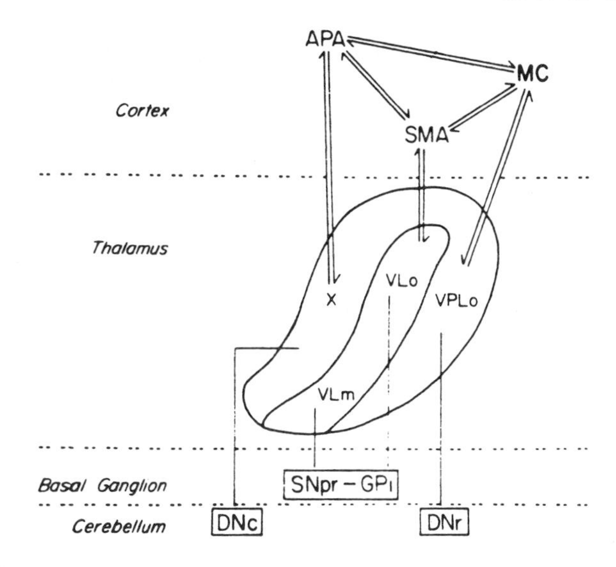

backwards to parietal areas. Area 5 also sends some fibres forwards to area 4. Thus, the primary motor area mainly receives its sensory input after prior processing in higher order regions of parietal cortex.

Sensory input also reaches areas other than the primary motor strip. The input to SMA is similar to that of area 4, coming principally from areas 2 and 5. Sensory inputs to premotor cortex arise from much wider regions of both parietal and visual cortex.

Outputs of Cortical Motor Areas. The major descending output of the cortical motor areas is the corticospinal tract (see Chapter 7), the bulk of which is made up of fibres from area 4. However, area 4 also has cortico-cortical connections to parietal areas 2 and 5, reciprocating the input from these regions. There are important outputs to basal ganglia, pontine nuclei (and thence to the cerebellum), red nucleus and reticular formation. The latter two pathways allow for the possibility of indirect cortical actions from area 4 on the spinal cord. Other areas of motor cortex (except MII, or 6B of Vogt and Vogt), contribute only a small proportion of the corticospinal tract. A substantial part of the output from supplementary and premotor areas projects to area 4, suggesting that they may function as 'higher order' motor regions. In addition, these areas have other indirect pathways to the spinal cord. The SMA sends efferents to the red nucleus, and the premotor area projects to the medial tegmental region of the brain stem. In addition, there are fibres going to the pons (and thence to cerebellum) and the basal ganglia.

Detailed Electrophysiology of Primary Motor (MI) Cortex

Motor Cortex Mapping

One question that has dogged physiologists for many years is precisely how the body is represented in the motor cortex. Are all the pyramidal tract cells which project to a given muscle aggregated together in a specific area, or do pyramidal tract cells which project to different muscles intermix fully within the cortex? If the former were true, then if one could stimulate the pyramidal tract cells within a very small area of cortex, this would produce contraction of only one muscle. If the latter were true, such stimulation might produce contraction of very many different muscles. The debate is often phrased as 'muscles versus movement'. (See relevant chapters of Phillips and Porter, 1977, for the best historical review of this topic.)

The way the cortex is mapped out has implications for the type of instructions that it might receive from higher centres. If each single muscle is mapped out discretely, rather like the keys on a piano, then the instructions which 'play' upon the motor cortex must be very well organised and highly detailed if accurate movements are to be made. If the representation of muscles is intermixed, so that certain elemental movements are represented, then the instructions to motor cortex would not have to be so detailed, and some of the organisation of movement could be performed by the motor cortex itself (Figure 8.5).

As shall be seen, one of the difficulties in answering the question of motor cortex representation is that a given area of cortex projects to spinal cord via several separate pathways. Not only is there a direct monosynaptic (or corticomotoneurone) projection from the pyramidal tract, but also polysynaptic pyramidal routes as well as cortico-rubro-spinal and cortico-reticulospinal pathways. The more indirect a route is, the more likely it is to be less specific in its final projection. The classical homunculi and simunculi of Penfield and Jasper (1954) and Woolsey, Erickson and Gilson (1979) do not help to resolve the question of whether muscles or movements are represented in the motor cortex. By the very nature of their technique, they mapped movements produced by the cortex rather than muscles activated. In addition, in order to produce visible movements in the operating theatre, the stimulating currents were strong, and probably activated an area larger than the minimal cortical outflow unit.

In order to examine the question of localisation in more detail, recordings were made in monkeys of the contractions of *single* muscles during stimulation of the cortical surface with short trains of impulses. Occasionally, points were found which produced contraction of only a single muscle. For any one muscle, these points lay together in a cluster, and stimulation within the cluster always provoked the largest muscle contraction. Stimulation outside these points also produced contraction of the muscle, but with longer latency and less force. The foci of different muscles never overlapped, but the fields did (Figure 8.6). Sometimes the cortical field of one muscle overlapped with the focus of another.

Figure 8.5: Possible Variations in the Organisation of Corticomotoneuronal Projections. Large circles denote two different spinal motor nuclei; small circles denote the areas of cortex which project to the two spinal motor nuclei. In D and E, the columns above the small circles represent the cortical network terminating on pyramidal tract cells. A shows a true mosaic representation of pryamidal tract cells, with no intermingling between projections to different muscles. This form of representation probably does not exist. B shows an overlapping organisation of pyramidal tract cells. C is similar to B except that some pyramidal tract cells are shown as having projections to more than one spinal motor nucleus. B and C are probably reasonable representations of the true organisation of motor cortex. In the second row, D and E illustrate two possible patterns of control of pyramidal tract cells. In D, the input selectively activates widespread pyramidal tract cells, which project only to one spinal motor nucleus. In this instance, the motor cortex might be said to be more interested in muscles than movement. In E, neighbouring pyramidal tract cells are activated without regard for their final destination. This case might be said to represent movements, not muscles (from Jankowska, Padel and Tanaka, 1975; Figure 11, with permission)

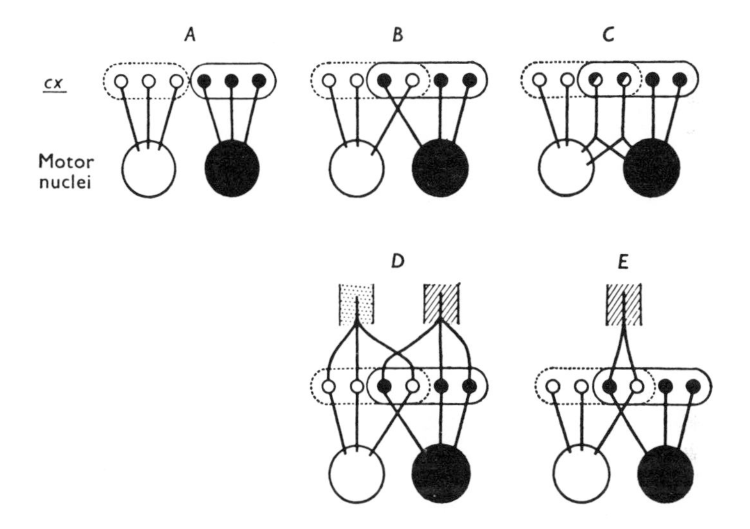

Thus, although there were points on the cortical surface which appeared to project only one muscle, there were areas away from these foci which projected to several different muscles.

These early localisation experiments did not investigate the route by which activity reached the spinal cord from the cortex. As noted above, trains of stimuli can activate several pathways other than the direct corticospinal pathway. Trains of impulses may also spread quite some way from the point of stimulation by activating intracortical interneurones beneath the stimulating electrode. The importance of the nonpyramidal pathways were demonstrated by Lewis and Brindley (1965) who mapped the motor cortex of baboons with 50Hz stimulation. They then sectioned the medullary pyramids and repeated the procedure. The cortical motor map remained unchanged, but the threshold for muscle activation was increased. It was later shown that repetitive stimuli were essential to produce activity via this route.

Figure 8.6: Hypothetical Distribution of Corticospinal Cells Activating Four Individual Muscles. The four rows of myograph records (M1, M2, M3, M4) show the size of the muscle twitch produced by short graded trains of stimuli applied at A and B on the cortical surface. Concentric circles around points A and B represent the area to which current spreads with liminal, medium and strong stimulation. Liminal stimulation at A or B produces contractions limited to muscles 1 or 2, respectively. As the intensity is increased, other muscles are recruited (from Ruch and Fetz, 1979; Figure 3.20, with permission (after Ruch, T. *et al.,* 1946 *Res. Pub. Ass. Nerv. Ment. Dis, 26,* pp. 61–83))

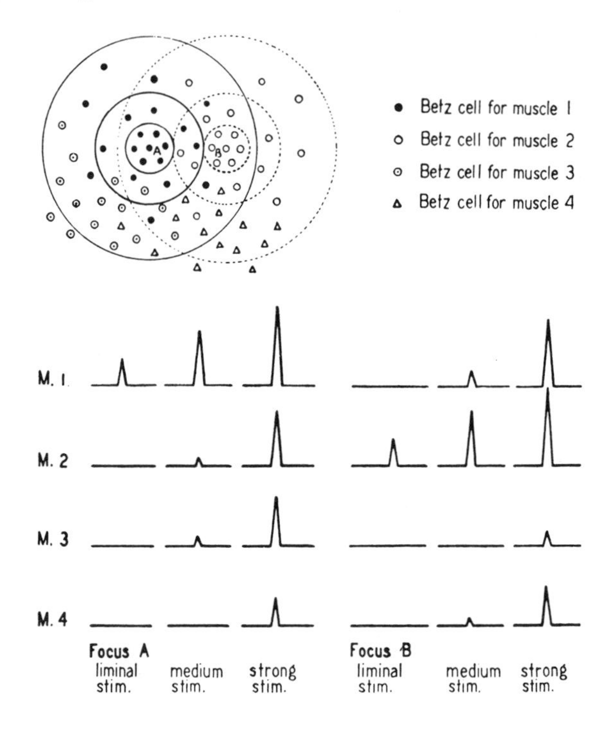

Corticospinal Localisation in MI. More detailed experiments have been carried out to exclude contributions of indirect pathways to localisation of function in MI. The techniques make use of the fact that only corticospinal axons make direct monosynaptic connections with α-motoneurones in the spinal cord. In monkey, baboon and man, a relatively large number of the large diameter axons of corticospinal tract have direct monosynaptic connections (or cortico-motoneurone connections) onto spinal alphamotoneurones, particularly those innervating the distal muscles of the hand. Thus, corticospinal activation can be identified by the presence of monosynaptic EPSPs in the impaled spinal motoneurones.

Phillips and his colleagues performed a series of experiments to map the distribution of corticomotoneurone cells within the primary motor cortex (see Phillips and Porter, 1977, for further reading). They used single pulses of surface anodal stimulation to activate the cortex. This means that a large plate

Figure 8.7: Top: A, Experimental Arrangement for Determining How Far Stimulating Current Spreads From a Focal Cortical Anode. A pyramidal tract axon is recorded and single stimuli (S) are applied at various points on the cortical surface. The size of the threshold stimulus necessary to produce activity in the cell is then plotted as a function of distance from the lowest threshold point. In B, two spinal motoneurones were impaled, one innervating a distal muscle and the other innervating a proximal muscle. A ball anode was then used to explore the cortex and find the best point for activating the motoneurones. With the electrode at this point, a graph was constructed of the size of the recorded EPSP in the motoneurone versus the stimulus intensity applied to the cortex. Maximal EPSPs are produced when all the corticomotoneuronal cells of a cortical colony are active. Thus, the spread of the stimulus intensity necessary to produce a maximal EPSP can be read off the graph in A, to give an estimate of the spatial extent of the cortical colony projecting to the impaled spinal motoneurone.
Bottom: Diagram Showing How the Size of the Monosynaptic EPSP from Corticomotoneurone (right) and Ia Muscle Afferents (left) Varies from Muscle to Muscle in the Baboon's Forearm and Hand. EDC, extensor digitorum communis; R, radial nerve innervated muscles of the forearm; Uh ulnar-innervated intrinsic hand muscles; Mh, median-innervated intrinsic hand muscles; FDS flexor digitorum superficialis; PL, palmaris longus. Circles represent motoneuronal pools in the spinal cord: thickness of the arrows is proportional to the size of the monosynaptic EPSP. (Top: from Ruch and Fetz, 1979; Figure 3.21 with permission. Bottom: from Clough, Kernell and Phillips, 1968; Figure 7 with permission)

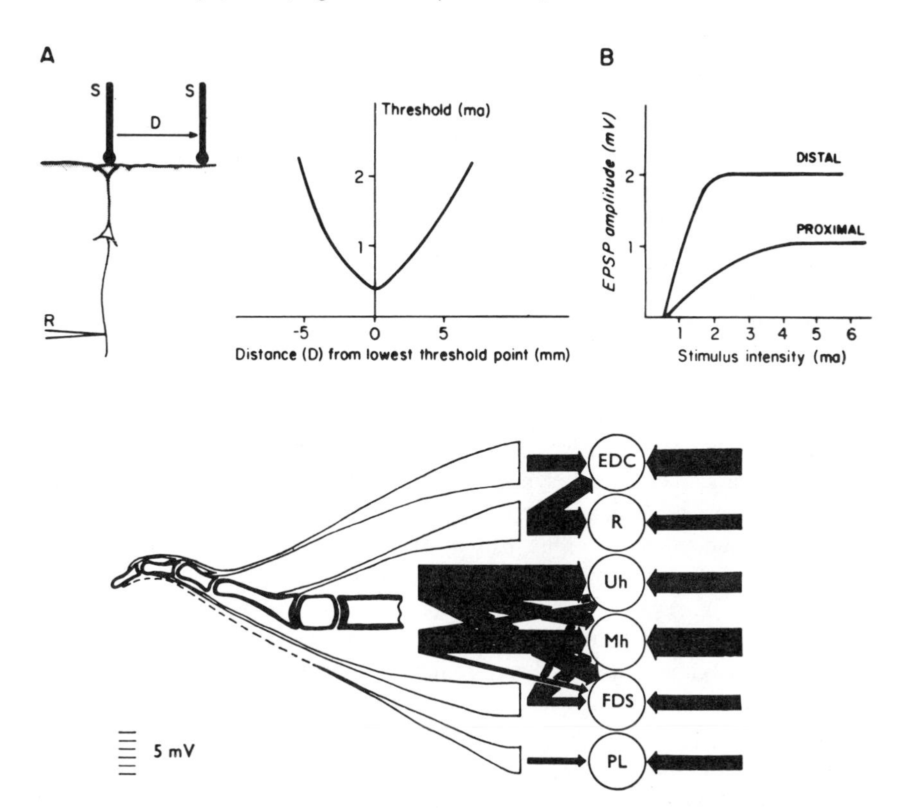

electrode attached to the skull serves as the cathode of the stimulator, while a small ball anode can be used to explore the exposed cortex. This method of stimulation preferentially stimulates pyramidal output cells in layer 5 of the cortex with little effect on cortical interneurones. To determine how far the current from a single pulse spreads in the cortex, single corticospinal tract axons were recorded in the lateral funiculus of the spinal cord. The intensity of stimulus needed to activate a given axon then was charted from different points on the surface. There was a single point of minimum threshold, with current intensity increasing beyond that point with the square of distance moved by the electrode (see Figure 8.7). From such curves, it can be seen that, for example, a 1mA pulse might activate any corticospinal neurones within a circle 6mm in diameter centred on the stimulating electrode.

After this, a single α-motoneurone was impaled in the spinal cord, and the size of the monosynaptic EPSP was measured following corticospinal stimulation. First, the best point for that cell was found — that is, the point of minimum threshold. Then, leaving the electrode on that point, the intensity of stimulation was increased until a maximum EPSP was recorded in the motoneurone. At this stage, all pyramidal tract neurones projecting to that cell must have been stimulated, and the spatial spread of the stimulus could be read off the previous graph. For example, in the motoneurone illustrated in Figure 8.7 the maximum EPSP in a distal motoneurone occurred with an intenstiy of 2mA, indicating that pyramidal tract cells within a circle diameter 10mm had been activated. This area represents the total cortical field for the motoneurone under study. The neurones within this area which project to the motoneurone are known as a cortical colony. Within the colony, there are 'hot spots' of localisation. Stimulation at these points at very low intensity produces activity in the motoneurones of only one muscle. However, the majority of the area occupied by the cortical colony consists of a mixture of cells. Some project to the motoneurones of the muscle under study, while others project to motoneurones of other muscles.

During these mapping experiments it was found that larger maximum EPSPs could be elicited in motoneurones of distal rather than proximal muscles (see Figure 8.7). Thus, all other factors being equal (for instance, motoneurone size, synaptic density and location, synaptic efficiency), more cortical cells project to spinal motoneurones which innervate distal muscles. Despite the larger number of corticomotoneurone cells, the *area* of the cortical colony projecting to each distal motoneurone was found, in contrast, to be smaller than that projecting to motoneurones innervating proximal muscles. Thus, for distal muscles, there are a large number of corticomotoneuronal cells concentrated in a small area of cortex. For proximal muscles, there are fewer corticomotoneuronal cells within a wider area.

Intracortical Microstimulation. The picture which emerges from experiments with surface stimulation of the cortex is of patchy 'hot spots' in which cortical cells project only to motoneurones innervating a single muscle. These 'hot spots'

Figure 8.8: Three Overlapping Projection Areas to Three Different Spinal Motor Nuclei, Each Containing Many Pyramidal Tract Neurones (Small Dots). The neurones projecting to the specific muscles are most densely packed in the centre of the projection areas. Circles around groups of pyramidal tract cells suggest that they may be aggregated together in units (columns), smaller than the total projection area. The unanswered question is whether all the pyramidal tract cells within a column project to the same spinal motor nucleus, or whether columns (for example, those in overlapping parts of the three projection areas) contain neurones which project to more than one muscle (from Phillips and Porter, 1977; Figure 6.3, with permission)

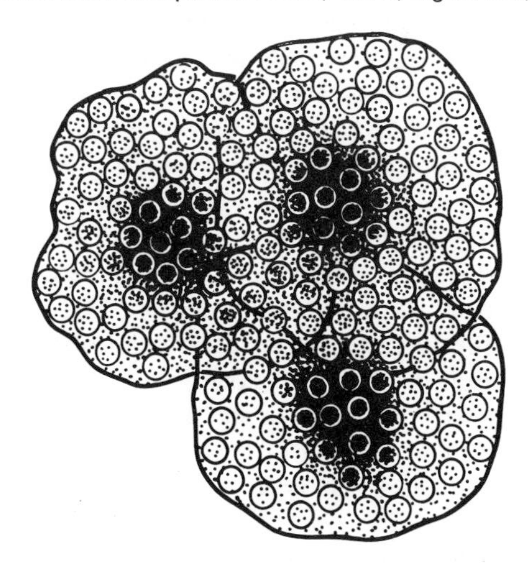

are surrounded by areas in which the representation of different muscles is intermixed. The question arises as to the nature of muscle representation within these areas of diffuse representation. Surface stimulation excites relatively large areas of cortex, even when small ball stimulating electrodes are used. It is a possibility that even within apparently diffuse areas of representation there is a very fine mosaic of individual muscle representation, rather than a true intermingling of cortical colonies. It has been proposed that the units of this mosaic are small groups of pyramidal tract cells aggregated together in a vertical *cortical column.* These columns would share common input and output connections as do the columns which have been demonstrated in the visual and somatosensory systems.

The mosaic concept is illustrated in Figure 8.8. Each dot represents a pyramidal tract cell, and the three large circles outline overlapping territories of the projection to three different muscles. The question of localisation can be put in terms of whether pyramidal tract cells are grouped together into columns (outlined by small circles) in which all cells project to the same muscle, or whether pyramidal tract cells intermingle freely.

If they are grouped together, this would mean that the output of each column controls only one muscle. Groups of columns innervating the same muscle might be found together under the surface 'hot spots'. Intermingling of columns might occur in the areas of diffuse innervation. The essence of the matter would be that individual muscles would still be represented separately at columnar level, even though there appeared to be an intermixed representation when surface stimulation was used.

The technique of intracortical microstimulation was introduced to examine the minimum unit of cortical organisation. Rather than stimulating the surface (which may be 1mm or more from the cortical efferents of layer 5), needle electrodes are inserted into the cortex in an effort to minimise spread of stimulating current. Using this method, individual muscles can be activated using single current pulses of less than 10 μA. Cathodal rather than anodal stimuli are most effective. When first used, the original investigators thought that such small currents would spread very little within the cortex. The effective radius of a 10 μA stimulus was calculated to be only 80–90μm. They found that if currents of 5–10μA were used, it was always possible to activate only a single muscle. Moreover, the lowest thresholds were in layer 5 and in a radial cylinder about this layer. It was suggested that muscles were indeed represented singly, within very small diameter cortical columns, and that these columns could intermingle within a very fine mosaic. The effects were mediated via the pyramidal tract since they disappeared on tract section.

Unfortunately, the extent of stimulus spread was underestimated. Later, it was found that cells within a diameter of 1mm were activated by microstimulation principally through the synaptic actions of cortical interneurones and pyramidal cell collaterals. There was no proof of the idea of a fine cortical mosaic of individual muscle representation, and the question remains unresolved (see Asanuma, 1981; Jankowska *et al.*, 1975; Phillips and Porter, 1977).

Stimulation of Motor Cortex in Man. In the past five years it has become possible to stimulate the motor cortex of man directly through the intact scalp. The technique uses a specially designed stimulator to deliver a short-duration, high-voltage (700V) electric stimulus to an electrode placed on the scalp. The high voltage appears to decrease the resistance of intervening tissues so that the current can penetrate to the brain. Stimulation over the motor area of the brain in man produces activation of muscles on the contralateral side of the body at very short latency. As expected from the high density of corticospinal projections found in the monkey, the responses are most easily obtained in the distal muscles of the hand and forearm. The responses are much larger if the subject voluntarily contracts the muscles under test. This is partly because of an increase in excitability of spinal α-motoneurones, making them more easily recruited by the descending volley.

With this technique it is possible to show that the descending volley has access to all the cells in the spinal motoneurone pools of distal hand muscles. This can be done quite simply by recording the size of a muscle twitch produced by a single electrical stimulus to the scalp. Under certain conditions, the isometric twitch force produced by the cortical stimulation is as large as the maximum isometric twitch produced by supramaximal stimulation of the muscle nerve at the wrist. Given that each motoneurone only fires once (otherwise it may be possible to get fused contractions, which are larger than a single twitch), this means that all the motoneurones can be accessed at short latency by the cortical volley.

The application of the technique is of some clinical use, for it provides the only way to monitor function in the central parts of the motor tracts. When peripheral conduction delays from spinal cord to muscle are taken into account, one is left with a measure of central motor conduction time from the brain to the segment of spinal cord under test. This is delayed in patients with multiple sclerosis or cervical spondylosis. Direct recording of descending impulses from an electrode inserted into the spinal epidural space can also provide a means of monitoring motor function of the spinal cord during surgical procedures on the exposed spinal column (see Marsden, Merton and Morton, 1983, for further information on stimulation of human motor cortex).

The Pyramidal Tract

Electrical stimulation studies have shown that there is a remarkable degree of localisation in the cortical representation of the body muscles. Yet considering the number of branches that an individual pyramidal tract cell may make on its journey to a spinal motoneurone, it is surprising that any detailed cortical organisation should occur at all. This is true even of the corticomotoneuronal portion of the pyramidal tract. For example, antidromic stimulation of pyramidal tract terminals within spinal motor nuclei has shown that one third of axons with terminals in cervical segments also project to lumbar levels. Such an arrangement would seem to guarantee absence of somatotopy. Possible explanations are that impulses are not normally conducted to the end of all branches of a particular axon, or that synaptic density is particularly high on the motoneurones belonging to a particular muscle. Axons innervating forelimb muscles also tend to branch significantly less than those to proximal muscles, which would maintain the discrete somatotopy in the cortex.

The electrophysiology of the corticomotoneuronal connection has been analysed in detail. All corticomotoneuronal synapses are excitatory. IPSPs may be produced by pyramidal tract or cortical stimulation, but their latency is about 1ms longer than the shortest latency EPSP. Early IPSPs are disynaptic and employ the Ia inhibitory interneurones of the spinal intermediate zone.

The size of the maximal EPSP in spinal motoneurones is greatest in motoneurones innervating distal muscles of the forearm, which correlates with the density of terminals in spinal motor nuclei. EPSPs are particularly large in the intrinsic muscles of the hand (see Figure 8.7). However, they are still quite small in absolute terms, being about 1–2mV, and are not capable of raising a resting motoneurone above its firing level (which usually takes about 10mV). This is compensated by a remarkable 'frequency potentiation' at the

Figure 8.9: Intracellular Records from a Baboon α-Motoneurone in the Cervical Cord Showing Frequency Potentiation at the Corticomotoneuronal Synapse. Upper pair: group Ia volleys at 200Hz (upper record) and monosynaptic EPSPs evoked by them (lower record) in the motoneurone. Lower pair: corticospinal volleys at 200Hz (upper record) and corresponding EPSPs in the motoneurone (lower record). The first EPSP is the same size as that to a Ia volley. Later EPSPs show remarkable frequency potentiation (from Phillips and Porter, 1964; Figure 21, with permission)

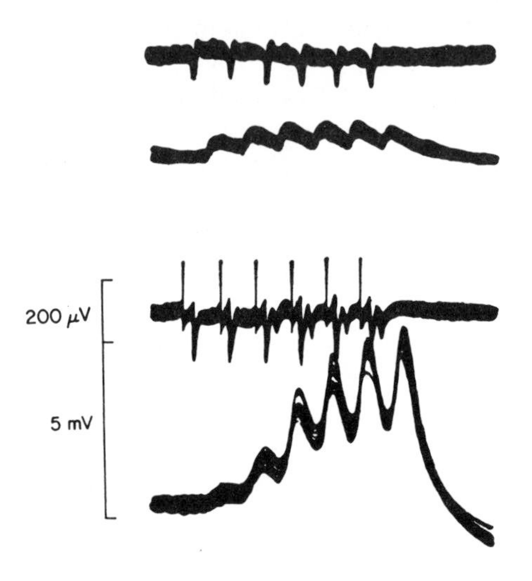

corticomotoneuronal synapse. Repetitive discharges in the pyramidal tract result in progressively increasing sizes of EPSP in the motoneurone (Figure 8.9). Such an effect is not seen at the synapses of Ia muscle afferents. It is a presynaptic action produced by augmenting transmitter release in the second and subsequent volleys of a train. The degree of potentiation depends on the frequency of activation, and is maximum at frequencies of about 500Hz. This is more than the maximum steady firing rate of pyramidal cells, which is around 100Hz. However, higher frequencies are observed at the onset of pyramidal cell firing, before the rate declines slowly to more tonic levels. Such a phasically decreasing discharge rate shows better frequency potentiation at corticomotoneurone synapses than steady high frequency firing. It is probably used most effectively in the initiation of rapid movements.

As well as their connections with α-motoneurones, there are also direct corticomotoneuronal connections with γ-motoneurones. The fibres originate in the same area of cortex as those which project to α-motoneurones innervating the corresponding extrafusal muscle. This organisation may indicate a degree of alpha-gamma coactivation in the pyramidal tract projection. Indeed, this may be the anatomical basis for the results from microneurographic recordings of muscle spindle firing during voluntary contractions in man (see Chapter 6).

The corticomotoneuronal connection represents only a small proportion of the total corticospinal projection. Most of the spinal terminations of the motor cortex component of the pyramidal tract are in the intermediate zone. There they may synapse onto spinal interneurones, such as the Ia inhibitory interneurones.

Finally, the pyramidal tract cells show extensive branching via axon collaterals within the motor cortex itself. These collaterals may spread within a radius of 30μm and extend vertically to layer 2 of the cortex. The action of these collaterals is complex. When a single motor cortical pyramidal tract cell is recorded and the pyramidal tract stimulated in the medullary pyramids with an intensity subthreshold for direct antidromic activation of that cell, two later phases of membrane potential change can be recorded: an initial depolarisation followed by a hyperpolarisation. The initial phase is thought to be caused by monosynaptic excitation from collaterals of pyramidal cells which were excited antidromically by the stimulus. The hyperpolarisation is probably due to inhibition transmitted via cortical interneurones. Repetitive stimulation greatly increases the size and duration of the hyperpolarisation, and it may be that during normal activation, a group of discharging pyramidal tract cells can cause 'surround inhibition' of neighbouring areas of cortex through this action of axon collaterals on cortical inhibitory interneurones (see Asanuma, 1981; Phillips and Porter, 1977).

Motor Cortex Cell Activity During Voluntary Movements

One of the most important advances in recent years has been the ability to record single cell activity from the brains of conscious animals while they perform particular learned movements. The technique was pioneered by Evarts in the 1960s (1966, 1968). A small chamber enclosing an hydraulic microdrive is fixed above a hole in the skull during an operation in an anaesthetised animal. The drive is then capable of inserting a tungsten microelectrode into parts of the underlying cerebral cortex at various depths in the behaving animal to record extracellular unit activity. This technique can be used to record activity in any part of the brain. When used to study motor cortex activity, another (nonmovable) set of electrodes is usually implanted within the medullary pyramids. These electrodes are used to stimulate the axons of pyramidal tract (PT) cells so that PT neurones in the cortex can be identified, and their conduction velocity measured.

The procedure to identify a motor cortex output cell is as follows: a unit is encountered in the motor cortex while moving the microelectrode. A stimulus is then given to the pyramid. This usually produces an impulse in the cell which may be an antidromic potential conducted along the axon of the cell, or a synaptic potential produced via activity in other cells. To distinguish these modes of activation, the motor cortical cell is itself stimulated through the recording electrode shortly before the arrival of the antidromic volley. If the latter is an antidromic impulse, it will collide with the descending (orthodromic) potential and will be unable to activate the cortical cell. In contrast, a synaptically conducted potential will still activate the cell. This test is known as the collision

Figure 8.10: Top: Timing of Discharge in Pyramidal Tract Neurones (PTN) before a Voluntary Movement in a Monkey. Upper traces are a series of 12 trials in which a monkey had been trained to react as rapidly as possible to a visual stimulus (presented at the start of each sweep) by extending its wrist. Recordings are from a single PTN neurone (upper traces) which was silent during flexion movements, EMG from wrist extensor muscles (middle traces), and onset of movement (bottom traces). The timing of the PTN discharge is more closely related to the timing of EMG activity than it is to onset of the visual stimulus. This is seen most clearly in the second and third trials of the first row. In the second, the PTN onset latency was short, and the EMG reaction time was also short; in the third, the onset of PTN discharge was long, as was that of the EMG activity. The duration of the sweeps was 500ms (from Evarts, 1966; Figure 2, with permission)
Bottom: Distribution of Onset Latencies of a Sample of Precentral (Upper Histogram) and Postcentral (Lower Histogram) Neurones in Relation to Abrupt Flexion and Extension Movements of the Wrist. Latencies have been measured in relation to the onset of movement (R). The reason for using movement, rather than EMG onset as a reference, was that responses to several different types of movement were combined in this graph, and it was not possible to discover which muscle started its activity first in each case. The y-axis plots the number of cells responding at a particular latency. Note that precentral neurones commonly changed their firing 60ms before movement, whereas postcentral neurones only began to discharge after movement. The cells which were sampled were both PTN and non-PTN neurones (from Evarts, 1972; Figure 5, with permission)

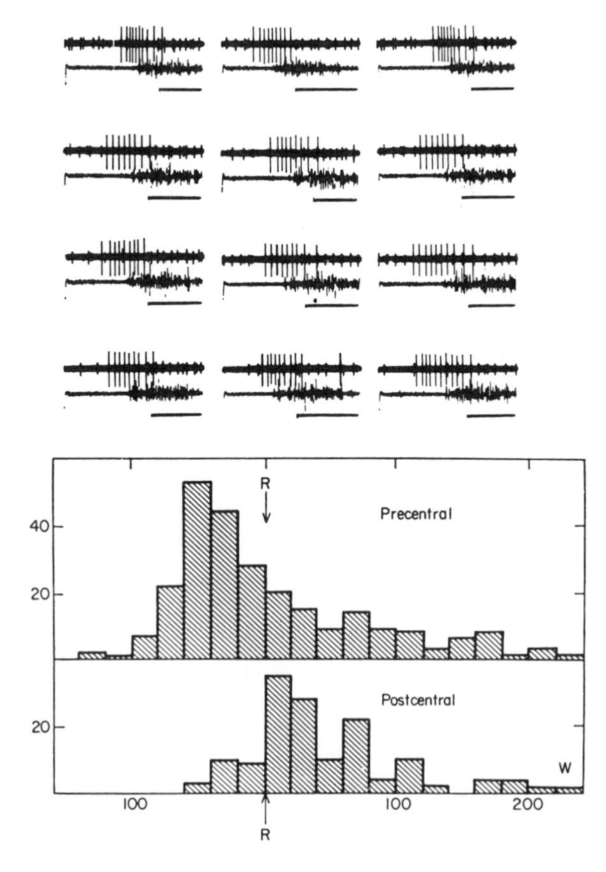

test. If it is positive, then the motor cortical cell which is being recorded from is a pyramidal tract (PT) neurone. If not, it is likely to be an intrinsic interneurone, or an output neurone which projects elsewhere than the pyramidal tract.

Most studies of motor cortex activity have concentrated on the relationship of pyramidal tract (rather than interneuronal) activity to movement.

Timing of Discharge

One of the first questions answered by this technique was the temporal relationship between motor cortical neuronal activity and muscle contraction. Monkeys were trained to perform reaction time movements of the hand to release a key as soon as possible after a visual stimulus. The time between the signal and the EMG response in wrist extensor muscles was of the order of 150ms. However, many motor cortex cells including PT neurones began to change their firing rate up to 100ms before the EMG activity (Figure 8.10). Because the timing of this activity was more closely correlated with EMG onset than with the time of the visual stimulus, it was suggested that it played a direct role in the production of movement.

The interval of up to 100ms before EMG activity in the agonist muscle is much longer than the minimum time taken for impulses to travel from motor cortex to the arm muscles in a monkey (5–10ms). The reason for the extra delay is that more than one pyramidal tract impulse is needed to bring spinal motoneurones to threshold and there is a short period of subthreshold facilitation prior to EMG onset. This can be detected using monosynaptic testing of the motoneurone pool at different times before movement onset. As the onset of EMG activity approaches, the size of the monosynaptic reflex in the agonist muscle gradually increases, consistent with a gradual increase in facilitation prior to movement. The remainder of the time taken between visual signal and muscle response is taken up with cortical processing and with the transformation of this information into motor commands.

The timing of motor cortical cell discharge can be contrasted with that of postcentral cells. In general, postcentral cells only fire after the onset of movement, whereas most precentral cells fire before (Figure 8.10) (see Evarts, 1966, 1981; Phillips and Porter, 1977).

Relationship of Discharge to Movement Parameters

The second question to be tackled with this technique was whether the firing of pyramidal tract cells was more closely related to the *force* of muscle contraction or to the *position* of the joint during the contraction. In effect, this type of experiment investigates the nature of the abstract command encoded by the discharge of pyramidal tract cells.

In the first experiments monkeys were trained to move their wrist through various positions of flexion and extension against a series of different loads. Pyramidal tract cells showed a closer relationship of their firing rate to the force

Figure 8.11: Relationship of Discharges in a Pyramidal Tract Neurone (PTN) and Wrist Flexion/Extension Movements Made Against Different Loads. Three experiments are shown: movements made against a load of 400g opposing flexion (and assisting extension) (HF), movements with no load (NL), and movements with a load of 400g assisting flexion (opposing extension) (HX). In each section, the traces are wrist position (lower trace: upwards deviation is flexion), PTN cell activity (middle traces) and a marker showing when the monkey had reached the end stops at the limit of the flexion and extension movements. In the 'no-load condition', the PTN can be seen to discharge before and during movement of the wrist into flexion. In the presence of an opposing load (HF), the discharge is greatly increased, and even continues during the extension phase of movement. In this instance, it may have been that extension was made by using the flexor muscles as a brake on the assisting extensor load. When a load assisting flexion was applied (HX), the cell was almost totally silent. Thus, even though the same wrist movements were made in the three experiments, cell discharge varied greatly and was related to the active force of movement rather than the position of the wrist (from Evarts, 1968; Figure 7, with permission)

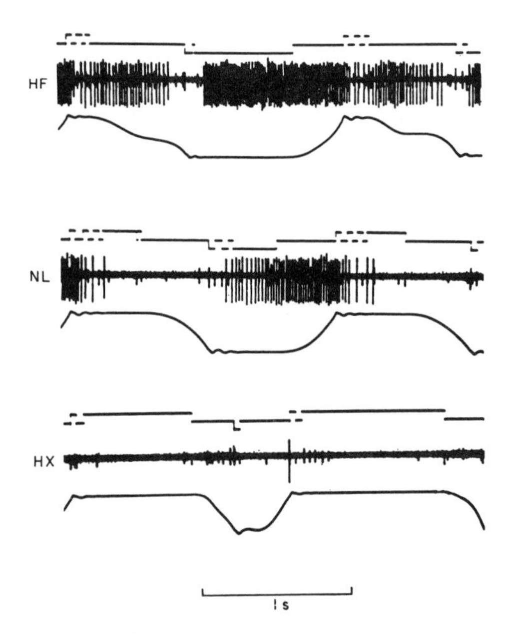

of contraction than to the position of the wrist (Figure 8.11). However, in subsequent experiments, it was found that if the monkey had to maintain a constant wrist position against either a flexor or extensor load, then the largest changes in discharge rate occurred when the direction of force shifted (that is, from flexion to extension or vice versa). Thus, it was argued that pyramidal tract activity might be more important in *selecting* the muscles to be activated, rather than in determining the level of activity in a given muscle. Other areas of brain or other descending tracts might be more involved in specifying these parameters.

There have been many subsequent experiments into the precise nature of the command encoded in pyramidal cell activity. The discrepancies seem now to

Figure 8.12: Relationship Between Firing Rate of a Corticomotoneuronal Cell and the Torque Exerted at the Wrist in a Monkey. The monkey was required to make wrist extension movements against spring loads (that is, the larger the movement, the larger the force needed to extend the spring) of various stiffness. Trials against three sizes of spring load are shown on the right. Spring stiffness decreases from a to c. Traces are instantaneous firing frequency of the neurone (top), wrist torque (middle) and wrist position (bottom). On the left, the graph plots the relation between wrist torque and both the phasic peak of activity of the cell during the movement, and the tonic plateau of activity recorded as the monkey held its wrist at the final end position. In this corticomotoneuronal cell, even the phasic activity recorded at the onset of movement is related to the torque exerted until saturation occurs at very high torques (from Ruch and Fetz, 1979; Figure 3.35, with permission)

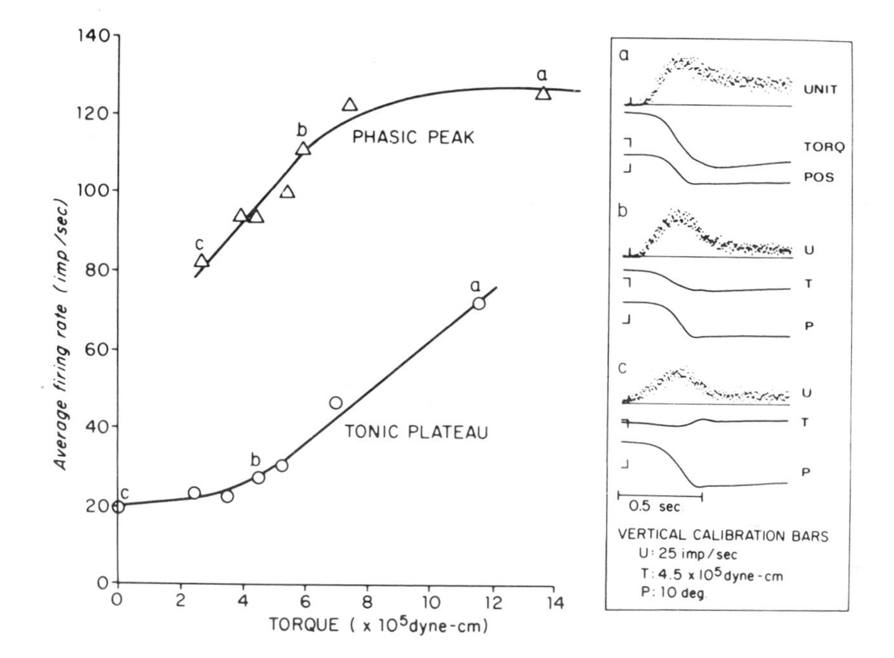

have been resolved by dividing the pyramidal tract cells into two groups. Those cells with monosynaptic connections to the motoneurones (corticomotoneuronal cells) have a firing rate which is well related to the force of contraction (Figure 8.12). At the wrist, this relationship is steeper for extensor coupled neurones than for flexors. Those pyramidal tract cells without monosynaptic connections to spinal motoneurones discharge phasically at onset of movement, but fail to show sustained levels of firing thereafter.

Not only have arm muscles been investigated in this way; the jaw-closing muscles also have been studied. During isometric biting, many precentral (but not, of course, pyramidal tract cells, since these are bulbar innervated muscles), cells showed a clear relationship between firing rate and force exerted. In contrast, most precentral cells failed to show any modulation of discharge during natural chewing movements. Presumably, such natural rhythmic movements are controlled by other centres of the brain.

A final addendum to this story is that the firing pattern of pyramidal tract motoneurones is affected by the type of movement that is made. Pyramidal tract neurones show greater changes in firing rate during finely adjusted movements than they do for large movements. This can be seen in Figure 8.14, where the discharge is not a great deal smaller prior to small movements than it is prior to large ballistic movements with a velocity 10 to 15 times larger. One conclusion from this could be that the motor cortex is preferentially involved in control of finely graded movements, rather than in gross movements of the limbs (see Evarts, 1968, 1981; Evarts, Fromm, Kroller *et al.*, 1983; Phillips and Porter, 1977).

Role of Small Pyramidal Tract Cells

Microelectrodes inserted into the cortex record perferentially from large cells. Thus most studies have analysed behaviour of the fastest conducting pyramidal tract neurones. Recently, attention has focused on small pyramidal tract neurones which contribute the majority of axons to the pyramidal tract. Analysis of firing behaviour of small and large cells has revealed differences which, in some ways, are reminiscent of the size principle of spinal motoneurone recruitment. Small pyramidal tract cells are more likely to maintain a steady firing rate in the absence of load than large cells. Their firing rate also tends to be related to muscle force over a wider range. Large cells may only be recruited above zero force levels, and have an 'S' shaped relation between firing rate and muscle force. Therefore, during very small movements, or small adjustments of force, small pyramidal tract cells change their firing frequency far more than large cells. Changes in firing rate of the latter are much more readily seen in movements requiring large amounts of EMG activity. The role of small pyramidal tract neurones may be principally in the control of fine movements at relatively low levels of force (Evarts *et al.*, 1983).

Sensory Input to Motor Cortex

The activity of pyramidal tract cells is influenced not only by central commands fed into motor cortex from other brain areas, but also by afferent input from the peripheral moving parts. Very short latency activation of precentral cells can be produced by electrical stimulation of cutaneous, muscle or mixed nerve trunks, or by natural stimulation of cutaneous or muscle receptors. Cutaneous input is directed mainly to rostral area 4, and muscle input to the caudal part, where it may lie buried within the central sulcus. There is a considerable overlap between these projection areas.

Controversy Over the Route by Which Afferent Input Reaches Motor Cortex

The short latency with which these sensory inputs reach motor cortex has generated some controversy over the anatomical route which they traverse to get there. At the present time there are two seemingly contradictory facts about this projection. Responses of precentral cells to peripheral stimulation are abolished by lesions of the dorsal columns, yet may remain after ablation of primary sensory cortex. The logical conclusion from these two experiments is that peripheral sensory input travels up the dorsal columns and reaches area 4 without first travelling to the postcentral cortex. Unfortunately, it has proved impossible to demonstrate this route anatomically. Dorsal column — medial lemniscal afferents terminate, as described above, in the VPL_c nucleus of the thalamus, and project only to primary sensory cortex. There is no dorsal column input to either of the thalamic nuclei VPL_o or VL_c, which project to motor cortex (see Figure 8.4).

The dilemma has not yet been resolved. Nevertheless, several reasons for the discrepancy between anatomical and physiological results are becoming apparent. One of these is that natural stimuli were used to evaluate the role of the dorsal columns in transmitting input to area 4, whereas electrical stimulation of peripheral nerve was used in experiments where sensory cortex was removed. It may be that natural stimulation, which provides a relatively weak and dispersed signal, overestimated the role of the dorsal column medial lemniscal system, whereas the synchronous and non-physiological afferent volley produced by electrical stimulation of nerve underestimated the role of sensory cortex. The present opinion is that there may indeed be a direct route for peripheral afferent signals to reach motor cortex, but that this is relatively small. Anatomically, it would have to use that part of the spinothalamic input to VPL_o thalamus which then projects to area 4. The main input to motor cortex is thought to travel via cortico–cortical connections from areas 3a and 3b via areas 1 and 2 to the precentral cortex. Consistent with this is the finding that peripheral receptive fields of motor cortex neurones are larger than those found in primary sensory cortex (see Schell and Strick, 1984; Weisendanger and Miles, 1982).

Relation between Sensory Input and Motor Output

Whatever the route the afferent signals take to reach motor cortex, their terminations within area 4 are carefully related to the pyramidal tract output of that region. Intracortical microstimulation has shown how tight this coupling can be. The precentral cortex is explored with a stimulating microelectrode and a 'hot spot' may be found which produces, say, flexion at a particular joint. Cortical stimulation is then stopped, and the same electrode is used to record the neuronal response of the nearby cells to peripheral sensory stimulation. Almost inevitably, the input is from receptors at or nearby the joint that moved during stimulation.

Cutaneous Inputs. The first experiments by Rosen and Asanuma (1972) suggested that the cutaneous receptive field was likely to be on a region of skin which would encounter obstacles during the movement produced by

Figure 8.13: Relationships Between Receptive Field of Afferent Input and Movement Produced by Intracortical Microstimulation (< 5 μA) at Various Sites in the Thumb Area of Monkey Motor Cortex. The diagram shows a reconstruction of electrode tracks (solid vertical line) and cell locations (small filled dots on the lines). Cortical spots stimulated at 5μA without producing motor effects are shown by small solid lines perpendicular to the electrode tracks. The symbols and diagrams connected to various points show the thumb movements produced and receptive fields recorded at each location. For example, in electrode track 12, the first cell encountered a cutaneous receptive field on the distal tip of the thumb. Microstimulation at this point produced a thumb extension movement. Further down the same track, the next cell had a cutaneous receptive field in the middle of the thumb about the interphalangeal joint. Stimulation here produced thumb flexion. UD, undefined input (from Rosen and Asanuma, 1972; Figure 2, with permission)

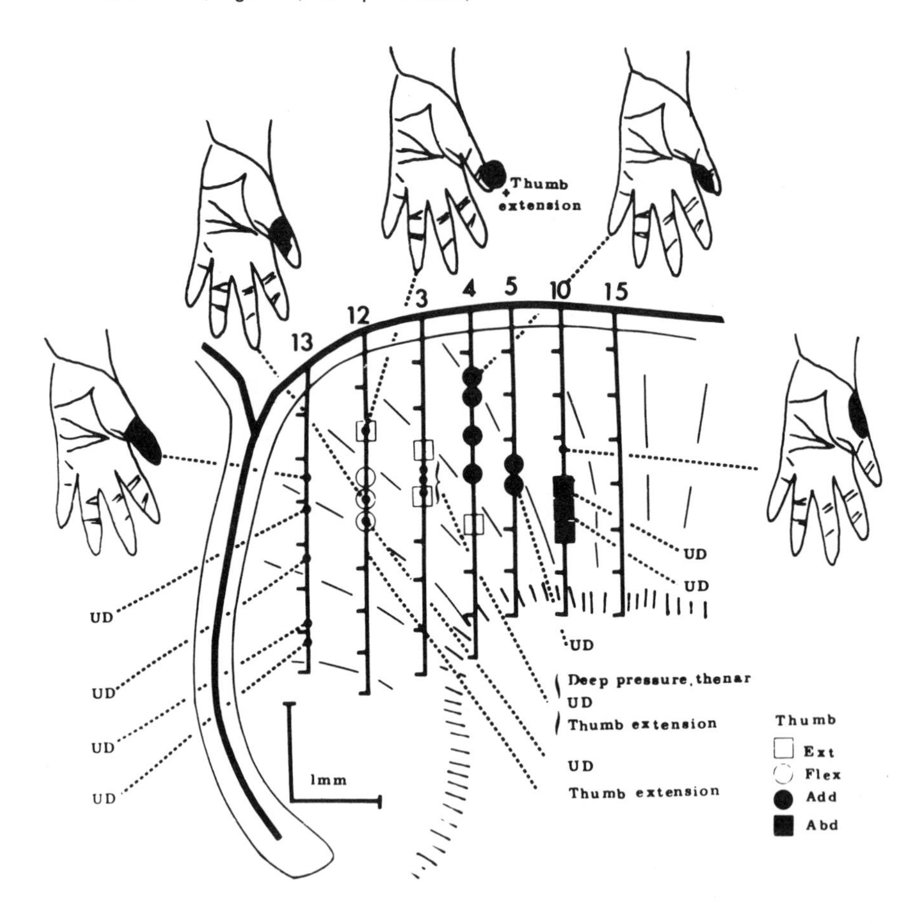

microstimulation (Figure 8.13). For example, in one experiment (not illustrated), they found a series of neighbouring cells, stimulation of which evoked, separately, extension, adduction and abduction of the thumb. The cutaneous receptive fields for each of these sites came, respectively, from the distal tip, medial surface, and lateral surface of the thumb — precisely the regions likely to be stimulated during the movements.

They speculated that such a tight input–output coupling could underlie the phenomenon of 'forced grasping' (see below) which sometimes is seen following lesions of prefrontal areas which project onto area 4. It was suggested that lesions might release the reflex circuit set up by these connections so that cutaneous input could drive selected pyramidal cells to maintain a grasp on an object.

Other authors have found, however, the coupling between input and output to be less tight, particularly for proximal muscles. Cutaneous receptive fields are sometimes found on the side of a limb opposite to that expected from the results of Asanuma and Rosen.

Deep Inputs. Muscle and joint afferent input is equally tightly coupled to motor cortical output. The predominant arrangement is that if microstimulation of cortex produces movement in one direction, then *passive* movement of the same joint in the *opposite* direction is likely to excite that area of cortex. The receptors responsible are the jont receptors, tendon organs and muscle spindles. For the spindles, this arrangement means that if a muscle is passively stretched, their afferent input goes to that area of cortex where it can excite cells which produce contraction of the same muscle.

The response of motor cortex cells to afferent input is enhanced during the performance of finely graded movements. This is shown in Figure 8.14 in which small torque pulses were given to a monkey during performance of simple wrist movements. The response of pyramidal tract cells was large if the pulses were given during or before a small movement. It was much smaller if pulses were given before or during a large, fast movement. The heightened sensitivity to afferent input, coupled with the increased discharge during small movements, suggests that motor cortex may be preferentially involved in muscular control at small force levels. It is not known what mechanism is used to change motor cortex responsiveness to afferent inputs. However, it is possible that motor cortex output itself can change the effectiveness of transmission through sensory pathways via its projections to the dorsal column nuclei and dorsal horn (see Evarts, 1981, for further references).

Transcortical Reflexes

The demonstration that muscle afferent input reaches the motor cortex and can affect the firing rate of pyramidal tract neurones, establishes a circuit between muscle afferent input and pyramidal cell output, which appears functionally to be very similar to the monosynaptic connection between Ia afferents and motoneurones at the spinal level. The question is whether this transcortical circuit can work as a reflex pathway in the same way as the monosynaptic spinal loop. Is the short latency excitation of pyramidal tract cells sufficiently strong to affect the firing rate of the spinal motoneurones to which they project? Does activity in the transcortical route evoke substantial reflex activity in muscle? One might expect perhaps that, since there are more synapses in the long loop pathway, it

Figure 8.14: Heightened Responsiveness of a Pyramidal Tract Neurone to Peripheral Inputs During Small, Carefully Controlled Movements. Upper row, during a wrist pronation task; lower row, wrist supination. The monkey was required either to hold the wrist in a constant position, to make a small pronation or supination movement, or to make a large fast movement. In random trials, a torque pulse was given by a motor in order to disturb the wrist position. In each row three displays are shown: (1) superimposed position traces; (2) histograms of unit discharge; and (3) rasters of unit discharge. The histograms plot the number of times the unit fired in a certain time interval over the whole series of trials. The rasters show the occurrence of each impulse in the PTN as a dot for each separate trial. Note: (1) Unperturbed small pronation (top left) is preceded by more prolonged unit discharge than unperturbed ballistic movement (top right); (2) The inhibitory effect of a torque pulse is greater when delivered during a small pronation (top, small + torque) than when delivered during holding (top, torque pulse holding); (3) There is a corresponding accentuation of excitatory effects of a supinating torque pulse when this is delivered during small movement (bottom, small + torque) compared with holding (bottom, torque pulse holding); (4) There is attenuation of the torque pulse response when the pulse is delivered immediately before a rapid wrist movement (preballistic torque) (from Evarts and Fromm, 1978; Figure 2, with permission).

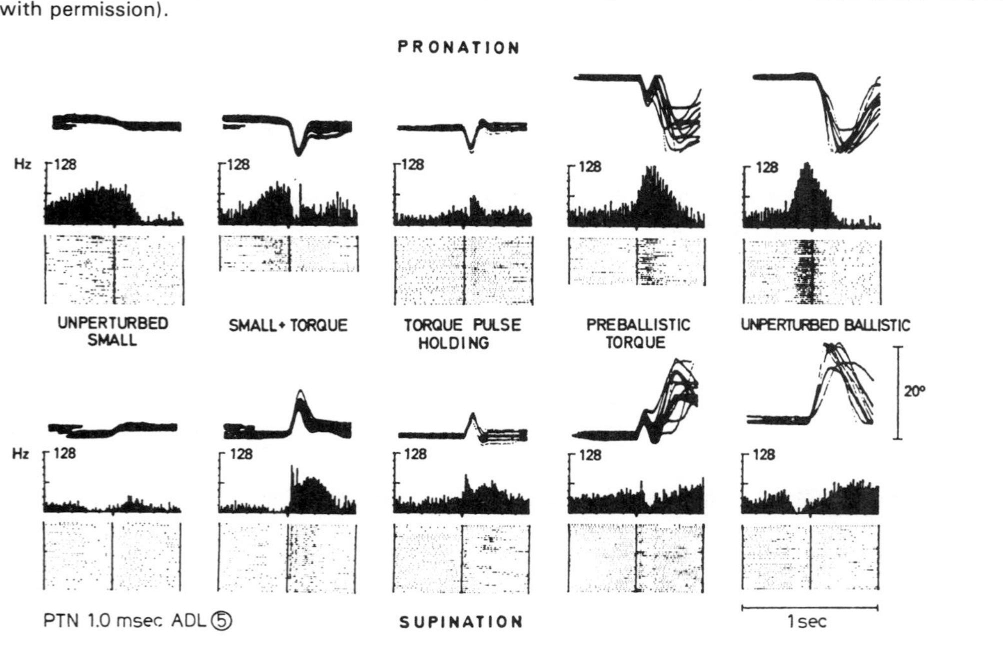

Figure 8.15: Demonstration that Peripheral Input Generated by Muscle Stretch Can Change the Firing Frequency of a Single Cortico-motoneuronal Cell in the Monkey Motor Cortex. The same cell could be shown to project monosynaptically to the motoneurones of the stretched muscle. Left: Average responses of the cortico-motoneuronal cell (top) and extensor digitorum muscle (second trace) to imposed flexion disturbances applied to the wrist during maintenance of a constant extensor contraction. The firing of the cortical cell is plotted as a histogram in which the number of impulses in each 2ms period have been counted for each of the 50 wrist disturbances. The EMG response to stretch consists of the usual short (M1) and long (M2) latency peaks of activity. In this animal, they had a latency of 10 and 24ms respectively. Cortical cell activity began to change some 14ms after stretch. Right: A method for proving that this cell had a monosynaptic connection to motoneurones of the extensor muscle under study. The technique involves recording the discharge rate of the cell and the small fluctuations in EMG activity in the muscle for long periods of time. An average picture of the muscle activity is then constructed, with the average triggered by each of the spikes in the train of cortical activity. The idea is that any muscle activity time-locked to the cortical spikes will emerge as a peak, and that randomly occurring changes in EMG activity will average out to a straight line. The technique is known as spike-triggered averaging. In this case, the cell produced a 'post-spike facilitation' of the extensor EMG with a latency of 10ms. Adding together 10ms with the 14ms delay between stretch and the change in cell firing shown on the left, gives 24ms. This is the precise timing of the M2 component of the stretch reflex. (Note that spike-triggered averaging only detects relatively strong monosynaptic connections. With disynaptic pathways, the extratemporal dispersion produced by an additional synapse makes any post-spike facilitation very difficult to detect) (from Cheney and Fetz, 1984; Figure 3, with permission)

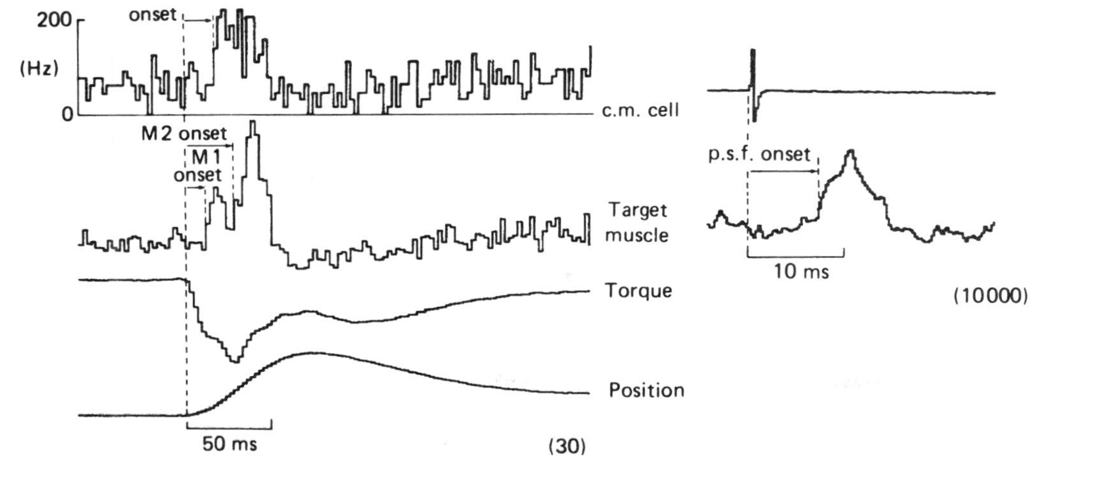

Figure 8.16: Dependence of Pyramidal Tract Cell Activity on Both Afferent Feedback from the Periphery, and on the Nature of the Voluntary Motor Command. The traces are raster displays of unit activity: each dot represents the time of occurrence of a spike in the cortical cell; the parallel horizontal rows are the responses to repeated trials. The monkey was trained to hold a push-pull handle in a centre zone. At the middle of the traces, the handle was moved in the 'pull' direction, stretching the triceps muscle. The firing of the neurone illustrated was linked to the activity in triceps, and was excited at short latency by this sudden lengthening of its muscle. The upper panel illustrates neuronal responses when the monkey was instructed to oppose the disturbance, and the lower when instructed to assist the disturbance. The intended discharge was highly dependent upon the instruction, remaining high when opposing the disturbance (and activating triceps) (from Evarts and Tanji, 1976; Figure 3, with permission)

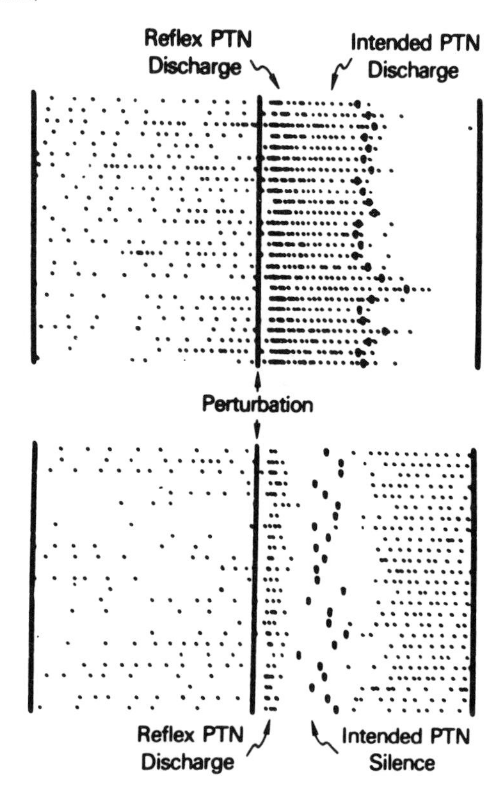

would be less effective than the shorter spinal route. The muscle input to the cortex might be used for more subtle control of motor cortical activity.

Experiments using the technique of single unit recording in conscious monkeys have shown that spindle input to motor cortex is powerful enough to change the firing pattern of pyramidal tract cells during the course of a movement, and that this activity can influence spinal motoneurone discharge. For example, a monkey can be trained to hold a constant position of the wrist by contracting the flexor muscles against a constant load. If the wrist is suddenly extended by increasing the load, the active (flexor) muscles are stretched. The spindle discharge produces at cortical level, a short latency (20ms or so) increase in firing rate of the pyramidal cells which project to the flexor muscles.

Alternatively, if the load is suddenly removed, thereby unloading the active (flexor) muscles, the activity of the pyramidal cells are followed by corresponding changes in EMG activity of the flexor muscles with a latency appropriate for the known conduction times from motor cortex to periphery (Figure 8.15). This activity is the long latency reflex described in Chapter 6. Pyramidal cell discharges are believed to be responsible for part, if not all, of the long latency reflex EMG responses. Thus the muscle spindle projection to motor cortex is strong enough to function as part of an active 'long-loop' reflex pathway which operates in parallel with the conventional spinal monosynaptic reflex (Cheney and Fetz, 1984; Evarts, 1973).

Pyramidal tract discharge is controlled by central command to motor cortex as well as by the reflex effect of sensory input. The two types of input can readily be dissociated by arranging an experiment where the sensory input for example, will increase cell firing, while the central command will be to decrease the firing rate. Monkeys can be trained to receive random stretches of, say, triceps and to react as rapidly as possible to the stretch by either extending or flexing the elbow as fast as possible. As shown in Figure 8.16, pyramidal tract cells which project to triceps modulate their firing in two stages following the disturbance. Initially, there is an increase in firing, as expected from the long-loop reflex projections from triceps muscle spindles to motor cortex. This activity is the same whether the monkey has been instructed to flex or to extend the elbow on perceiving the disturbance. However, some 40ms later, the discharge comes to depend critically on the prior instruction given to the monkey. If instructed to extend, then firing remains high. If instructed to flex, there is a sudden decrease in firing. This later part of the response is due to the central input to pyramidal tract neurones from the voluntary motor command (Evarts and Tanji, 1976).

Other Motor Areas of Cortex

Premotor Cortex

Resumé of Anatomical Connections. In contrast to the wealth of information available on primary motor cortex, very little is known about other cortical motor areas. The premotor cortex, in particular, has suffered because of arguments over precisely where it is. As discussed above, the best definition probably is that part of area 6 lying on the lateral surface of the brain in front of the representation of axial and trunk muscles of the primary motor cortex. This region sometimes is termed the arcuate premotor area. The remaining, mesial part of area 6 constitutes the supplementary motor area. Remarkably, while areas 4 and 6 of cortex are of approximately equal size in monkey, in man area 6 is approximately *six times larger* than area 4. Although little understood, this region of brain may therefore prove to be particularly important in the control of human movement.

The inputs to premotor cortex come via thalamic nucleus X from the cerebellum, and via cortico-cortical connections from parietal, visual and prefrontal cortex. A major output is to the primary motor cortex. In addition there is a small direct projection to spinal cord via the pyramidal tract, and an indirect connection via brainstem tegmental areas. The latter, which projects particularly to axial and proximal muscles may indicate a possible role of premotor cortex in control of posture. Premotor cortex also has strong reciprocal connections with mesial area 6, the supplementary motor area.

Effects of Electrical Stimulation. Electrical stimulation of premotor cortex has given few clues as to its function. The results are very similar to stimulation of the supplementary motor area, and of the frontal eye fields in area 8. No response occurs until some seconds after stimulus onset (50Hz alternating current surface stimulation), and may outlast offset by some time. Movement usually consists of deviation of the eyes, head and trunk towards the side opposite stimulation, and may persist, if high stimulation intensities are used, when are 4 is removed. Such movements, which involve many muscles bilaterally, may therefore be produced via the direct pyramidal tract projection, or via the indirect tegemental reticular connection. They are termed *ad*versive movements, a confusing term since the movement is *away* from the stimulated side. However, it is towards the side to which the eyes deviate. Epileptic seizures with a focus in premotor cortex often begin with similar adversive movements. No somatotopic representation of movement has been described (see review by Wiesendanger, 1981).

Electrical Recording and Lesion Studies. Recording and lesion studies have generated several useful hypotheses about the function of premotor cortex. An important concept that has directed the most recent research efforts is the division between movements that are triggered by sensory input from the periphery, and self-paced movements that are produced by internal mechanisms. Because of its inputs from visual areas of cortex, the premotor area is believed to be particularly important in control of visually guided movements.

In a typical experiment, neuronal activity will be recorded from premotor neurones during the following type of task. Monkeys are trained to move their arm rapidly from a rest position and align their fingers with a target light. The signal to move is dimming of the target light. However, before dimming it is first moved into the correct position for the next movement to instruct the animal in which direction to move (see Figure 8.17). Under such conditions three classes of neurones can be distinguished: stimulus-related, set-related, and movement-related. Movement-related cells behave like precentral neurones, with discharge onset some 100–150ms (that is, slightly before precentral cells) before movement. The timing of their discharge is temporally correlated with onset of EMG. Stimulus-related cells respond transiently at short latency (60ms) following either visual signal (the instruction or the trigger), and are temporally coupled to the onset of the stimulus. There are very few cells of this type in primary motor cortex.

Set-related neurones, which comprise over half of the cells recorded in premotor cortex, are the most interesting. They fire in response to the warning stimulus and maintain their discharge until onset of movement. Many of these cells are directionally specific. For example, a cell may increase its discharge if the warning stimulus indicates a forthcoming movement to be made to the right hand side, but will decrease its discharge if the forthcoming movement is to be to the left. If a rightward instruction is given, and then replaced 0.5s later by a leftward instruction, the cell will initially be excited and will then be inhibited after onset of the second warning signal. These cells do not fire at all if no movement is required.

Exactly what these set-related cells are doing in the interval between warning stimulus and trigger stimulus is, at the moment, speculation. There are several possibilities:

1. They may be involved in focusing attention towards the expected direction of the trigger stimulus;
2. They may be part of a plan of movement (specifying direction), which is preformulated in premotor cortex and set to be released by the trigger stimulus. However, it should be noted that no neurones have been recorded in this area which might specify other parameters of movement such as speed or extent;
3. They may preset the activity of other systems in preparation for the forthcoming movement.

There is one piece of evidence favouring the latter possibility. Monkeys with lesions of premotor cortex and supplementary motor area have difficulties in performing visually guided movements. They are capable of reaching directly towards a target, but cannot reach around a transparent obstacle to obtain the reward. Instead they persevere in trying to reach directly towards the object. It has been suggested that the set-related cells in premotor cortex may normally inhibit primitive brainstem circuits that mediate direct reaching towards a target.

Human studies have pointed to two possible roles of premotor cortex. Patients with relatively pure lesions of premotor cortex have weakness of shoulder and hip muscles during tasks involving support of the arm or leg. There is no distal weakness, as expected if the precentral cortex was lesioned, and not all combinations of girdle muscles were affected. The interpretation of these results is that the deficits are due to interruption of indirect projections to spinal cord via the brainstem tegmental system, which is directed preferentially towards axial and proximal muscles. If the lesion is more widespread, patients may have difficulties similar to the monkeys described above: visuomotor ataxia and perseveration (which are also symptoms of damage to anterior prefrontal regions, and are not specific to premotor areas).

Other human studies point to a role in visually guided movements. Blood flow through any area of brain can be monitored in intact human subjects by radioisotope scanning. Blood flow increases to active areas of brain, and can be seen to increase particularly in premotor cortex during the performance of movements requiring directional guidance from sensory information (see Freund and Hummelsheim, 1984; Haaxma and Kaypers, 1975; Weinrich *et al.*, 1984; Wise and

Figure 8.17: Classification of Neuronal Response Types in the Premotor Area of Monkeys. The diagram on this page shows the experimental arrangement. The monkey faces a target screen, the top half (P) of which gives a visual display of the position of its arm. The bottom half (T) is reserved for target and warning lights. Starting at the top of the figure, the monkey is at the end of a previous trial, waiting in the inter-trial interval. In B, the lower light is illuminated fully and moves to the right. This is the warning instruction to the monkey that the next movement should be made to the right. After a random delay (from B to C), the light dims: this is the reaction signal for the monkey to move. In D, the monkey has moved to the right and the task is completed. Panels E and F show the procedure when the monkey does not need to make a movement. The results opposite show examples of the responses of each class of cell: signal-related, movement-related and set-related. In each panel, the responses of a single neurone are displayed as response histograms (upper) and as rasters of responses in repeated trials (lower). The responses have been aligned in time to the appearance of the warning signal (IS) or to the onset of movement (Mvt). Signal-related cells change their firing shortly after the warning signal. There is a later change when the target signal (that is, dimming of the light) is given. This is seen best in the rasters, where the large dots indicate the time at which the target signal was given with respect to the time of the warning signal. After each large dot, there is a transient increase in firing. Set-related cells change firing with the warning signal and maintain a high rate until the onset of the trigger signal (see rasters). Movement-related cells do not show any activity related to the warning signal (upper right), although the rasters show an increase in discharge after the trigger signal. In the lower right, these traces have been realigned to the onset of movement, to show the pronounced increase in firing about the time that movement begins (from Weinrich, Wise and Mauritz, 1984; Figures 1 and 2, with permission)

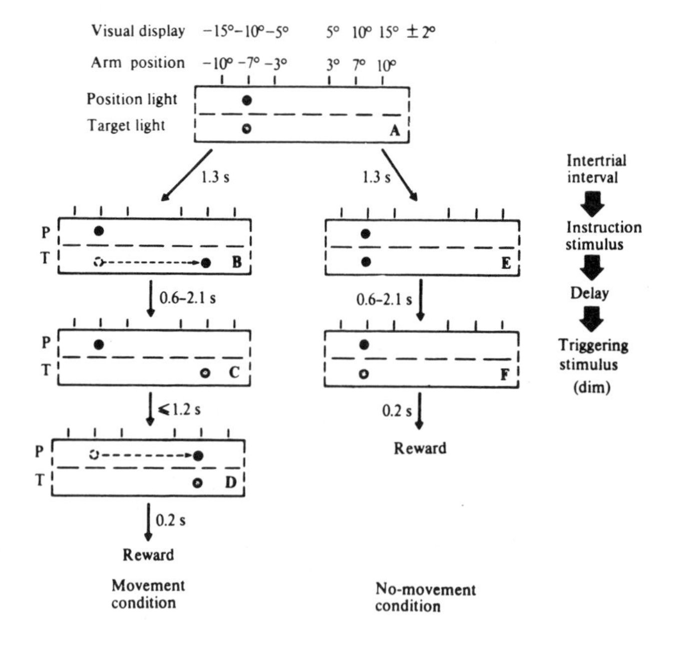

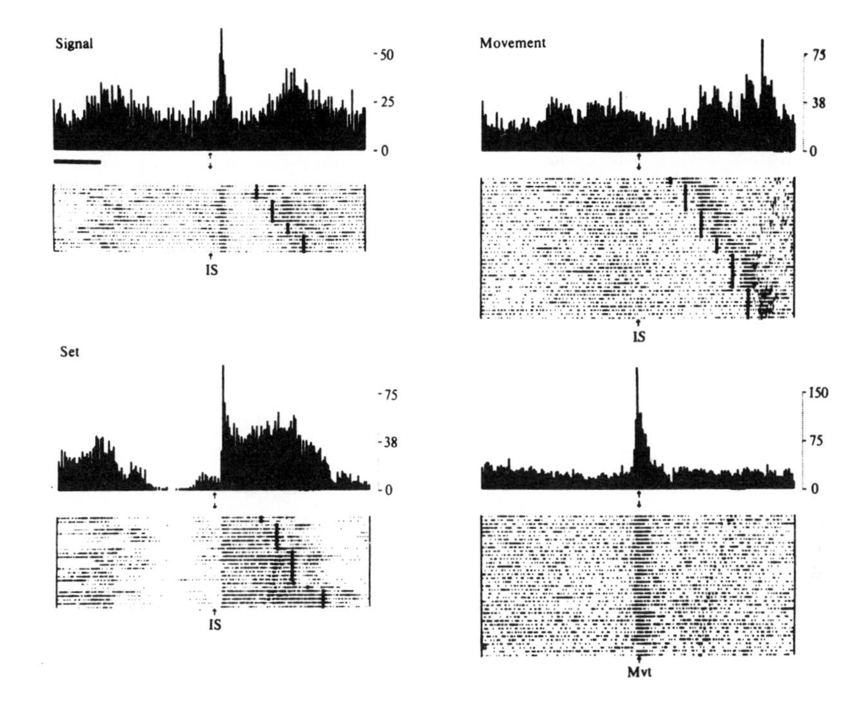

Mauritz, 1985, for further information on the work covered in this section).

Grasp Reflexes and Spasticity. Two other symptoms of premotor area lesions have been reported, but the extent to which they represent pure dysfunction of premotor cortex or damage to other structures is debated. They are the appearance of a grasp reflex and spasticity. A grasp reflex is seen in normal human infants, but disappears within a few months of birth. It may reappear after lesions of area 6 in posthemiplegic patients. There are three different types of grasp reflex and, in any one patient, one or all three of them may be present.

The classical grasp reflex of neurology consists of involuntary prehension elicited by a tactile stimulus to the palm. It is usually accompanied by a traction response, which is a heightened stretch reflex in the finger flexor muscles, such that attempts to remove the object from the hand by extending the flexor tendons results in an increased force of grip. This traction response is known sometimes as forced grasping, to indicate the difficulty which the patient or animal has in releasing his grip on an object under these conditions. The traction response can be influenced by the posture of the patient. When lying on one side, the reflex is heightened in the upper limbs, and decreased on the opposite side.

The final version of the grasp reflex is rather more complex, and known as the instinctive grasp reflex. It is an involuntary closure of the hand elicited by tactile stimulation on *any* part of the hand or fingers. The hand is first oriented towards the object so as to bring it into the palm. This stimulus then secondarily elicits a tactile grasp reflex to close the fingers on the object. The initial exploratory orientation of the hand in the instinctive grasp reflex was considered by Denny-Brown (1966) to be the opposite of the tactile avoiding reaction, in which the hand is withdrawn from a stimulus. Lesions involving the premotor cortex produce the instinctive grasp, whereas lesions of the parietal cortex produce the avoiding reaction.

Unfortunately, despite the wealth of clinical description of the grasp reflex, it is not known which is the precise region of cortex which must be lost in order for the reflex to appear. Many authors consider it equally well to be a sign of damage to supplementary motor area.

There is also disagreement about the role of premotor cortex lesions in production of spasticity. The old idea was that there was a 'suppressor' strip just anterior to area 4 ('strip area 4s') which had to be lesioned for spasticity to occur. This idea is now discredited. It is probably safest to conclude that if a lesion is made in area 6 (and this includes both premotor and supplementary motor areas), then the larger it is, and especially if accompanied by lesions of area 4, the more likely spasticity is to ensue (see Wiesendanger, 1981).

Supplementary Motor Area

Resumé of Anatomical Connections. The supplementary motor area, which is the mesial portion of area 6 lying above the cingulate sulcus, appears to be, like the premotor cortex, a region which relays high level command signals to primary motor cortex. Cytoarchitecturally, the two areas are the same and share much of the same thalamic inputs from VA/VL nuclei. However, the SMA receives a far greater proportion of input from basal ganglia (via VL_m and VL_o) than the premotor area. Other differences are in the intracortical connections. Premotor cortex receives processed visual input from extrastriate visual cortex, whereas the supplementary motor area does not. However, like the premotor cortex, it does receive somatosensory input from parietal regions. The outputs of the two areas are similar, going to spinal cord, brain stem and (bilaterally in the case of the SMA) to primary motor cortex.

Effects of Electrical Stimulation. Electrical stimulation of SMA has produced conflicting results. In anaesthetised monkeys, Woolsey and his colleagues (1952) reported a fairly discrete somatotopic map, and drew a second motor simunculus on the mesial surface of the hemisphere (see Figure 8.3). Forelimb areas were represented rostral to hindlimb areas. In contrast, studies by Penfield and Jasper (1954) in man led them to suggest a rather different localisation in which the leg area was located deep within the SMA, along the cingulate sulcus, and the arm area was on the dorso-medial bank of the SMA. In Woolsey's studies, stimulation intensities were high, and other authors have suggested that the effects were due to current spread to area 4. (In the monkey, the SMA is only 1cm long antero-posteriorly.) No somatotopic map can be found after ablation of primary motor cortex, which reinforces this suggestion.

Electrical stimulation of SMA following ablation of area 4 produces effects very like those seen after stimulation of premotor cortex. These are deviation of the head and eyes to the contralateral side, and raising of the contralateral arm. In humans, speech arrest or vocalisation may occur, the latter said to be a quite characteristic vowel cry, the 'cri de l'aire motrice supplementaire'. As with stimulation of premotor area, these complex synergies are seen only with prolonged repetitive surface stimulation. Epileptic seizures which begin in the SMA involve very similar reactions: adversive movements of the head and eyes, and elevation of the contralateral arm, together with speech arrest or vocalisation.

Recent experiments have changed our view of somatotopy in the SMA. Three approaches have been used:

1. Electrodes have been implanted chronically along the SMA in some human patients for relief of chronic epilepsy. Stimulating through these electrodes produces a tonic arm response in the posterior portion of the SMA, vocalisation in the middle portion and deviation of the head and eyes most anteriorly;
2. The corticospinal projection from SMA has been reinvestigated in monkeys using retrograde HRP labelling. Injection of horseradish peroxidase (HRP) into cervical and lumbar enlargements of cord produced small clusters of labelled neurones in SMA. Those projecting to cervical levels were concentrated mainly in the mesial part of SMA, with those from lumbar levels straddling them on either side, on the dorsolateral convexity of the hemisphere and on the bank of the cingulate sulcus. This arrangement is similar to that found by Penfield in man, but different from the chronic stimulation studies above;
3. Microstimulation has also been used to define the SMA somatotopy. With this technique, the motor map is the same as that found with HRP labelling, although intensities of stimulation needed to produce movement were higher than those for precentral cortex. The conduction velocity of the neurones was slow, at about $10ms^{-1}$, and therefore it might be expected that their role might be in producing postural adjustments rather than rapid movements.

Why is there a difference in the somatotopy found with microstimulation and that reported by Woolsey? One possibility is that microstimulation preferentially investigates the corticospinal projection from SMA. However, the SMA may influence motoneurones via many parallel pathways, including the projection to precentral cortex. Since all routes are activated by surface stimulation, there may be different somatotopies for each route. Indeed, the anatomical projection from SMA to primary motor cortex has a rostro-caudal pattern (face, forelimb, leg) similar to that expected from Woolsey's map, and different to that seen with microstimulation and HRP mapping (see Macpherson, Wiesendanger, Marangoz *et al.*, 1982; Wiesendanger, 1981).

Electrical Recording and Lesion Studies.

Electrical Recording. Few single cell recordings have been made in the SMA of awake, behaving animals. In those that have, neurones have been found which discharge prior to movement at both proximal and distal joints. Initial studies suggested that there was a substantial proportion of neurones related to movements of the ipsilateral limb. However, more recent studies with monkeys carefully trained to move only one limb on request show a clear contralateral predominance in the SMA. Such recordings confirm the rostro-caudal representation of the forelimb–hindlimb neurones, as suggested by Woolsey *et al.*, (1952). The changes in neuronal firing were more variable both in size and in latency than those of precentral neurones recorded in similar tasks, suggesting that the SMA is more remote from the motor apparatus than area 4.

A possible role of SMA in programming motor performance was suggested by reaction experiments in which monkeys had to press a key in response to either an auditory or tactile stimulus, but not both. Before the reaction stimulus, the monkey was given a warning signal to instruct the animal which type of stimulus to respond to. Of the neurones in the SMA related to the task, nearly half changed their firing rate following the instruction signal. Most of them maintained this new discharge rate until appearance of the triggering stimulus (whether or not a non-triggering stimulus intervened). One half of these cells responded differently to the two types of instruction: they might, for example, have increased their discharge rate following a signal warning the monkey to respond to auditory stimulus, whereas the rate would be unchanged, or increase much less, following an instruction to react to tactile stimulus. There were far fewer cells in the precentral motor cortex which were related to the instruction signal, and none of these had a differential response to the two types of instruction. The conclusion is that the SMA plays an important role in the preparatory processes leading to movement, perhaps being involved in initiating the correct type of response, or suppressing inappropriate responses.

The phenomenon of the release of the grasp reflex after lesions involving the SMA (see below) prompted a different approach to the problem of the role of the SMA in movement. Stimulation here in intact animals reduced the responses of cells in primary motor cortex to somatosensory disturbances of movement. As already seen, these inputs to motor cortex can generate reflex changes in firing of area 4 neurones. It may be that one function of the SMA is to modulate the gain of these reflexes. The grasp reflex might then be seen as the consequence of their release from SMA control (see Tanji and Kurata, 1985, for more information on SMA electrophysiology).

Blood Flow. Some of the most exciting studies on the function of the SMA come from the measurement of cerebral blood flow in humans performing different motor tasks (Figure 8.18). Such studies use computerised tomographic techniques to monitor the washout of radioactive ^{133}Xe injected or inhaled into the blood. During simple repetitive tasks like tapping the finger, cerebral blood flow increases to area 4, but not the SMA. Assuming that blood flow increases to those parts of the brain which are active in a task, this indicates preferential involvement of primary motor area in this type of movement. If the task was made more complicated, such as requiring sequential opposition of the thumb to each of the

Figure 8.18: Areas of Increased Cerebral Blood Flow (CBF) in Human Brain During Different Types of Finger Movement. In A, the subject was required only to make a sustained isometric contraction of the contralateral thumb and index finger. CBF increases in pre- and postcentral areas (numbers indicate size of the increase in flow). In B, the task was to touch each finger in turn with the thumb of the same hand. The right and left sides of the brain are shown, but contralateral movements were studied in each case. Note the more widespread increase in CBF to include the SMA and other areas. In C, the task was to rehearse in the mind the movements performed in B. In this case only the SMA shows an increase in CBF (from Roland, Larsen, Lassen, *et al.,* 1980; Figures 11, 7, 9, with permission)

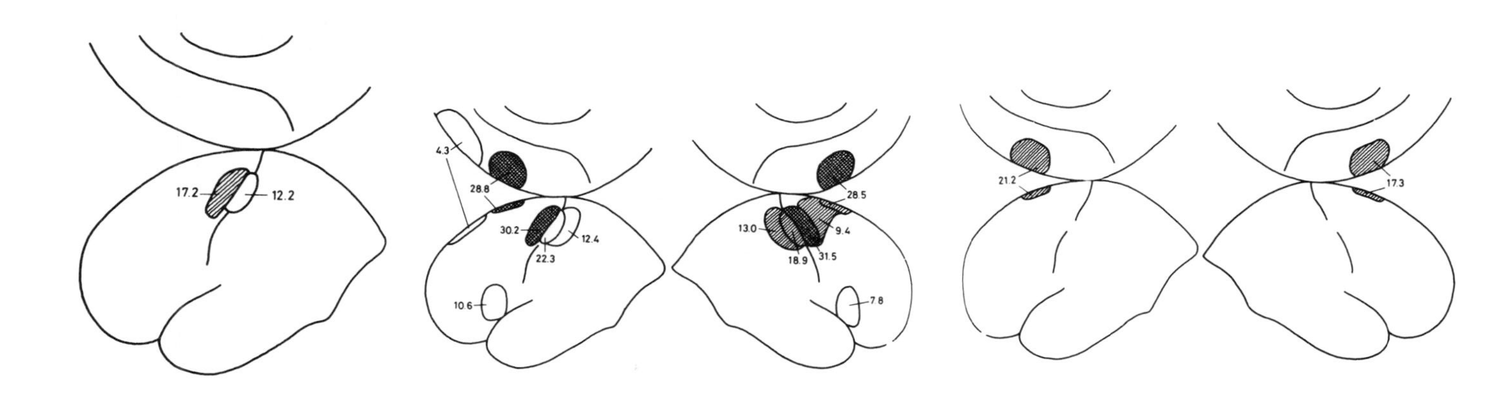

forefingers in turn, the blood flow increased in both the SMA and area 4. Remarkably, if this sequence was just rehearsed in the mind, but not actually performed, then SMA blood flow was still increased, but not that to the primary motor cortex. Exciting as these experiments are, they do not reveal any information about the timing of cortical activity with respect to movement, because of the time taken to accumulate the information to prepare the picture for each section of the brain scan. However, they provide clear evidence for a role of SMA in the 'higher order' programming of movement (see Roland, Larsen, Lassen, *et al.*, 1980).

Lesion Studies. Lesion studies have revealed little about the function of SMA. Like lesions of premotor area, early studies showed that lesions of SMA could lead to the appearance of spasticity and the appearance of a grasp reflex. The extent of the lesion needed to produce either effect was never clear. In man, a grasp reflex is prominent in the early stages after a lesion, but, unlike lesions of premotor cortex, there is no weakness unless area 4 also is involved.

Of more interest is the description of three patients who had their SMA removed unilaterally because of intractable epilepsy. All three patients had an initial three-week period of contralateral akinesia and paucity of verbal expression, although there was no weakness of the muscles. The face was motionless and there was difficulty in lateral deviation of gaze. Such akinesia is remarkably similar to that seen in parkinsonism, and it is possible that it is a consequence of lesioning one of the main targets of basal ganglia outflow via VL_m and VL_o thalamus to SMA.

Figure 8.19: Effect of Unilateral SMA Lesion on the Performance of a Complex Bimanual Task in the Monkey. Left: Normal monkey pushing from above to collect a bait lodged in a hole in a Perspex sheet uses the index finger of the preferred hand while the non-preferred hand is cupped underneath, anticipating catch of the falling bait. Right: Reversal of hand position and identical behaviour of both hands five months after ablation of the right SMA. The non-preferred hand is now above the plate, and both hands are used in pushing with the index finger. Even with full visual control, the movements are not corrected (from Brinkman, 1981; Figure 1, with permission)

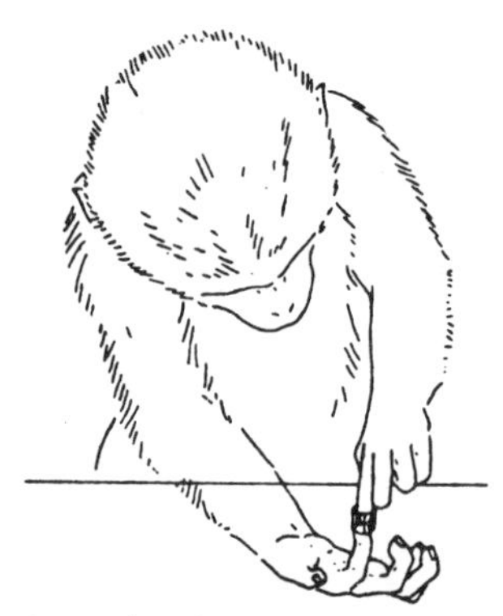

Recovery was almost complete, except that the patients were left with a permanent reduction in the speed at which they could perform rapid alternating movements of the hands. As with the blood flow monitoring this sequence of events suggests a role in 'higher level' control of movement (Laplane, Talairach, Meininger, *et al.*, 1977).

Recent lesion studies in monkeys have revealed another subtle and more enduring deficit than the grasp reflex and spasticity. After removal of the SMA on one side of the brain, monkeys were left with difficulties in tasks requiring bimanual co-ordination (Figure 8.19). In retrieving food pellets from a well in a Perspex board, monkeys had to push a currant down through the hole with a finger from one hand while they held the other hand cupped underneath to catch it when it fell. After the lesion, there was a lack of co-ordination between the hands. Rather than sharing the task between each hand, the monkey might try to use both hands to push the currant out of the hole. It may that, in the normal animal, the SMA on one side of the brain provides command signals only for the ipsilateral motor cortex. In the lesioned animals, the remaining SMA might influence precentral areas bilaterally and result in mirror movements of the hands (Brinkman, 1984).

Motor Functions of Parietal Cortex

Primary and Secondary Sensory Areas. In their original mapping studies of the excitable cortex, Woolsey and his collegues (1952) found that electrical stimulation of postcentral cortex could elicit contralateral movements of the body. They believed that there was a direct outflow from postcentral cortex to spinal cord. Because of this, they proposed the terms SmI and SmII for primary and secondary sensory areas to indicate their combined sensory and motor functions.

The idea that there is direct motor outflow from postcentral cortex to spinal cord has been modified in the recent years. The threshold for motor effects using repetitive surface stimulation of brain is much higher than for precentral cortex, and no muscular contractions can be observed using the technique of intracortical microstimulation. Such studies do not disprove the idea of a direct corticospinal projection, but they indicate that if it is present it is weak, and requires the spatial and temporal summation of activity in many neurones to be activated. These are also the characteristics to be expected of indirect pathways to spinal cord. It is now believed that the motor effects produced by stimulation of postcentral areas are mediated by indirect projections via, for example, the precentral cortex and pontine nuclei, both of which receive very strong projections from postcentral areas.

Two lines of evidence support the assumption of indirect motor pathways. One is the finding that after ablation of precentral areas, or after interruption of cortico-cortical connections in the white matter between post- and precentral cortex, it is sometimes impossible to produce movements by stimulating sensory cortex. The other line of evidence comes from single cell recording in conscious monkeys. There are some cells in postcentral cortex that change their firing rate before movement, rather like the cells of precentral cortex. This behaviour is completely abolished by deafferentation, which has no effect on the early discharges of precentral cells. The early change in discharge of postcentral cells may have been caused by postural adjustment prior to movement.

The present view that there are no direct motor pathways from postcentral sensory cortex to spinal cord is supported by anatomical studies. The pyramidal tract outflow from sensory areas terminates in the base of the dorsal horns of the spinal cord, rather than in the intermediate zone or ventral grey. It is more likely to function in the control of afferent input than motor output. This is not to say that postcentral cortex has no motor function at all. In conscious monkeys, cooling of the forelimb area of precentral cortex produces weakness of the contralateral wrist. Only if the cooling is applied also to the postcentral cortex is the wrist completely paralysed. Indirect motor output from the postcentral cortex (via, for example, the pons) may be responsible for this effect.

Parietal Areas 5 and 7 (see reviews by Hyvarinen, 1982; Wiesendanger, 1981).

Lesion Studies. Although parietal areas 5 and 7 are usually thought of in terms of their sensory functions, lesions in this part of brain also produce characteristic motor deficits in man. These include a reluctance to use the contralateral arm and neglect of that side of the body. Objects in the contralateral visual field are ignored. If patients can be persuaded to use their contralateral arm, deficiencies in reaching for, grasping and manipulating objects become evident. The arm frequently over-reaches its target, and the fingers and thumb fail to orient themselves so that the shape of the hand is not properly prepared to grasp the object. Patients have difficulties in knowing how to manipulate familiar objects (apraxia) and cannot recognise objects placed in the hand when blindfolded. They may be able to distinguish whether an object is hard or soft, but be unable to synthesise the various impressions gained by manipulation into a three-dimensional concept of the object (astereognosis). Such deficits sometimes are said to be most marked if the lesion is in the nondominant (usually the right) hemisphere.

Monkeys with parietal lesions in these areas have very similar deficits. Unilateral cooling of area 5 in monkeys resulted in misreaching in all directions by the contralateral arm. Cooling of area 7 led to misreaching by either arm when movements were made in the contralateral visual field. The observations can be summarised by saying that areas 5 and 7 have a role in the sensory guidance of movement. Because of their connections, area 5 probably is involved in tactile guidance, and area 7 in visual guidance.

In many respects, areas 5 and 7 lie at a point midway between sensory input and motor output. Area 5 receives processed somatosensory information from the primary sensory areas, and also a smaller input from area 6. The output is directed forward to area 6 (that is, the premotor and supplementary motor areas) and to area 4 with a backward projection to area 7. Thus, area 5 has fairly tight connections with both somatosensory and motor areas of cortex. There is also is a pyramidal tract projection to spinal cord, although like that of the primary sensory area, this is directed towards the dorsal horn, and probably functions in corticofugal control of sensory input.

Area 7 is not intimately related to motor cortex. Its input comes mainly from visual peristriate cortex, nonspecific thalamic nuclei, and from limbic structures. The output is directed forwards to the prefrontal cortex, cingulate cortex and temporal lobe. Only the most rostral portion of area 7 projects to premotor cortex. Thus, the influence of area 7 on movement must be indirect. Possible pathways include cortico-cortical connections from prefrontal to motor areas, or the subcortical projections to basal ganglia or to the pontine nuclei (and thence to cerebellum).

Electrical Recording. The first studies of single unit activity in posterior parietal cortex were made by Mountcastle, Lynch, Georgopoulos, *et al.* (1975). Among the many types of cell that they encountered, they identified a group which apparently was not activated by any peripheral sensory inputs, but which discharged in association with active movements made by the animal. These neurones had the remarkable property that their discharge was not tightly linked to a particular movement such as arm extension, but was seen only if a given movement occurred in a particular behavioural context. A typical example was the discharge recorded when the animal reached out its arm to obtain food. No discharge was recorded if the same movement was performed spontaneously.

Two hypotheses were put forward to explain the role of these cells. One was that they were 'command' neurones working at a high level in the command process leading to movement. The other was the possibility that the neurones were receiving a corollary discharge from precentral areas. The first possibility might be related to the neglect of contralateral limbs in lesioned animals or man. The second might be related to the pheomenon of astereognosis.

The situation is still far from clear. Many of the 'command' neurones later were found to be responsive to external stimuli which had not been tested in the previous experiments. Some of those in area 7 discharged in response to relatively large visual stimuli rather than small spots of light that were routinely used to test visual fields. However, studies on deafferented monkeys revealed that there were substantial numbers (14 per cent) of cells in area 5 which still discharged before a movement made in response to an auditory signal, even though no peripheral sensory input was available. These cells did not respond to the auditory signal alone. Finding such neurones in area 5 is in great contrast to the absence of similar cells in primary sensory cortex following deafferentation. Some of the area 5 cells fired before the earlier precentral neurones, and hence might be true 'command' neurones. Others fired after precentral changes, and might have been in receipt of corollary inputs from motor cortex (Kalaska, Caminiti and Georgopoulos, 1983; Seal and Commenges, 1985).

Pathophysiology of Motor Areas of Cerebral Cortex

Epilepsy and Myoclonus

Jacksonian Seizures. The movements produced by abnormal cell activity in supplementary and premotor areas during the onset of epileptic seizures are described above. In contrast with such complex synergies involving the activation of many muscles in different parts of the body, seizures with onset in primary motor cortex have much simpler characteristics. They were first described by Hughlings Jackson several years before the discovery of the electrical excitability of the cerebral cortex. Such fits, now known as Jacksonian seizures, begin with focal twitching of a small group of muscles in one part of the body. Usually, this is in the hands or the feet or the corner of the mouth. From this point, the twitching spreads proximally to involve more and more muscle groups, over the space of several seconds. For example, it may progress from one hand to the forearm, elbow, shoulder on the same side; finally a full tonic-clonic convulsion of the whole body may ensue.

At cortical level, the seizure begins as focal activity in one part of the motor cortex, and because of their relatively large representation this frequently involves the hand, foot or mouth. Activity spreads to neighbouring areas of cortex, to recruit neurones projecting to neighbouring muscle groups, according to the somatotopic cortical representation. A full grand mal convulsion follows when (and if) the seizure spreads beyond the motor strip to the rest of the cerebral cortex.

Cortical Myoclonus. A less devastating form of increased cortical excitability can be seen in patients with cortical myoclonus. Myoclonus refers to muscular twitching in any part of the body and may involve one or many muscles of the body. Hypnic jerks that are experienced by many people on falling off to sleep are a form of physiological myoclonus.

In contrast to these whole body jerks, cortical myoclonus usually involves muscles in only one part of the body. The jerks are brief, and look very like the muscle twitch produced by a single electrical stimulus to a peripheral nerve. The jerks may occur at rest, during voluntary activation of the affected part, or may be provoked by sensory stimuli, particularly from the affected body part. Jerks which occur at rest and are repetitive, regular and persist continuously for many days, disappearing only during sleep are known as *epilepsia partialis continua*. In some patients, epilepsia partialis continua is associated with Jacksonian seizures if the activity in the cortical focus spreads to other areas.

The reflex variety of cortical myoclonus has been studied in some detail. It is the result of hyperactivity in the cortical reflex loops from peripheral skin or muscle receptors to motor cortex and back down to muscle (Figure 8.20). This can be demonstrated in many patients quite easily by recording the muscle and brain activity which is produced by the somatosensory stimuli that evoke myoclonic jerks. Brain activity in man cannot be recorded via intracortical microelectrodes, as in a monkey. Instead, distant electrodes have to be attached to the surface of the scalp over the area of brain of interest. Electrodes placed over the sensorimotor areas normally record small waves of activity with onset 18–20ms after electrical stimulation of the median nerve at the wrist. These potentials are known as somatosensory evoked potentials (SEPs).

Figure 8.20 shows results from a patient who had reflex myoclonus of the left hand. Electrical stimulation of the median nerve at the wrist on the right (normal) side evoked normal scalp potentials and no myoclonus. Electrical stimulation of the left side produced myoclonic jerks which had a latency of about 50ms in the first dorsal interosseous muscle. In addition, it produced giant evoked potentials over the contralateral central region of scalp. The initial part (N1) of this potential was normal, but the later components (from P1 onwards, at 25ms) were greatly enlarged. The hypothesis is that in this patient, a normal afferent volley reached somatosensory cortex, but that later stages in processing were abnormal. Increased excitability produced an abnormal SEP which may have driven cells in motor cortex to discharge and produce a visible myoclonic jerk. The latency between onset of the enlarged SEP and the muscle EMG response is the same as the corticospinal conduction time to hand muscles in man.

Patients with spontaneous jerks can be studied in the same way. In this case, spike activity over contralateral pre/postcentral areas can be recorded some 20ms before spontaneous muscle jerks in the arm. In such cases, it may be that an excitatory focus is discharging spontaneously in the cortex (see Obeso, Rothwell and Marsden, 1985, for further details of cortical myoclonus).

Lesions of Descending Pathways

Capsular Lesions. The most common cause of damage to the cortical motor system occurs in capsular hemiplegia. Interruption of cortical efferent fibres in the internal capsule gives rise to a variety of symptoms, the severity of which depends upon the extent and position of damage. For many years, the symptoms were considered to be due to destruction of pyramidal tract fibres, and the symptoms referred to as pyramidal tract symptoms. However, the cortical outputs to red nucleus, pontine nuclei and brainstem tegmental areas are also interrupted and contribute to the symptoms.

Hemiplegia produced by a fairly complete motor stroke in man is characterised by the following sequence of events. Immediately after the stroke there is a period of complete paralysis. The contralateral limbs are flaccid and no reflexes can be obtained. This is the initial or shock stage. The first reflexes reappear after 10 to 12 hours. The plantar reflex, elicited by firm stroking of the lateral border of the sole of the foot, usually occurs before the others. However, it is now inverted (Babinski's sign), and the big toe dorsiflexes rather than plantarflexes, as is normal (see Chapter 6). A little later, tendon jerks reappear and, over a week or so, gradually become extremely brisk. The patient becomes hyper-reflexive and clonus may appear in some muscle groups.

At the same time, there is a gradual increase in muscle tone which is unevenly distributed. It affects preferentially the extensors of the legs and the flexors of

Figure 8.20: Cortical Reflex Myoclonus in Man, Evoked by Somatosensory Peripheral Stimulation. Left: Postulated transcortical pathway involved in the reflex muscle jerks (from Rothwell, Obeso and Marsden, 1986; Figure 4, with permission)
Right: Recordings from a patient with reflex myoclonus of the right hand. A and B: evoked potentials to electrical stimulation of the median nerve at the wrist recorded from electrodes over the cervical spinal cord (C5) and the somatosensory hand area of cerebral cortex (LHA, RHA). Stimulation of the normal (right) hand gives the usual complex of evoked responses at the spinal cord and scalp (N1, P1, N2; negative deflection upwards). Stimulation of the abnormal (left) hand gives normal early responses over the cervical cord and a normal N1 response at the scalp. However, the later P1, N2 responses are greatly enlarged. (The enlarged later waves of the C5 lead are due to pick-up of remote activity from the scalp, and not to activity in the spinal cord.) C: EMG recordings from the first dorsal interosseous muscle on a longer time sweep. Stimulation of the normal hand produces no EMG response in the relaxed muscle on that side. Stimulation of the abnormal hand generates a myoclonic jerk, seen in the EMG as increased activity with a latency of about 50ms after stimulation (from Rothwell, Obeso and Marsden, 1984; Figure 1, with permission)

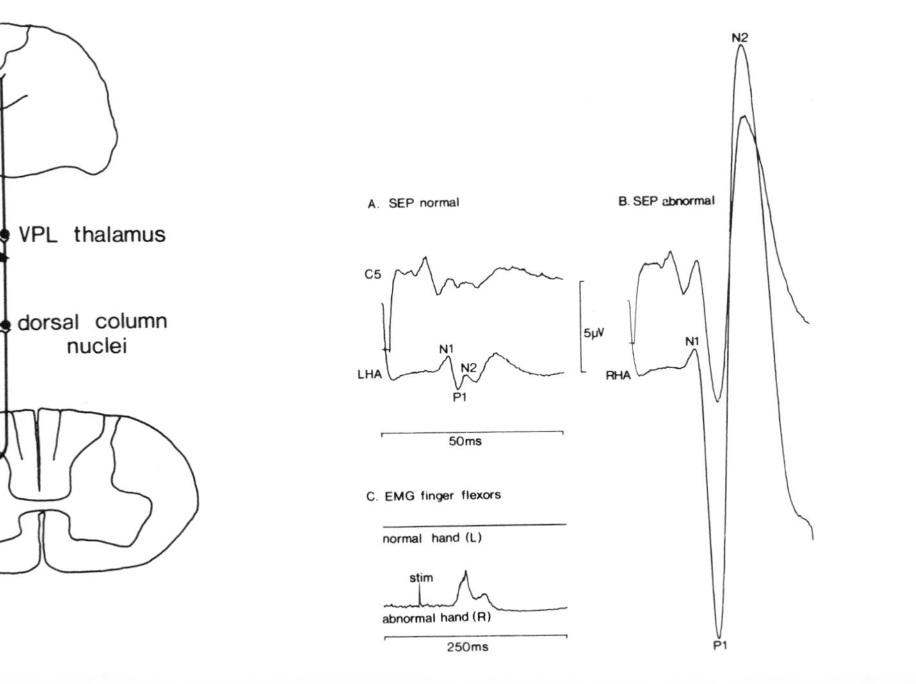

the arms. Such spasticity is accompanied by a clasp-knife reflex (see Chapter 6). Two reflexes do not reappear: the abdominal and cremasteric reflexes. The former consists of contraction of the abdominal muscles during stroking of the abdomen. The latter consists of retraction of the testicle following stroking the inner surface of the ipsilateral thigh.

In most cases, there is a gradual recovery of voluntary movement. Recovery begins with crude synergic movements of all the muscles in a limb. In the leg these are the extensor and flexor synergies: extension of the knee, for example, is always accompanied by abduction (usually, although sometimes adduction) of the hip and dorsiflexion of the ankle and toes. Flexion of the knee is accompanied by flexion at the hip and plantar flexion of the ankle and toes. In the arm, comparable flexor and extensor synergies occur. Independent movements of the digits are the last to appear, if at all (see Brodal, 1973, for the patient's viewpoint).

The Effect of Pyramidal Tract Lesions. Pure lesions of the pyramidal tract in man are rare since, except in the medullary pyramids, the fibres of the tract are contaminated throughout their length by other systems. Nevertheless, a few cases with lesions in this area have been described. All the patients were described as being weak on the contralateral side, but muscle tone was affected by varying amounts, ranging from flaccid to severe spastic paralysis. Differences between patients were probably caused by varying degrees of involvement of the medial reticular formation and leminiscal systems.

In the late 1950s and 1960s, neurosurgeons experimented with pedunculotomies to relieve abnormal movements, due, for example, to chorea (Chapter 10). They sectioned the middle third of the peduncle, through which it was believed that the majority of pyramidal tract fibres ran. In one patient at autopsy, three and one half years later, it was found that 83 per cent of fibres within the pyramids had degenerated. In contrast to the results above, this patient was said to have retained 'an almost normal volitional control' of the contralateral limbs although the involuntary movements had been abolished. There was no spasticity (Bucy, Keplinger and Sequiera, 1964).

From these reports in man, it would seem as though lesions of the pyramidal tract have very little effect on either movement or tone. More controlled experiments have been conducted on monkeys and chimpanzees, but only recently with trained animals and with quantitative techniques. As in man, there is surprisingly little gross deficit. Immediately after bilateral pyramidotomy, monkeys can sit, run and climb in their cage with their head erect. However, despite this appearance of normality in such gross movements, careful observation reveals substantial deficits in the control of fine independent finger movement. For example, although the monkeys can use their hands in a power grip to hold onto and swing from the bars of their cages, they cannot pick up food with their hands from the floor. The critical difference in these tasks is that in picking up food, the individual fingers have to move independently, rather than in concert, as in the power grip. Recovery occurs with time, but the animals are left with a deficiency in performing individual movements of the fingers. The best test is to have the animals retrieve food pellets from a shallow depression in a food board. Normally, this is done using the thumb and forefinger together in a 'precision grip', while at the same time, the other fingers are held flexed out of the way. Pyramidotomised monkeys can no longer perform the task in this way. Instead, they use only a 'hooking' movement of all the fingers together, which usually proves inadequate to retrieve the pellet.

Residual muscle weakness is slight and in recent experiments is said to be zero. The only deficit is that the speed of reaction in these monkeys is slower than normal. This is due to a delay and slow rise in the onset of EMG activity in the muscles.

The muscle tone of monkeys with sectioned pyramids was never found to be enhanced. In fact, there is a slight hypotonus caused by decreased stretch reflexes. This is probably due to a diminished gamma drive to muscle spindles as a result of interruption of cortico-fusimotor fibres in the pyramidal tract.

These experiments on animals suggest that few of the symptoms of hemiplegia are caused specifically by damage to corticospinal fibres. The conclusion is that the corticospinal tract superimposes speed and fractionation upon the movements produced by other descending systems. It fits well with the anatomical findings which show an increasing number of corticospinal tract terminals in the spinal motor nuclei, and a gradual increase in importance of the corticomotoneurone connection in monkey, chimpanzee and man. It is speculated that these direct connections from brain to α-motoneurones provide the basis for independent control of finger muscles in higher primates.

A thorny problem has been, and still is, the role of the pyramidal tract in the production of Babinski's sign. Pyramidotomy in chimpanzees produces upgoing big toes after stroking of the plantar surface of the foot, and abolishes the abdominal and cremasteric reflexes. There are no such clear-cut data in man. Reports on the correlation between the presence of pyramidal tract damage, confirmed at autopsy, and the sign of Babinski are contradictory. The most recent suggestion is that the upgoing plantar response probably occurs after damage to the monosynaptic component of the corticospinal projection to the extensor muscles of the toe (see Lawrence and Kuypers, 1968, for classical studies of pyramidotomy in monkeys).

Effects of Lesions of Other Descending Tracts. Lawrence and Kuypers (1968) provided the most complete description of the deficits following transection of descending pathways in the Rhesus monkey. There are no comparable data available in man.

As described in Chapter 7, Lawrence and Kuypers divided the descending pathways into pyramidal tract and brainstem pathways group A and B. Group A comprises the interstitiospinal, tectospinal, vestibulospinal and reticulospinal tracts; group B is mainly the rubrospinal tract. In order to evaluate the contribution of these pathways to movement control, a pyramidotomy was first performed. This removed any pyramidal tract contribution to control of proximal and distal muscles.

Bilateral transection of the brainstem pathways in the upper medullary medial tegmental field (the group A pathways), together with bilateral pyramidotomy produced animals with severe postural deficits. Even after some weeks recovery, animals tended to slump forward when sitting and had great difficulties in avoiding obstacles when walking. There was severe impairment in righting reactions. In contrast to the immobility of axial structures, the hands and distal parts of the limbs were very active. Manipulative ability returned to that seen in pyramidotomised animals; the monkeys could retrieve food from a board with their hands and bring it to their mouth. However, the animals would not reach out to grasp food. They would wait until food was in reach, and bring the food to the mouth mainly by flexion of the elbow.

Animals with pyramidotomy and lesions of the group B descending brainstem pathways looked very different. Their posture appeared normal. They sat up in their cages, walked and climbed the bars. Limbs would be used relatively normally in gross movements of climbing and walking. However, when seated, they held their arms limply from the shoulder. In reaching for food the hand was brought to target by circumduction at the shoulder, with little movement at the elbow. Finger movement was poor, and food usually was obtained by simultaneous flexion of all fingers as part of a total arm movement.

The conclusion from these experiments is that group A pathways are used chiefly in control of gross postural movements. Arm movements only occur as part of a postural synergy. These tracts travel in the ventromedial parts of the spinal cord and terminate on motoneurones of proximal and axial muscles and on long propriospinal neurones which interconnect many segments of cord. The group B pathways terminate on motoneurones innervating distal muscles and on short propriospinal interneurones which interconnect only a small number of spinal segments. These group B pathways appear to give the animals the capability to make arm movements, particularly at the elbow and wrist, which are independent of gross postural synergies. In addition to these methods of control, the corticospinal tract superimposes the ability to fractionate movements even further. This is seen particularly well in finger movements.

Apraxia

Lesions of the cerebral cortex do not only produce the primary disorders of movement described above, that is, a disorder of the effector regions of motor cortex, but also can produce secondary disorders of movement in which there is no weakness or damage to the primary motor areas. Such secondary disorders of movement provide some insight into the mechanisms used to control the final output of motor cortex. Diseases of the cerebellum or basal ganglia provide good examples of secondary disorders of movement, and are described in the following chapters. At a cortical level, an example of this type of disturbance is *apraxia*.

Apraxia may be defined as an impairment in producing an appropriate movement in response to visual or auditory instruction. For true apraxia to be considered, the incorrect movement cannot be due to weakness (that is, damage to effector structures), or to akinesia or bradykinesia (as in basal ganglia damage), ataxia (as in cerebellar damage), or to impairment in comprehension or sensation. Furthermore, the patient must be attentive and co-operative. The reason for this fairly strict definition is to define a syndrome in which there is a disorder of movement control in the absence of deficits in either effector or perceptual mechanisms.

The most common form of apraxia is *ideomotor* apraxia. If a patient is asked to hold out his tongue, he may be unable to do so, or will make an inappropriate movement such as chomping the teeth. However, although unable to use the tongue on command, the same patient may be able to move it quite normally in automatic movements of licking his lips or speaking. While ideomotor apraxia refers to the inability to perform single movements, *ideational* apraxia (less common) refers to an inability to perform correctly a sequence of movements, such as taking a cigarette, lighting it and using an ashtray. Each movement may be performed well, but, for example, in the wrong order.

Apraxia usually affects movements of the face or limbs. Axial apraxia is rare. An example of apraxia of facial movements is an inability to blow out a match. Patients may fail to round their lips to perform the task, and may instead shake the match in their hands to put it out. This type of error clearly shows that the patient knows what to do; he is simply unable to transform the idea of the movement into the appropriate action.

The question is, where in the brain does this transformation occur? The answer is not yet clear. Apraxia is a symptom of lesions in several parts of the brain. It is particularly common in patients with left hemisphere lesions, who often have aphasia as well, and in patients with lesions of the anterior part of the corpus callosum. Both types of lesions have been interpreted as producing 'disconnection syndromes', disconnecting areas of the brain which receive the instruction to move or formulate the idea of the movement, from the final effector areas of the motor system.

For example, one particular patient with a lesion of the anterior part of the corpus callosum could use both legs normally in walking, turning and so on. Yet when asked to demonstrate how he would kick a ball, he would only use his right leg; he was unable to demonstrate any movement with his left leg. If the patient was asked to imitate a visual instruction, performance was normal with both legs. The interpretation is as follows. The left side of the brain was able to interpret the auditory command and to feed this forward to premotor centres on the left so that movements of the right side were normal. However, the lesion of the corpus callosum disconnected the frontal areas on the right side of the brain, making them unable to receive information about the verbal command. Hence, movements on the left were apraxic. In contrast, visual information to the right side of the brain was sufficient to produce appropriate movements on the left.

With left hemisphere strokes, there may be paralysis of the right limbs and apraxia of the left, often associated with aphasia. The explanation is the same as above. There is damage to the areas of the left hemisphere which send commands to the right side of the brain. This also explains why axial apraxia is rare. Cortical control of axial muscles is mainly via brainstem reticular pathways, which are bilaterally organised.

Is there, then, a 'centre' in the left hemisphere responsible for transforming commands into appropriate plans for movement? One clue to the answer comes from the observation that lesions of the lower parts of areas 5 and 7 in the left hemisphere can lead to pronounced bilateral apraxia of the limbs (but not face). Geschwind and Damasio (1985) have suggested that this may be a 'master control system' receiving inputs from all sensory modalities and projecting to frontal motor areas on both sides of the brain (see Geschwind and Damasio, 1985, for review of apraxia).

Premovement Potentials. An indication of the widespread areas of cortex involved in planning and executing a single self-paced movement can be gained by recording the premovement potential (Bereitschaftspotential) before voluntary movements in man. The procedure involves placing EEG electrodes on the scalp and recording the average electrical activity preceding a simple voluntary (usually finger extension) movement. The important aspect is that the movements must be self-placed, so that no extraneous events can trigger unwanted brain activity. An example is shown in Figure 8.21. There are two remarkable features about this

Figure 8.21: Premovement (Bereitschaftspotential) Potentials Recorded Simultaneously From Several Scalp Locations During a Self-paced Voluntary Extension Movement of the Right Index Finger. The subject was instructed to move in his own time every 5–10s, and 128 trails were averaged. The EMG responsible for the movement can be seen in the first trace. About 1.5s before onset of EMG, a slowly rising (negative polarity is upwards) wave can be seen in the EEG records, culminating in a peak just after the EMG has begun. This wave of activity is known as the premovement potential (courtesy of Dr J.P.R. Dick)

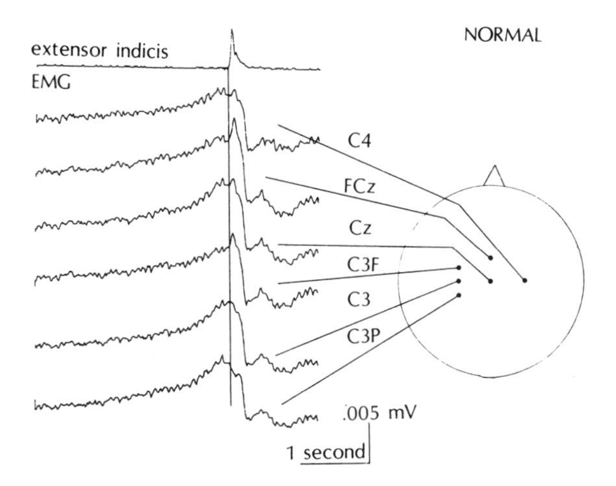

activity. First, its duration: it begins up to 1.5s before movement. Second, its topography: the waves are extremely widespread, indicating an involvement of wide areas of cortex in the preparation for movement. As movement onset approaches, the activity becomes steeper over the contralateral sensorimotor areas.

Perhaps the most interesting fact about the premovement potentials is that they appear to start before one is aware of taking the decision to move. In a fascinating experiment, Libet, Gleason, Wright *et al.* (1983) showed that if subjects estimated the time at which they made the decision to move, this was always some 0.5s later than the onset of the premovement potential. Estimation of the time of other types of past event such as sensory stimulation was, in contrast, quite accurate. The reason for the interest in this fact is purely philosophical. It means that consciousness must be secondary to activity of cerebral neurones. The idea that humans decide to move and then somehow set in train the physiological processes to accomplish the task is quite wrong. The decision to move is made first: only later do we become consciously aware of it (Libet *et al.*, 1983).

References and Further Reading

Review Articles and Books

The following chapters in V.B. Brooks (ed.), *Handbook of Physiology,* sect. 1, vol. II, part 2, Williams and Wilkins, Baltimore:

Asanuma, H. (1981) 'The Pyramidal Tract', pp. 703–33

Porter, R. (1981) 'Internal Organisation of the Motor Cortex for Input-Output Arrangements', pp. 1063–82

Evarts, E.V. (1981) 'Role of Motor Cortex in Voluntary Movements in Primates', pp. 1083–120

Wiesendanger, M. (1981) 'Organisation of Secondary Motor Areas of Cerebral Cortex', pp. 1121–47

Denny-Brown (1966) *The Cerebral Control of Movements,* University Press, Liverpool

Geschwind, N. and Damasio, A.R. (1985) 'Apraxia' in J.A.M. Fredriks (ed.) *Handbook of Clinical Neurology,* vol. 1(45), Elsevier, Amsterdam, pp. 423–32

Hyvarinen, J. (1982) 'Posterior Parietal Lobe of the Primate Brain', *Physiol. Rev., 62,* pp. 1060–129

Phillips, C.G. and Porter, R. (1977) *Corticospinal Neurones. Their Role in Movement,* Academic Press, London

Rothwell, J.C., Obeso, J.A. and Marsden, C.D. (1986) 'Electrophysiology of Somatosensory Reflex Myoclonus' in S. Fahn *et al.* (eds.), *Advances in Neurology,* vol. 43, Raven Press, New York, pp. 353–66

Wiesendanger, M. and Miles, T.S. (1982) 'Ascending Pathway of Low-threshold Muscle Afferents to the Cerebral Cortex and Its Possible Role in Motor Control', *Physiol. Rev., 62,* pp. 1234–70

Original Papers

Brinkman, C. (1981) 'Lesions in Supplementary Motor Area Interfere with a Monkey's Performance of a Bimanual Co-ordination Task', *Neurosci. Lett., 27,* pp. 267–70

——— (1984) 'Supplementary Motor Area of the Monkey's Cerebral Cortex: Short- and Long-term Deficits After Unilateral Ablation and the Effects of Subsequent Callosal Section', *J. Neurosci., 4,* pp. 918–29

Brodal, A. (1973) 'Self-observations and Neuroanatomical Considerations After a Stroke', *Brain, 96,* pp. 675–94

——— (1981) *Neurological Anatomy in Relation to Clinical Medicine,* Oxford University Press, Oxford

Bucy, P.C., Keplinger, J.E. and Sequiera, E.B. (1964) 'Destruction of the "Pyramidal Tract" in

Man', *J. Neurosurg.*, *21*, pp. 385–98
Cheney, P.D. and Fetz, E.E., (1984) 'Corticomotoneuronal Cells Contribute to Long-latency Stretch Reflexes in the Rhesus Monkey', *J. Physiol.*, *349*, pp. 249–72
Clough, J.E.M., Kernell, D. and Phillips, C.G. (1968) 'The Distribution of Monosynaptic Excitation from the Pyramidal Tract and from Primary Spindle Afferents to Motoneurones of the Baboon's Hand and Forearm', *J. Physiol.*, *198*, pp. 145–66
Evarts, E.V. (1966) 'Pyramidal Tract Activity Associated With a Conditioned Hand Movement in the Monkey', *J. Neurophysiol.*, *29*, pp. 1011–27
——— (1968) 'Relation of Pyramidal Tract Activity to Force Exerted During Voluntary Movement', *J. Neurophysiol.*, *31*, pp. 14–27
——— (1972) 'Pre- and Postcentral Neuronal Discharge in Relation to Learned Movement' in T. Frigyesi, E. Rinvik and M.D. Yahr (eds.), *Corticothalamic Projections and Sensorimotor Activities*, Raven Press, New York, pp. 449–58
——— (1973) 'Motor Cortex Reflexes Associated with Learned Movement', *Science*, *179*, pp. 501–3
——— and Tanji, J. (1976) 'Reflex and Intended Responses in Motor Cortex Pyramidal Tract Neurones of Monkey', *J. Neurophysiol.*, *37*, pp. 1069–80
——— and Fromm, C. (1978) 'The Pyramidal Tract Neuron as a Summating Point in a Closed-loop Control System in the Monkey' in J.E. Desmedt (ed.), *Prog. Clin. Neurophysiol.*, vol. 4, Karger, Basel
———Fromm, C., Kroller, J., *et al.* (1983) 'Motor Cortex Control of Finely Graded Forces', *J. Neurophysiol.*, *49*, pp. 1199–215
Freund, H.-J. and Hummelsheim, H.J. (1984) 'Premotor Cortex in Man: Evidence for Innervation of Proximal Limb Muscles', *Exp. Brain, Res.*, *53*, pp. 479–82
Friedman, D.P. and Jones, E.G. (1981) 'Thalamic Input to Areas 3a and 2 in Monkeys', *J. Neurophysiol.*, *45*, pp. 59–85
Haaxma, R. and Kuypers, H.G.J.M. (1975) 'Intrahemispheric Cortical Connections and Visual Guidance of Hand and Finger Movements in the Rhesus Monkey, *Brain*, *98*, pp. 239–60
Jankowska, E., Padel, Y. and Tanaka, R. (1975) 'Projection of Pyramidal Tract Cells to α-motoneurones Innervating Hindlimb Muscles in the Monkey', *J. Physiol.*, *249*, pp. 637–67
Kalaska, J.F., Caminiti, R. and Georgopoulos, A.P. (1983) 'Cortical Mechanisms Related to the Direction of Two-dimensional Arm Movements: Relations in Parietal Area 5 and Comparison with Motor Cortex', *Exp. Brain. Res.*, *51*, pp. 247–60
Laplane, D., Talairach, J., Meininger, V., *et al.* (1977) 'Clinical Consequences of Corticectomies Involving the Supplementary Motor Area in Man', *J. Neurol. Sci.*, *34*, pp. 301–14
Lawrence, D.G. and Kuypers, H.G.J.M. (1968) 'The Functional Organisation of the Motor System in the Monkey, Parts I and II', *Brain*, *91*, pp. 1–14 and 15–36
Lewis, R. and Brindley, G.S. (1965) 'The Extrapyramidal Cortical Motor Map', *Brain*, *88*, pp. 397–406
Libet, B., Gleason, C.A., Wright, E.W. and Pearl, D.K. (1983) 'Time of Conscious Intention to Act in Relation to Onset of Cerebral Activity (Readiness Potential)', *Brain*, *106*, pp. 623–42
Marsden, C.D., Merton, P.A. and Morton, H.B. (1983) 'Direct Electrical Stimulation of Corticospinal Pathways Through the Intact Scalp in Human Subjects', in J.E. Desmedt (ed.), *Advances in Neurology*, vol. 39, Raven Press, New York, pp. 387–91
MacPherson, J., Wiesendanger, M., Marangoz, C., *et al.* (1982) 'Corticospinal Neurones of the Supplementary Motor Area of Monkeys. A Single Unit Study', *Exp. Brain. Res.*, *48*, pp. 81–8 (see also *45*, pp. 410–16)
Mountcastle, V.B., Lynch, J.C., Georgopoulos, A. *et al.* (1975) 'Posterior Parietal Association Cortex of the Monkey: Command Functions for Operations within Extrapersonal Space', *J. Neurophysiol.*, *38*, pp. 871–908
Obeso, J.A., Rothwell, J.C. and Marsden, C.D. (1985) 'The Spectrum of Cortical Myoclonus', *Brain*, *108*, pp. 193–224
Penfield, W. and Jasper, H.H. (1954) 'Epilepsy and the Functional Anatomy of the Human Brain', Little Brown, Boston
Phillips, C.G. and Porter, R. (1964) 'The Pyramidal Projection to Motoneurones of Some Muscle Groups of the Baboon's Forelimb' in J.C. Eccles and J.P. Schade (eds.), *Progress in Brain Research*, vol. 12, pp. 222–42
Roland, P.E., Larsen, B., Lassen, A., *et al.* (1980) 'Supplementary Motor Area and Other Cortical

Areas in Organisation of Voluntary Movements in Man', *J. Neurophysiol., 43,* pp. 118-36 (see also pp. 137–50)
Rosen, I. and Asanuma, H. (1972) 'Peripheral Afferents to the Forelimb Area of the Monkey Motor Cortex: Input-output Relations', *Exp. Brain, Res., 14,* pp. 257–73 (see also pp. 243–56)
Rothwell, J.C., Obeso, J.A. and Marsden, C.D. (1984) 'On the Significance of Giant Somatosensory Evoked Potentials in Cortical Myoclonus', *J. Neurol. Neurosurg. Psychiat., 47,* pp. 33–42
Ruch, T.C. and Fetz, E.E. (1979) 'The Cerebral Cortex: Its Structure and Motor Functions' in T. Ruch and H.D. Patton (eds.), *Physiology and Biophysics,* vol. II, W.B. Saunders, Philadelphia, pp. 53–122
Schell, G.R. and Strick, P.L. (1984) 'The Origin of Thalamic Inputs to the Arcuate Premotor and Supplementary Motor Area', *J. Neurosci., 4,* pp. 539–60
Schiebel, M.E., Davies, T.L., Lindsay, R.D. *et al.* (1974) 'Basilar Dendrite Bundles of Giant Pyramidal Cells', *Exp. Neurol., 42,* pp. 307–19
Seal, J. and Commenges, D. (1985) 'A Quantitative Analysis of Stimulus- and Movement-related Responses in the Posterior Parietal Cortex of the Monkey', *Exp. Brain. Res., 58,* pp. 144–53
Tanji, J. and Kurata, K. (1985) 'Contrasting Neuronal Activity in Supplementary and Precentral Motor Cortex of Monkeys', *J. Neurophysiol., 53,* pp. 129–41 and 142–52
Weinrich, M., Wise, S.P. and Mauritz, K.-H. (1984) 'A Neurophysiological Study of the Premotor Cortex in the Rhesus Monkey', *Brain, 107,* pp. 385–414
Wise, S.P. and Mauritz, K.-H. (1985) 'Set-related Neuronal Activity in the Premotor Cortex of Rhesus Monkeys: Effects of Changes in Motor Set', *Proc. Roy. Soc. B., 223,* pp. 331–54
Woolsey, C.N., Erickson, T.C. and Gilson, W.E. (1979) 'Localisation in Somatic Sensory and Motor Areas of Human Cerebral Cortex as determined by Direct Recording of Evoked Potentials and Electrial Stimulation', *J. Neurosurg., 51,* pp. 476–506
——— Settlage, P.H., Meyer, D.R. *et al.* (1952) 'Patterns of Localisation in Precentral and "Supplementary" Motor Areas and Their Relation to the Concept of a Premotor Area', *Res. Pub. Ass. Nerv. Ment Dis., 30,* pp. 238–64

9 THE CEREBELLUM

There are more neurones in the cerebellum than there are in the whole of the rest of the brain, yet surprisingly little is known about the function of this remarkable organ. Lesions of the cerebellum do not produce muscle weakness nor disorders of perception. However, they do produce disturbances of co-ordination in limb and eye movements, as well as disorders of posture and muscle tone. Despite the lack of any direct efferent connexions between cerebellum and spinal or brainstem motor nuclei, it is a structure which is undoubtedly concerned with the control of movement.

Cerebellar Anatomy

The cerebellum consists of a convoluted outer mantle of grey matter covering an inner core of white matter. The structure is folded over upon itself so that the anterior and posterior ends meet beneath, above the surface of the fourth ventricle. The conspicuous features of the human cerebellum are two lateral hemispheres which lie either side of a narrow midline ridge known as the vermis (Figure 9.1). Deep within this folded mass are three pairs of nuclei, arranged either side of the midline. In animals they are known as the medial, interposed (subdivided into an anterior and posterior part), and lateral nuclei. In man, they have different names: fastigial, globose and emboliform, and dentate. The globose and emboliform are sometimes spoken of together as the interposed nucleus, as in animals. These nuclei receive a large proportion of the output from the cerebellar cortex and distribute efferents to other regions of the CNS.

The input and output of the cerebellum travels via three pairs of large fibre tracts which connect it to the brainstem: the superior, middle and inferior cerebellar peduncles, which are known also as the brachium conjunctivum, the brachium pontis and restiform body, respectively. There are far more fibres entering the cerebellum than there are leaving it. The ratio of afferents to efferents has been put as high as 40:1. Most of the afferents travel in the middle cerebellar peduncle (20 million fibres), and come from the pontine nuclei. The other afferents which come from the olive and via direct spinocerebellar pathways, are found mainly in the inferior peduncle (0.5 million fibres), with a small contribution in the superior peduncle. There are also inputs from visual and vestibular systems. Most of the efferent fibres leave the cerebellum through the superior peduncle (0.8 million fibres) and pass to the red nucleus, thalamus and brainstem.

Figure 9.1: Surface Views of the Superior (Upper) and Inferior (Lower) Aspects of the Human Cerebellum. In the lower diagram, the tonsil and the adjoining part of the biventral lobule on the right side have been removed (from WIlliams and Warwick, 1975; Figures 7.72 and 7.75, with permission)

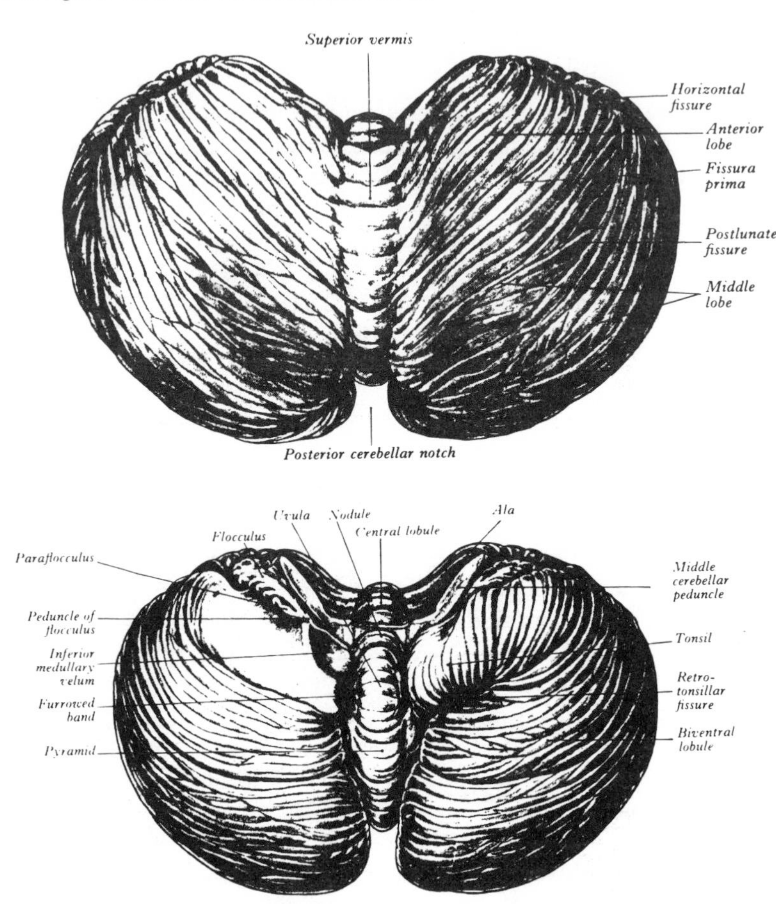

Subdivisions of the Cerebellum

The cerebellar cortex has basically the same neuronal organisation throughout (see below), so that cytoarchitectonics, which was so useful in describing the anatomy and functional subdivisions of the cerebral cortex, cannot be applied here. Various other schemes have been proposed which describe the structure in terms of gross anatomy, phylogeny and afferent and efferent connections.

Gross Anatomical Divisions. The surface of the cerebellum is divided into a series of ridges or *folia*, separated by deep fissures. These ridges are parallel and run transversely to provide the basis for a gross morphological nomenclature. Two of the fissures are deeper than the rest and divide the cerebellum into three *lobes*. Of these, the first to appear phylogenetically and ontologically is the *posterolateral fissure,* which in man is buried deep beneath the hemispheres. This divides the flocculus and nodule (flocculonodular lobe) from the rest of the cerebellum, which is termed the *corpus cerebelli*. The *primary fissure,* which develops later, and is clearly visible on the surface of the cerebellum divides the corpus cerebelli into two further lobes, the *anterior* and *posterior* lobes. These two lobes are further subdivided by smaller fissures into a number of *lobules*.

Different names are given to these lobules in man than in other mammals. A comparison is shown in Figure 9.2 (top). In addition, there is a third system of classification which uses a numerical terminology. This divides the cerebellum into 10 lobules numbered from I to X. Five of the lobules are in the anterior lobe (although only three are visible in man), four in the posterior lobe and the last (X) is the flocculonodular lobe. This numeric system is now in common use.

Phylogenetic Divisions (Figure 9.2 (bottom)). It has already been mentioned that the posterolateral fissure is the first to develop phylogenetically. This separates off the oldest portion of the cerebellum, the *archicerebellum* (that is, the flocculonodular lobe). The most recently acquired portions are the lateral hemispheres, and the middle portion of the vermis (lobules VI and VII) which are known as the *neocerebellum*. The rest of the cerebellum, that is, the vermal part of the anterior lobe, the pyramis, uvula and paraflocculus are known as the *paleocerebellum*.

Divisions Based on Input and Output Arrangements

Afferents. The afferent connections to the cerebellum follow these divisions roughly but not precisely. Vestibular inputs are found mainly in the archicerebellum, spinal cord inputs in the paleocerebellum, and pontine inputs (principally from the cerebral cortex) in the neocerebellum. Because of this pattern of inputs, these regions of the cerebellar cortex have often been referred to as the vestibulo-, spino- and ponto-cerebellum. However, it should be noted that this division of inputs is not precise. For example, direct spinal cord inputs (via spinocerebellar tracts) do not terminate in all parts of the paleocerebellum: for example, they occupy only a limited region of the paraflocculus. In contrast, they project densely to the paramedian lobule, which is part of the neocerebellum. Additionally, there is extensive overlap between all inputs. Even the archicerebellum is now known to receive afferents from a wide variety of sources such as the neck, visual and oculomotor systems.

Efferents. Efferent fibres of the cerebellar cortex project either to the cerebellar nuclei, or the vestibular nuclei, or both. Three main *longitudinal* divisions can be recognised: a medial zone (vermis) which sends fibres to the fastigial nucleus, an intermediate zone which sends fibres to the interposed

Figure 9.2: Top: Subdivisions of the Human Cerebellum. On the left half are the classical names of the various lobules in man. On the right are the names used for mammals in general. Longitudinal subdivisions into vermis, intermediate and lateral parts are shown by dotted lines. In the vermis, the Roman numerals I to X represent the numbering system now commonly in use. Li, lingula; L.c., lobulus centralis; Cu, culmen; De, declive; F.v. and T.v., folium and tuber vermis; P, pyramis; U, uvula; N, nodulus (from Brodal, 1981; Figure 6.1 with permission). '
Bottom: Evolutionary Subdivisions of the Monkey Cerebellum (from Gilman, Bloedel and Lechtenberg, 1981; Figure 2A, p.5, with permission)

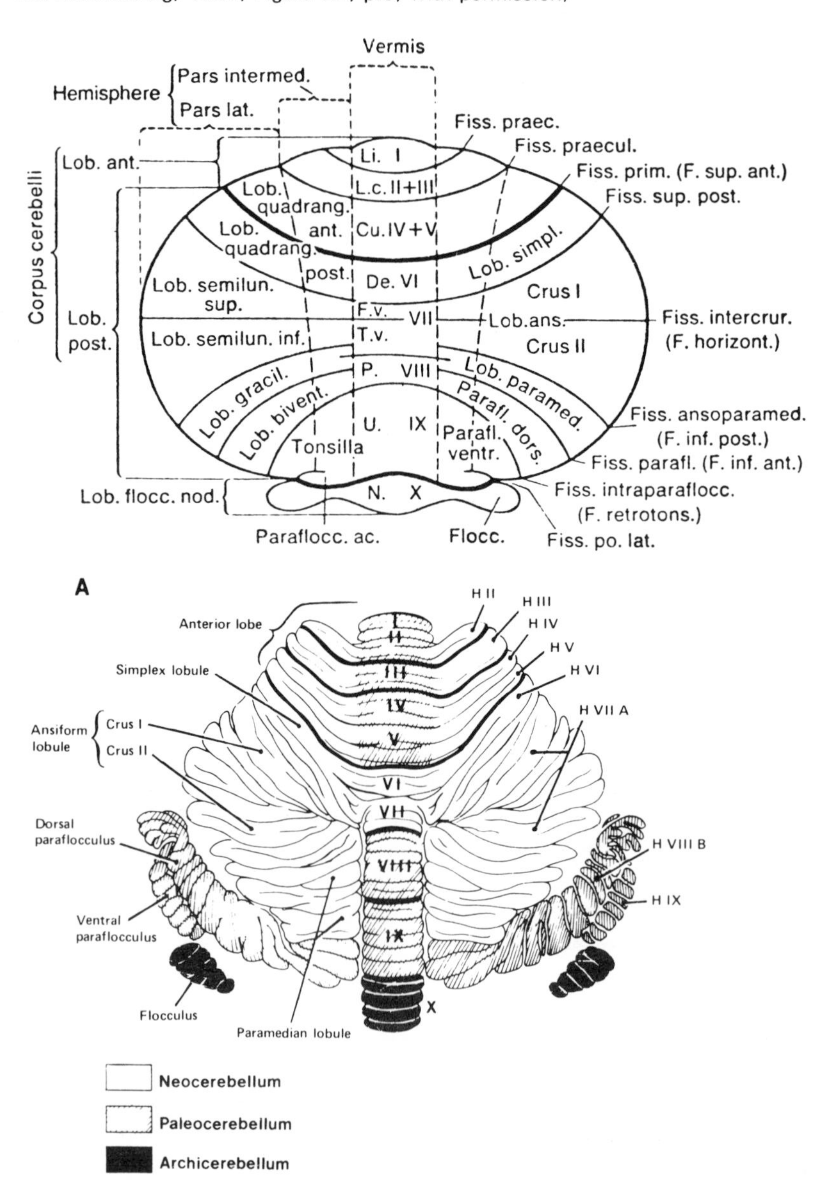

Figure 9.3: Simplified Diagram of the Longitudinal Subdivisions of the Cerebellum Based on the Pattern of Efferent Projections to the Cerebellar and Vestibular Nuclei. Similar shading of cerebellar cortex and nuclei indicates an efferent connection between the two. (from Jansen and Brodal, 1958; Figure 183, with permission)

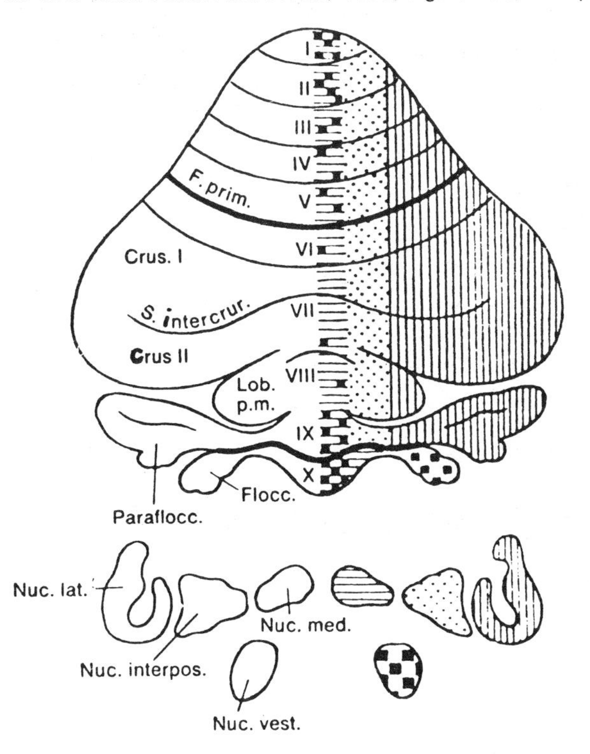

nucleus, and a lateral zone which sends fibres to the dentate nucleus (Figure 9.3). The vestibular nuclei receive afferents from the flocculonodular lobe and from regions of the vermis in the anterior and posterior lobes.

More detailed work on the microscopic anatomy of the cerebellar cortical projections has led to further refinement of these longitudinal subdivisions. Staining for myelin sheaths reveals that the white matter is divided into a series of longitudinal compartments which are distinguished by the number of thick and thin fibres within them. The axons in these compartments come from longitudinal strips of cerebellar cortex which run perpendicular to the long axis of the lobules. Each strip projects to a specific part of the cerebellar nuclei, and receives its climbing fibre input from a particular region of the inferior olivary nuclei (Figure 9.4).

At one time it was common to consider the inputs and outputs of the three major longitudinal zones as comprising three quite separate cerebellar circuits:

1. Lateral: receiving input primarily from cerebral cortex and projecting via dentate nucleus back to motor areas of cortex;
2. Intermediate: receiving input primarily from spinal cord and also from

Figure 9.4: Fine Subdivision of Longitudinal Zones of Cat Cerebellum According to the Pattern of Olivocerebellar Projections and the Projections from Cerebellar Cortex onto the Cerebellar Nuclei. As indicated by the different shading, different regions of the inferior olivary complex (top right) project to longitudinal strips of cerebellar cortex. These strips in turn project in an orderly fashion onto the cerebellar nuclei. F, fastigial nucleus; IA, IP, anterior and posterior interposed nuclei; D, dentate nucleus; Dei, lateral vestibular nucleus of Deiters (from Groenewegen, Voogd and Freedman, 1979; plate 16, with permission)

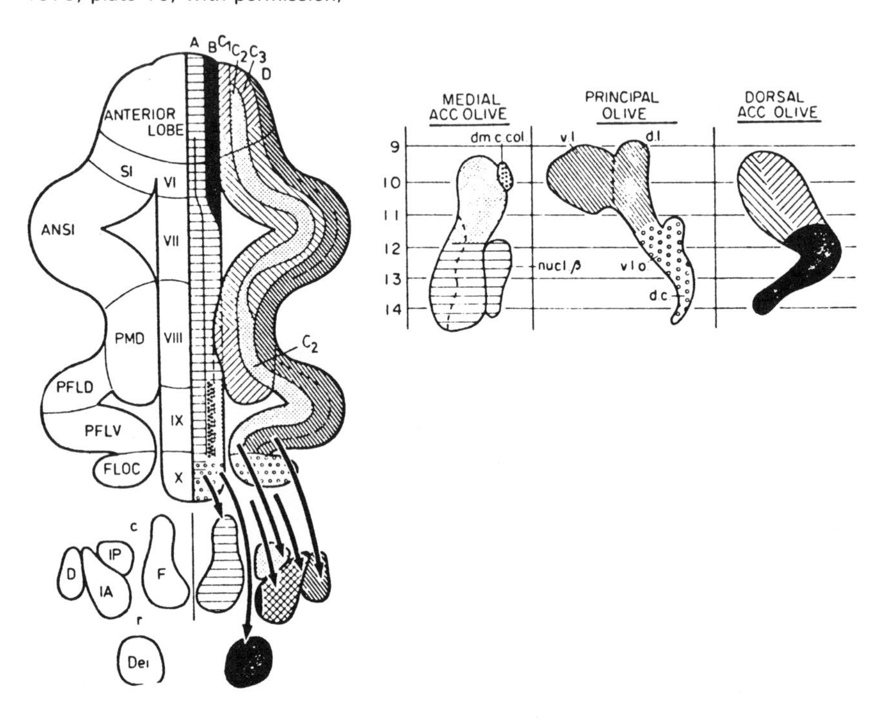

cerebral cortex, projecting via interposed nuclei to red nucleus (and hence the rubrospinal tract) and also back to cerebral cortex;

3. Medial: receiving spinal cord and vestibular input, projecting via fastigial nucleus to the reticular formation and vestibular nuclei, and acting on spinal cord via the reticulospinal and vestibulospinal tracts.

These subdivisions inspired a good deal of electrophysiological experiments on the cerebellum, and are still useful in interpreting results of these experiments as we shall see below. However, it is now becoming more common to stress the *interconnections* between these loops. As detailed below, *all* cerebellar zones receive direct or indirect input from both the spinal cord and cerebral cortex, and all project to both cord and to cortical structures.

Afferent Connections of the Cerebellum

The cerebellum receives primary, or direct, afferent fibres from two sources, the spinal cord and the vestibular apparatus. Other afferents project first to other

Figure 9.5: Highly Simplified Diagram Showing Main Sources of Afferent Input to the Cerebellar Cortex and Nuclei. Note that both mossy fibre (left) and climbing fibre (right) inputs project to both cortex and nuclei.

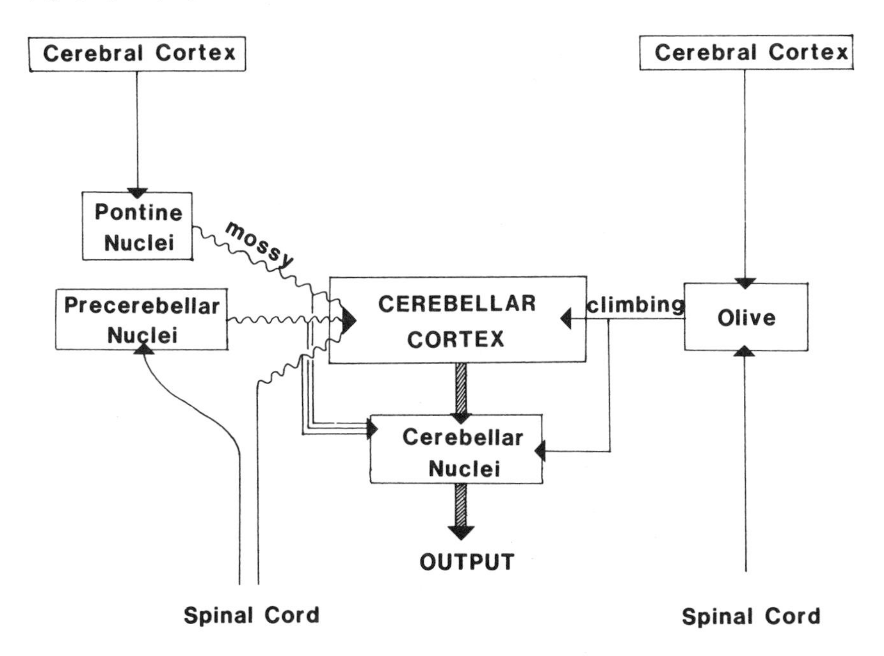

nuclei in the pons or medulla. These *precerebellar relay nuclei* receive convergent input from many different centres. Figure 9.5 gives a very simplified diagram of the main sources of cerebellar afferents.

It is thought that all afferents to the cerebellum terminate in both the cortex and the cerebellar nuclei. This arrangement has given rise to the idea that the cerebellar cortex functions as a modulatory 'side-loop' in the direct pathway from afferent input to nuclear output. However, too much weight should not be given to this speculation since the relative strength and importance of the inputs to cortex and nuclei are in most instances, not known (see Bloedel and Courville, 1981, for detailed summary of cerebellar afferent systems).

Direct Spinal Input (see also Chapter 7). Direct input from the spinal cord travels in the spinocerebellar tracts (Figure 9.6). The dorsal spinocerebellar tract and its forelimb analogue, the cuneocerebellar tract, both transmit input from large diameter sensory fibres with restricted peripheral receptive fields. Their terminations correspond approximately with the hindlimb and forelimb representations of the two somatotopic body maps (see below) in the anterior and posterior lobes. The ventral and rostral spinocerebellar tracts contain much smaller diameter fibres, and terminate more diffusely and bilaterally in the cerebellar cortex. Terminals of fibres in the rostral tract, which transmits

Figure 9.6: Schematic Diagram of Direct Spinocerebellar Pathways (Except the Rostral Spinocerebellar Tract) and Their Areas of Termination in the Cerebellar Cortex of a Cat. Note that all project to interposed and fastigial nuclei as well as the cerebellar cortex.

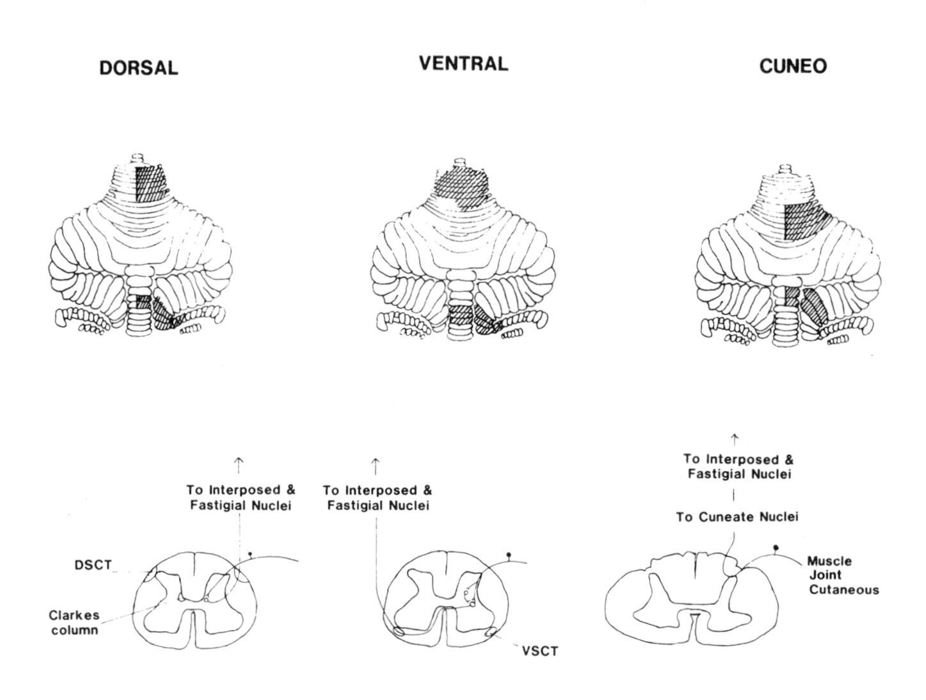

information only from the forelimb, can be found bilaterally in both forelimb and hindlimb areas of the somatotopic maps. The dorsal and cuneocerebellar tracts travel in the inferior cerebellar peduncle, the ventral spinocerebellar tract in the superior peduncle, and the rostral spinocerebellar tract in both superior and inferior peduncles. Spinal inputs are also found in the fastigial and interposed nuclei, and perhaps in the dentate.

Direct Vestibular Input. Primary vestibular afferents terminate in the nodulus, flocculus and the ventral part of the uvula (the vestibulo-cerebellum). A smaller number of fibres supply the whole of the vermis. The fibres also end in the fastigial nucleus.

Inputs from Precerebellar Reticular Nuclei. There are three precerebellar reticular nuclei which are usually considered separately from the pontine nuclei referred to below. They receive convergent input from more than one source. The best studied of these nuclei is the lateral reticular nucleus, which lies just lateral to the inferior olive. Its main input comes from spinal afferents with large receptive fields, which may cover all four limbs of the body. Fibres from lumbar regions end in the parvocellular part, fibres from the cervical region end in the magnocellular part. These then project via the inferior cerebellar peduncle to the hindlimb and forelimb areas of the ipsilateral cerebellar somatotopic map. Because of their large receptive fields, such inputs will be responsible for blurring the somatotopic representation in awake animals. Other inputs to the lateral reticular nucleus arise in the contralateral sensorimotor cortex, magnocellular red nucleus, fastigial nucleus, lateral vestibular nucleus and the trigeminal system.

The nucleus reticularis tegmenti pontis is situated dorsal to the pontine nuclei proper, and is separated from them by the medial leminiscus. It receives no afferents from the spinal cord and projects via the middle cerebellar peduncle to the contralateral vermal regions VI and VII, and the flocculus. These are the regions which receive visual input and are concerned with the control of eye movements. Its main inputs come from the cerebellar nuclei themselves, ipsilateral frontal cortex, vestibular nuclei, oculomotor nuclei, superior colliculus and pretectum.

The final precerebellar reticular nucleus is the paramedian reticular nucleus, which lies medially in the medulla slightly above the level of the hypoglossal nuclei. It receives afferents from spinal cord, vestibular nuclei, fastigial nuclei and cerebral cortex. It projects to all areas of the cerebellar cortex, bilaterally, via the inferior peduncle.

Inputs from the Pontine Nuclei. The pontine nuclei are the main relay station for cortical inputs to the cerebellum. The bulk of the fibres arise in the ipsilateral frontal and parietal lobes, Most of them are not collaterals of pyramidal tract fibres but constitute a specific corticopontine projection. Other inputs are from the cerebellar nuclei and the colliculi. The pontocerebellar projection is massive, and forms the bulk of the middle cerebellar peduncle. Terminals are found in almost all parts of the contralateral cerebellum (except the nodulus and ventral uvula), as well as in the interposed and dentate nuclei. However, the main areas of termination are in the lateral and intermediate zones of the cerebellar cortex.

The organisation of the pontine input is very different from that of the spinal inputs considered above. Injections of HRP into small areas of a cerebellar lobule reveal retrogradely labelled cells in the pons which are aggregated together into longitudinal columns. Each one of these columns has a divergent output which innervates small regions in two or more lobules. Conversely, each lobular region of the cerebellar cortex shows convergence of input being innervated from more than one pontine column. The inputs to the pontine nuclei follow a similar pattern. For example, a given area of cerebral cortex may project to more than one spatially separate column. Such a complex pattern of convergence and divergence within this system has the result that, although the sensorimotor inputs to the pons are somatotopically organised, it is very difficult to follow this pattern in the cerebellar projection.

Inputs from the Inferior Olivary Nuclei. The infererior olive is a relay station for a large variety of different convergent inputs. These arise in the spinal cord (travelling via the spino-olivary tract), motor cortex, superior colliculus, red nucleus, vestibular nuclei, trigeminal nuclei and pretectum. The spinal inputs are the best studied. They consist of fibres of the flexor reflex category with large receptive fields, and terminate in a somatotopic pattern.

Olivary afferents travel in the inferior cerebellar peduncle to end in all regions of the contralateral cerebellar cortex and nuclei. Those from the principal olive terminate mainly in lateral cerebellum and those from the dorsal accessory olive in the medial and intermediate zones. In the cortex, they are believed to terminate exclusively as climbing fibres. Each Purkinje cell receives one climbing fibre input (see below) from a single olivary neurone. Since there are more Purkinje cells than olivary axons, this means that each axon contacts seven to 15 Purkinje cells.

The pattern of termination within the cerebellar cortex is extremely precise (see Figure 9.4). Each region of the olive innervates a separate longitudinal zone of cortex running perpendicular to the axis of the folia. Within these six to seven major zones, microzones can be distinguished: certain sections are innervated by particular groups of cells within the olivary subregion. However, some olivary cells do not show such specific projections since the Purkinje cells which they contact sometimes are in widely separate regions of a zone.

The zonal regions defined by the olivary input send their efferents to separate regions of the cerebellar nuclei. These nuclear regions receive direct input from the same cell groups in the olive as innervate the cortical regions above them. In addition, to close the circuit completely, these nuclear regions project back to the very same cells of the olive.

Circuitry of the Cerebellar Cortex (Figures 9.7 and 9.8)

Anatomy. Because of its many convolutions, the cerebellar cortex of man covers a large surface area: it has a total antero-posterior length of over one metre. Yet throughout this area it maintains the same basic neuronal structure. Viewed in sections cut vertical to the surface, three neuronal layers can be distinguished. From the surface down these are the molecular layer, the Purkinje layer and the granule layer. The only output cell of the cortex is the Purkinje cell; cortical inputs terminate either as mossy or climbing fibres.

Purkinje cells are the main neuronal elements of the middle layer of the cerebellar cortex. These flask-shaped cells are the largest neurones in the brain and form a nodal point in cerebellar circuitry. Their axons leave the cortex and travel to the cerebellar or vestibular nuclei. Their dendrites stretch upward throughout the depth of the molecular layer (400 μm) forming a dendritic tree with a total length of some 4mm per cell. The arrangement of the dendrites is extremely precise. They are oriented in a single narrow plane perpendicular to

Figure 9.7: A: Three-dimensional Diagram of the Principal Circuitry Within a Cerebellar Folium. Input enters as mossy and climbing fibres: the only output is via Purkinje cell axons. B: Structure of the glomerular junction between mossy fibre axons, granule cell dendrites and Golgi cell axons (from Ghez and Fahn, 1985; Figure 39.3, with permission)

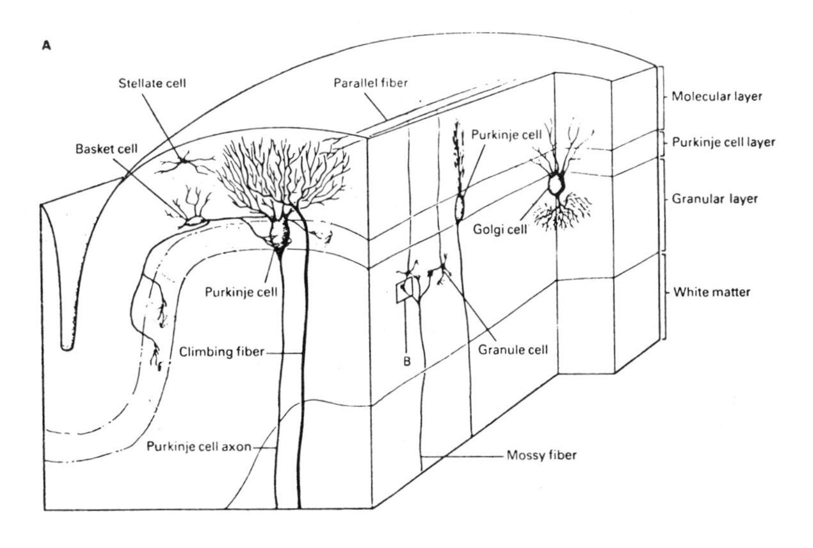

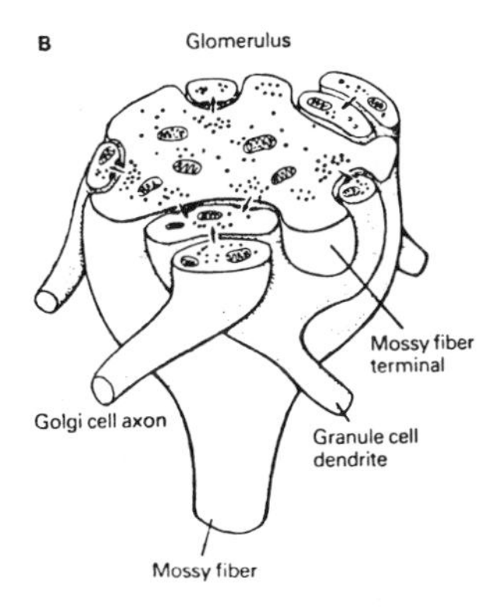

the long axis of the folia. Rows of Purkinje cells appear to have their dendritic trees stacked on top of each other like plates.

The two types of input to the cortex show extremes of convergence and divergence in their contacts with the Purkinje cells. Climbing fibres, which originate in the inferior olive, synapse directly with Purkinje cells. Each Purkinje cell is only contacted by one climbing fibre, which forms 150 to 200 synaptic contacts on the proximal dendrites. This synaptic arrangement is very powerful: a single impulse in the preterminal axon always produces firing of the Purkinje cell (see below). Each climbing fibre may, however, contact seven to 15 Purkinje cells.

Figure 9.8: Schematic Diagram of Neuronal Circuitry and Possible Synaptic Transmitters in the Cerebellar Cortex. Inhibitory cells and synapses are black; open cells and synapses are excitatory. The major efferents are excitatory: the climbing fibres (CF) from the inferior olive (IO) and the mossy fibres (MF) from the pontine nuclei (PN) and elsewhere. Inhibitory inputs are very much smaller in number and arise in the locus coeruleus (LC) and raphe nucleus (RP). The dotted circle represents the cerebellar nuclei, which receive inputs from all classes of afferent. PF, parallel fibres; GO, Golgi cell; GR, granule cell; BA, basket cell; PC, Purkinje cell; ST, stellate cell. NA, noradrenaline; AS, aspartate; TA, tachykinin (from Ito, 1984; Figure 3, with permission)

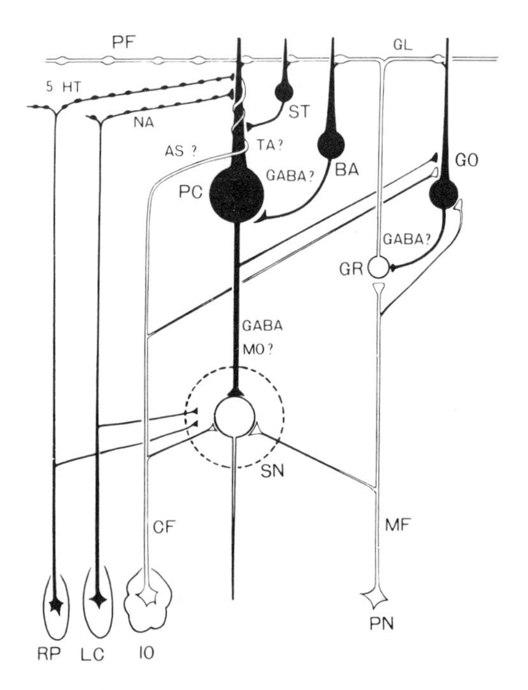

Mossy fibres, which do not contact the Purkinje cells directly, come from many sources and have axons of larger diameter than those of the climbing fibres. Their axons end in the granule layer. This is made up of an enormous number of small densely packed cells (10^{10} to 10^{11}) with very scanty cytoplasm. Each granule cell has four or five small dendrites which end in claw-like expansions known as rosettes. Rosettes from up to 28 different granule cells are clustered together to form a structure known as a glomerulus. It is at these sites that synapses are formed with the incoming mossy fibres and also with axons of Golgi cells (see below). A single mossy fibre may branch to innervate many glomeruli in more than one cerebellar lobule. Since the mossy fibres contact many granule cells in one glomerulus, it has been calculated that each mossy fibre may contact on average over 400 separate granule cells.

The axons of the granule cells ascend vertically through the Purkinje layer to the molecular layer, where they bifurcate in a 'T' junction. The branch on each of the arms of the 'T' is known as a parallel fibre, and runs along the axis of the folia so as to contact the maximum number of Purkinje cells, whose planar dendritic trees are stacked perpendicular to their course. Along its 2–3mm length, a single parallel fibre passes through the dendritic trees of 400,000 or so Purkinje cells. One, or at most two, synapses are formed with any Purkinje cell, and frequently none at all. A Purkinje cell has up to 80,000 parallel fibre synapses on its distal dendritic tree. It is the end point of a remarkable degree of divergence from a single mossy fibre input.

Both parallel and climbing fibre inputs are excitatory to Purkinje cells, and synapse at special sites known as dendritic spines. The other inputs are from cortical interneurones. They are inhibitory and contact the smooth surface of the cell soma or dendrites.

There are three types of cortical interneurone. Two of them, the stellate and basket cells have cell bodies in the molecular layer. These cells receive input from parallel fibres and direct their inhibitory output to the Purkinje cells. Stellate cells synapse with the middle and distal parts of the dendritic tree, while the basket cells have axon terminals which interweave, basket-like, around the soma and proximal dendrites of the Purkinje cell. Each Purkinje cell may receive input from 20 to 30 basket cells. Basket cells inhibit Purkinje cells which lie 'off the beam' of the parallel fibres which excite them.

The final interneurone of the cerebellar cortex is the Golgi cell, which has its cell body in the granule layer. Its dendrites ramify up into the molecular layer, where they receive input from parallel fibres. In addition, they receive input from mossy fibres at the glomeruli in the granule layer. Unlike the Purkinje, stellate and basket cells, their dendrites are distributed to form a cylinder rather than a plane. They inhibit granule cells, and may perhaps be involved in a feedback inhibition of the mossy fibre input.

Electrophysiology. Because of its size, the Purkinje cell has been intensively investigated with electrophysiological techniques. It contains the usual voltage sensitive Na^+ and K^+ conductances in the soma and axon membranes which are responsible for the production of rapid action potentials. In addition, there are other voltage sensitive channels which give the cell some unusual firing characteristics.

If the soma is depolarised by passing a just threshold steady current through an intracellular microelectrode, then the cell will fire a repetitive series of impulses. Increasing the intensity of the depolarising current produces a different behaviour. The onset of repetitive firing occurs earlier and, rather than giving a continuous train of impulses, rhythmic bursts of impulses are seen. The end of each burst is signalled by a reduction in the amplitude of the individual spikes of the burst.

Two main conductances underly this behaviour. A non-inactivating Na^+ conductance in the soma membrane is responsible for the repetitive firing of the cell during prolonged depolarisation, A voltage-sensitive Ca^{2+} conductance in the dendritic membrane is responsible for the bursting behaviour. This latter can

Figure 9.9: Intracellular Potentials Recorded from Cerebellar Purkinje Cells. Dots mark responses evoked by climbing fibre connections: complex spikes. The shorter action potentials are simple spikes caused by mossy fibre input (from Martinez, Crill and Kennedy, 1971; Figure 1A-D, with permission)

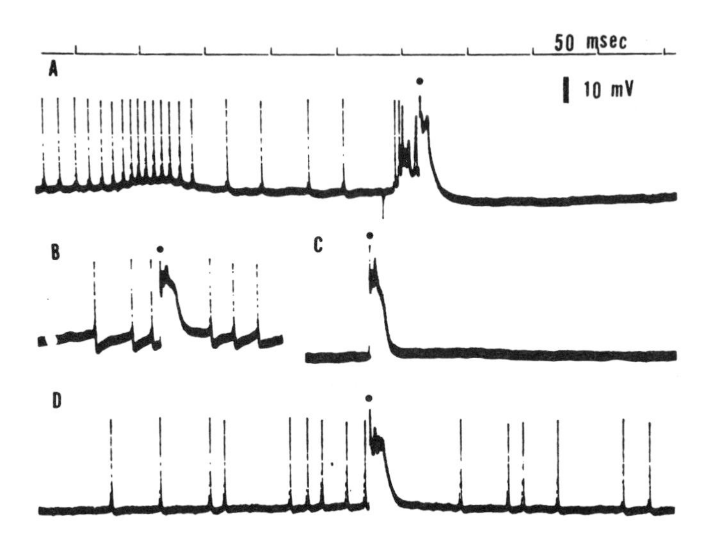

be demonstrated by blocking the Na^+ channels with tetrodotoxin (TTX), which abolishes the fast spikes, leaving a late, slow-rising burst potential intact. This potential is blocked by removal of Ca^{2+} channels (such as cobalt or manganese).

Under normal circumstances, the Purkinje cell is activated either by parallel or climbing fibre input. Activation of the climbing fibre input generates a very large (25mV) EPSP in the Purkinje cell. It is the most powerful synapse in the CNS, and is a consequence of the large number of synaptic contacts made by each climbing fibre. It causes a burst of spikes followed by a long repolarisation phase, which may have additional spikes superimposed upon it. This type of activity is produced by activation of both the dendritic Ca^{2+} conductance and the Na^+ and K^+ conductances of the soma and axon. The spikes are known as *complex* spikes. A single action potential in a climbing fibre always produces a complex spike in a Purkinje cell. It is an obligatory synapse with all-or-none properties.

Parallel fibre input is much weaker. It generates a graded EPSP in the Purkinje cell which may or may not generate a single rapid action potential. These are known as *simple* spikes and probably do not involve activation of the dendritic Ca^{2+} conductance. Due to its different types of spiking, the Purkinje cell is one of the few places in the CNS where the morphology of the response indicates the mode of afferent excitation (Figure 9.9). Only if an exceptionally synchronous activation occurs via the parallel fibre system can complex spikes be produced.

Inhibitory inputs from basket cells also have been studied in some detail. Activation of these cells produces graded inhibition of Purkinje cells which lie on either side of an activated bundle of parallel fibres. Recordings from Purkinje cells 'on-the-beam' of the active fibres show an EPSP followed by an IPSP. Purkinje cells on either side show only the IPSP. Basket cells thus produce lateral inhibition about the active parallel fibres.

There is another inhibitory input to the Purkinje cells whose function is unknown. Catecholaminergic terminals arising from cells in the locus coeruleus can evoke a large hyperpolarisation of Purkinje cells (see Llinas, 1981; Ito, 1984, for further information on microscopic structure and physiology of cerebellar cortex).

Efferent Pathways of the Cerebellum

The Purkinje cells are the only output cells of the cerebellar cortex. They project to the cerebellar nuclei and the vestibular nuclei and in all instances their action is inhibitory. The efferent fibres from these nuclei are the only route whereby the cerebellum can influence other parts of the brain and spinal cord.

Projections to the Cerebellar Nuclei. As already seen (Figure 9.3), it is possible to subdivide the cerebellar cortex longitudinally into three major zones. These zones send fibres to the three ipsilateral cerebellar nuclei. The medial zone (vermis) sends efferents to the fastigial nucleus, the intermediate zone to the interposed nucleus and the lateral zone to the dentate nucleus. With the discovery of further longitudinal subdivisions, based upon myeloarchitectonics and the pattern of termination of olivary afferents in the cerebellar cortex, more detail has been added to this scheme. However, the pattern of cortical projection to the cerebellar and vestibular nuclei remains very much as in the original tripartite scheme (see Figure 9.4). Although the zones are quite clear in the anterior lobe the situation is not so clear in the posterior lobe. It seems likely that the borders within the posterior lobe may undergo revision in the near future.

Projections to the Vestibular Nuclei. It is possible to divide the cerebellar cortical projections to the vestibular nuclei into those innervating the lateral vestibular nucleus (Deiter's), and those innervating the other vestibular nuclei. Projections to Deiter's nucleus come from the lateral zone of the anterior lobe vermis. These projections are somatotopically organised and have direct access to the lateral vestibulospinal tract. In addition, there are indirect projections from the medial zone of the anterior lobe vermis to the fastigial nucleus and thence to Deiter's nucleus.

Other regions of the vestibular nuclei receive afferents directly from the flocculonodular lobe and the uvula. These parts of the cerebellum do not send fibres to Deiter's nucleus and are involved mainly in the control of eye movements.

Efferent Pathways of the Cerebellar Nuclei. The dentate and interposed nuclei

Figure 9.10: Diagram of the *Non-thalamic* Outputs of the Dentate and Interposed Nuclei. On the left are the descending projections (only the dentate is shown here, for clarity: the interpositus connections are the same) to the nucleus reticularis tegmenti pontis (NRTP), the V and VIth cranial nerve nuclei, the raphe nuclei (R), reticular formation (RF) and inferior olive (IO). On the right are the ascending projections to the red nucleus. The dentate innervates the rostral and anterior caudal parts of the nucleus: the interposed (mainly the anterior region) projects to the caudal part. The rostral (or parvocellular (PC)) red nucleus projects to the ipsilateral inferior olive. In the monkey, the caudal (or magnocellular (MC)) region is the origin of the rubrospinal tract, whereas in the cat, some contribution is also seen from the rostral part (dotted line) (from Gilman, Bloedel and Lechtenberg, 1981; Figure 3, p. 125, and Figure 4, p. 128)

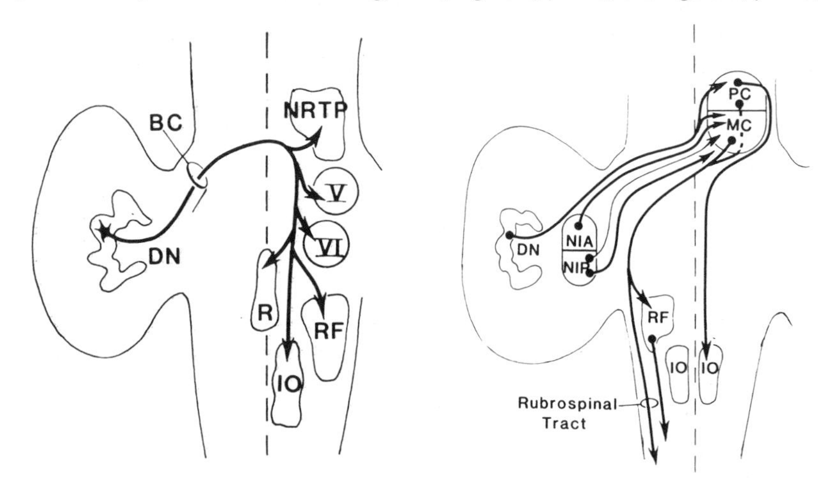

are the main output nuclei of the cerebellum. They send fibres rostrally to innervate the red nucleus and thalamus, and caudally to innervate nuclei in the pontine and medullary reticular formation. These connections provide the cerebellum with access to the following descending motor tracts: reticulospinal, rubrospinal and, via the thalamocortical projection, the corticospinal tract.

Efferents of the dentate and interposed nuclei leave the cerebellum in the ipsilateral superior peduncle, and cross to the contralateral midbrain tegmentum. At this point the tract divides into an ascending and descending component, with many of the fibres branching to provide axons in both components (Figure 9.10). The descending axons innervate nuclei in the pontine and medullary reticular formation. The ascending axons innervate the contralateral red nucleus, the VPL_o and VL_c thalamus and the neighbouring thalamic nucleus X.

The two portions of the red nucleus receive fibres from different parts of the cerebellar nuclei. The anterior interposed nucleus sends fibres somatotopically to the caudal, or magnocellular, portion of the red nucleus. Thus this part of the cerebellar nuclei has direct access to the rubrospinal tract which has its origin in the large cells of the red nucleus. There is also a small projection from posterior interposed to the medial portion of the red nucleus. Fibres from the dentate end in the rostral, or parvocellular part of the red nucleus. In man this region is much larger than the magnocellular part. From here, fibres descend in the ventral tegmental tract to innervate the ipsilateral inferior olive. The olivo-cerebellar projection completes a cerebello-rubro-olivo-cerebellar loop.

The cerebellar projection to the thalamus travels through the red nucleus, and hence is difficult to study in isolation. The main areas of termination (in VPL_0 and VL_c) are in those parts of the thalamus which project strongly to the motor cortex. This is the basis for a direct cerebellar influence on motor cortical activity. Within the VPL_0 and VL_c, there is considerable overlap between terminals of dentate and interposed fibres, which contrasts with the very precise pattern of projection from thalamus to motor cortex (that is, VPL_0 to caudal area 4 in which distal muscles are represented; VL_c to rostral area 4 where proximal muscles are represented). Despite this, it has been suggested that the outputs of the interposed and dentate nuclei may be directed preferentially to proximal and distal muscles respectively.

The projection from cerebellar nuclei to the thalamic nucleus X has only been discovered relatively recently. It forms the basis for a possible cerebellar influence on the activity of lateral area 6, the premotor area. At present, there have been no physiological studies on this pathway.

The efferent connections of the fastigial nuclei are much more complex than those of the dentate and interposed nuclei, although they constitute a much smaller proportion of the cerebellar output. Briefly, in the monkey, descending fibres from the fastigial nucleus leave the cerebellum in the ipsilateral inferior and superior peduncles and terminate in the lateral and medial vestibular nuclei; crossed fibres leave mainly in the hook bundle of Russell, which bends around the superior peduncle and terminate in the lateral and medial vestibular nuclei and in the reticular formation. Ascending fastigial afferents terminate in the superior colliculus and ventrolateral thalamus.

In addition to these connections with other areas of the brain, there are also projections from all three cerebellar nuclei back onto the cerebellar cortex. In general, these connections reciprocate the corticonuclear projections and end as mossy fibres. An exception to this rule in primates is that from the dentate, which sends fibres to the vermis as well as to the lateral areas of the cerebellar cortex (Brodal, 1981; Gilman *et al.*, 1981).

Electrophysiological Studies of the Cerebellum

Background Levels of Discharge

At rest in either decerebrate, lightly anaesthetised or awake animals, Purkinje and nuclear cells both discharge at high rates of between 40 and 80Hz. This is far higher, for example than the maximum 20Hz seen in pyramidal tract cells of relaxed animals and suggests that the cerebellum, even at rest, may be providing some tonic input to other structures. Indeed, removal of the cerebellum can produce a loss of muscle tone (see below) which would be consistent with this idea. The Purkinje cells and probably the nuclear cells as well are driven to fire at such high frequencies by sustained mossy fibre input. Recordings of complex spike activity show that climbing fibre input is infrequent, having a discharge rate of only about 1Hz. Even during active movement, when Purkinje rates may exceed 400Hz, climbing fibre input remains at approximately the same rate.

In general, nuclear and Purkinje cells change firing frequency in phase with each other. For example, increases in Purkinje cell discharge usually are accompanied by increases in nuclear cell firing. This is consistent with the anatomical findings that input to cerebellum projects both to the intracerebellar nuclei and to the cortex. Since the output of cerebellar cortex is then mainly to the nuclei, it may be that the cortical discharge *modifies* the nuclear firing. One idea is that the cortical output may *sculpt* the firing pattern of the nuclei. This might happen either in terms of space or time. Thus the inhibition from the cerebellar cortex may allow only certain groups of nuclear cells to respond to mossy fibre input. Alternatively, the inhibitory cortical output might cut short the nuclear discharge and sharpen the temporal firing pattern of the cells. Unfortunately there is rather little data on this point. Most studies on movement-related firing of cerebellar cells have concentrated on nuclear, rather than Purkinje cells. In fact, the relationship between cortical and nuclear firing is one of the least understood features of cerebellar physiology.

Afferent Input to the Cerebellar Cortex

Almost all classes of peripheral afferent input reach the cerebellar cortex, with a major input from group Ib and II fibres. Purkinje cells respond best to phasic rather than sustained stimuli and, like some cells in the motor cortex, are sometimes responsive preferentially to particular directions of joint rotation.

In addition to spinal and cortical afferents, the cerebellum receives inputs from visual, auditory and vestibular systems. These are distributed mainly to the flocculonodular lobe, and parts of the vermis in the posterior lobe. They will not be dealt with in this chapter. One of their main roles is in the control of the vestibulo-ocular reflex.

Afferent Input to the Cerebellar Nuclei

As noted anatomically, many, if not all, of the afferents to the cerebellar cortex also project directly or via collaterals to the cerebellar nuclei. The threshold for excitation is similar to that needed to evoke activity in the cerebellar cortex. Many inputs to the nuclei produce an initial excitation followed by a longer period of inhibition. The inhibition is abolished by cooling the cerebellar cortex, suggesting that it is produced by activity in Purkinje cells excited by the same afferent systems as project to the nuclei. This sequence of excitation and inhibition is again in agreement with the hypothesis that impulses in the cerebellar cortex can function as an inhibitory 'side-loop' which modulates transmission in the direct nuclear pathway (Brooks and Thach, 1981; Harvey, Porter and Rawson, 1979).

Figure 9.11: Differences in Cerebellar Somatotopy in Awake and Anaesthetised Cats. On the right are the three classical somatotopic representations produced by electrical stimulation of various parts of the body in anaesthetised animals. The study on the bottom left appears to confirm this. The diagram shows the size and distribution of responses in the anterior lobe of an anaesthetised animal after electrical stimulation of the saphenous nerve in the right leg. The potentials are localised in the ipsilateral 'leg' area of the anterior lobe. In contrast, the upper diagram shows the distribution and size of evoked responses after natural stimulation of the forepaw in an unanaesthetised decerebrate cat. Similar findings of widespread, bilateral activity also have been seen in unanaesthetised intact animals (from Combs, 1954; Figures 8 and 9, with permission)

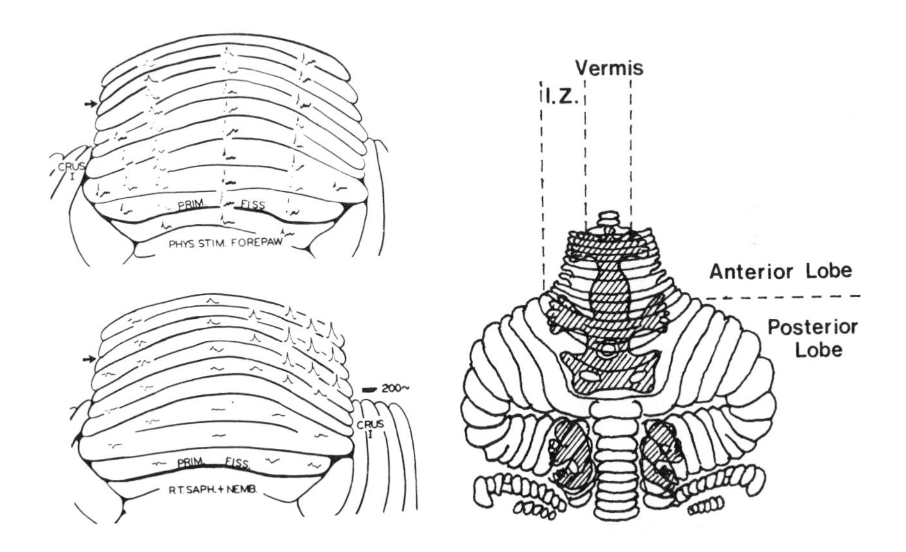

Somatotopy (Figure 9.11)

Anaesthetised. Afferent input from all parts of the body reaches the cerebellar cortex. In anaesthetised preparations, two complete somatotopic representations of the body can be found, one in the anterior lobe and one in the paramedian lobule and pyramis of the posterior lobe. The representations are inverted with respect to one another, and have axial structures represented in the midline. In the middle part of the posterior lobe vermis are found visual and auditory inputs. Overlying the peripheral somatotopic maps is another input from cerebral cortex. Stimulation of somatosensory cortex gives a topographic map on the surface of the cerebellar cortex in which stimulation of, for example, leg areas of the cerebral cortex evokes activity in the same cerebellar regions as receive direct spinal input from the leg.

Awake. Somatotopy disappears in awake animals. Stimulation of one arm can give responses over a wide area of cerebellar cortex. Such divergence can be seen at the level of single cells. For example, peripheral stimuli may arrive at a Purkinje cell via mossy fibres from one part of the body and via climbing fibres from another. The same can be seen when cerebral cortex is stimulated.

The reason for the differences is that in anaesthetised animals, only the most direct pathways remain active. The more indirect routes, containing more synapses, are more likely to be affected by anaesthesia. Thus input from the leg may arrive directly in the dorsal spinocerebellar tract. However, information from the same part of the body may also reach cerebellar cortex in ventral spinocerebellar tract and via relays in the olive and precerebellar reticular nuclei. These latter pathways, which are active in the awake animal probably are responsible for obscuring the somatotopy seen in anaesthetised preparations (Bloedel and Courville, 1981; Gilman *et al.*, 1981).

Activity of the Cerebellar Neurones During Active Movement

Many of the neurones of the cerebellar cortex and deep nuclei change their discharge rate in association with movements of the body. Unlike the cells of the motor cortex, cerebellar neurones have no direct connections with either spinal or bulbar motoneurones, so that their discharge is less likely to be related to specific patterns of muscle activity. Instead, their discharge may provide insight into how a different level of the CNS is involved in the control of movement.

Initiation and Control of Rapid Movements. The advantage of examining rapid movements is that they have a well-defined onset which can be used to define the time relationships of cerebellar firing to muscle activity. In the first experiments, neurones were recorded in the cerebellar cortex, dentate and interposed nuclei, together with neurones in motor cortex and muscle, while monkeys made rapid wrist movements in response to a visual signal. There was considerable overlap in the time at which a change in neuronal firing was seen at each of these sites. However, on average, the earliest changes were seen in the dentate nucleus and motor cortex. This was followed by activity in interpositus and muscle (Figure 9.12). In those studies in which Purkinje cells were recorded, they increased firing at about the same time as the nuclear cells to which they projected.

The anatomical projections show that the major output of dentate is to motor cortex, and it is possible that dentate activity may contribute to onset of activity in motor cortical cells. In fact, in some studies, dentate cells were found to start firing an average of 15 to 30ms *before* motor cortex. Two further lines of evidence support the idea that dentate activity may be involved in initiation of movement via motor cortical projections:

1. Electrical stimulation of the dentate can produce activity in motor cortex at latencies as short as 4ms, and even produce overt movement of the body in intact animals;
2. Cooling of the dentate increases a monkey's reaction time in similar tasks and delays the occurrence of motor cortical discharge. The motor cortical delay was not due to a removal of background facilitation provided by dentate since cooling

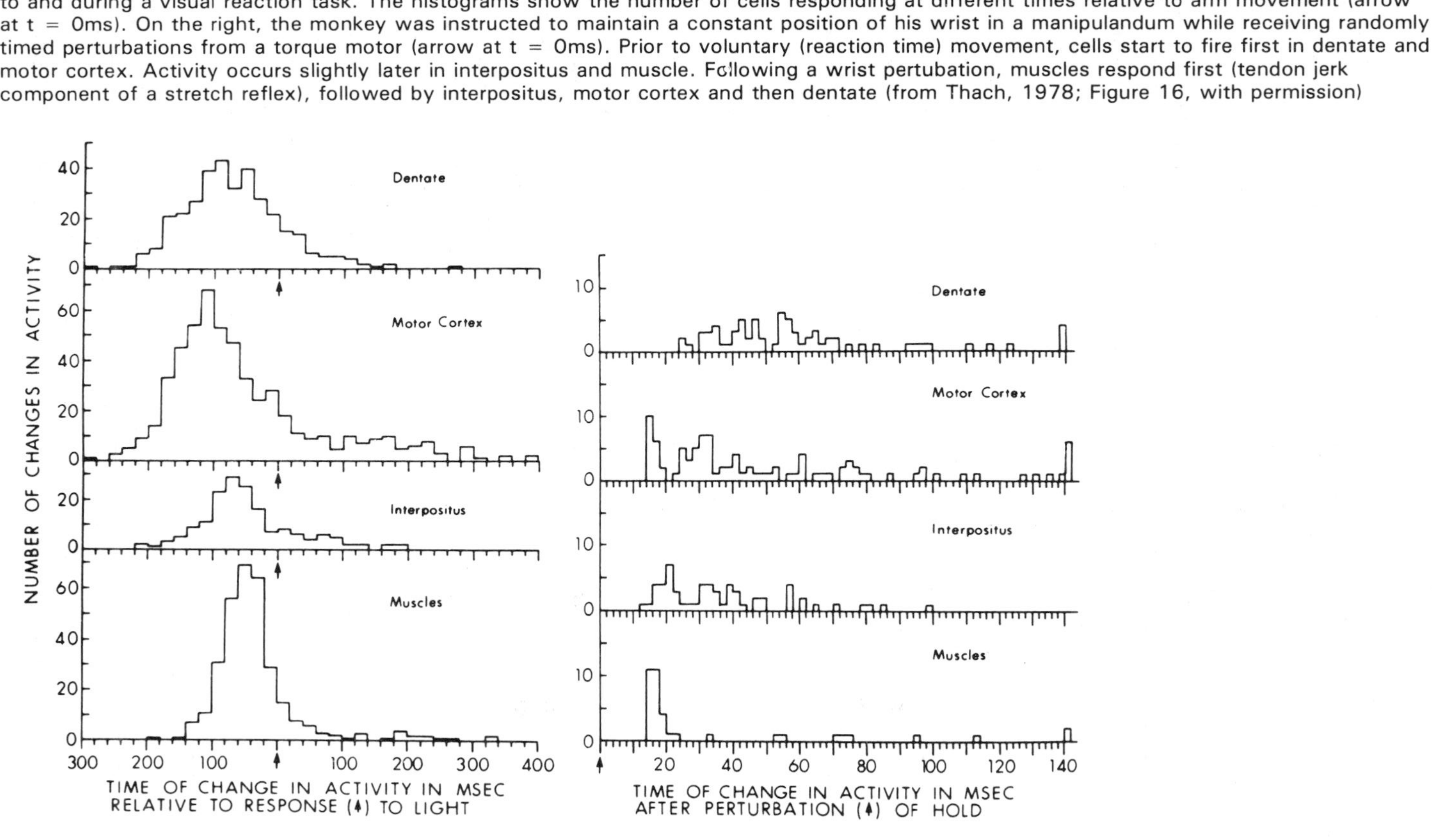

Figure 9.12: Timing of Changes in Cell Activity in Dentate and Interpositus Nuclei, Motor Cortex and Muscle. On the left are responses prior to and during a visual reaction task. The histograms show the number of cells responding at different times relative to arm movement (arrow at t = 0ms). On the right, the monkey was instructed to maintain a constant position of his wrist in a manipulandum while receiving randomly timed perturbations from a torque motor (arrow at t = 0ms). Prior to voluntary (reaction time) movement, cells start to fire first in dentate and motor cortex. Activity occurs slightly later in interpositus and muscle. Following a wrist pertubation, muscles respond first (tendon jerk component of a stretch reflex), followed by interpositus, motor cortex and then dentate (from Thach, 1978; Figure 16, with permission)

did not appreciably change the resting discharge in motor cortex cells.

Other experiments suggest that the situation is more complicated. For example, cooling the ventrolateral thalamus, which is thought to be the synaptic relay from dentate to motor cortex, does not produce any delay in reaction time. Similarly, cooling of motor cortex itself has little effect on reaction time in the monkey in comparison with dentate cooling. The movements are much smaller but they appear to begin at short latency. A possible explanation of this is that output from dentate can in certain circumstances produce movement via subcortical pathways (for example, its projection to the reticular formation), although this may not be the normal mode of activity (see section on electrical stimulation below).

Whatever the final explanation of these experiments, it is clear that the timing of changes in changes in the dentate nucleus allow for the possibility that it may have some role in the initiation of movements. In contrast, the late onset of change in interpositus activity suggests that it can only play a role in the control of ongoing movements. However, such studies provide little information as to the role of Purkinje cell discharge in producing these changes in nuclear activity. For example, since Purkinje cells tend to change frequency at the same time as the nuclear cells, their function may be to switch off nuclear discharge, rather than to initiate it.

A final piece of evidence supports these timing arguments. Rather than have a monkey flex its elbow in response to a target light, monkeys were trained to hold their wrist in a constant position around the middle of joint movement (Figure 9.12, right; Figure 9.13). At random intervals, the elbow was suddenly extended or flexed by application of an external force. The monkeys could be instructed in advance on how to react to the disturbance. On some occasions the animal was told to assist the disturbance by moving the elbow rapidly in the direction of the force. In others, the monkey had to assist the perturbation.

The EMG response consisted of an early reflex component followed by a later voluntary response which was dependent upon the prior instruction. In these circumstances, interpositus neurones changed their firing pattern before dentate neurones and were related to the direction of the perturbation. In contrast, dentate firing was more closely related to the direction of the voluntary response and could occur even if there was no preceding elbow perturbation. Again, dentate neurones appeared to play a role in the initiation of voluntary movement, whereas interpositus neurones may have been more involved in the automatic control of such movements.

These ideas are consistent with the anatomical relationships of the two nuclei. The dentate nucleus receives its major input from two sources, the overlying lateral cerebellar cortex and from direct pontine afferents. Both regions are supplied primarily by cerebral cortical fibres, perhaps providing input concerned with the command to move. The interpositus nucleus receives inputs from intermediate cerebellar cortex and from spinocerebellar and pontine afferents. It is thus receiving both peripheral and central input, and it has been

Figure 9.13: Responses of Human Muscle and Neurones in the Dentate Nucleus of a Monkey to Perturbations of the Arm and Their Dependence on Prior Instruction. A: Muscle responses recorded from a *human* subject. The task was to maintain a constant wrist position against a torque motor which could apply random supination/pronation perturbations. Imposed pronation stretched the biceps. In some trials the subject had to oppose the displacement and in others he had to assist it. The responses from biceps are shown in two ways: in the upper traces is an average rectified EMG, in the lower traces are raster displays of the responses recorded on each single trial. On the left, the biceps was stretched and the instruction was to oppose the stretch. In the EMG there is a stretch reflex starting about 20ms after stretch (short-latency response) with a later component just visible (long-latency response) starting at about 50ms. Following these reflex responses is the voluntary muscle activity at 100ms. On the right, the same stretch was given but the instruction was to assist the disturbance. In this case, the stretch reflex response remains, but the voluntary activity disappears. B: Responses of a dentate neurone recorded in an awake *monkey* are shown in the same way. In this instance, the monkey held a push/pull handle within a target zone. The handle could be moved by a motor so as to extend triceps (stim. away) by moving away from the monkey. When the instruction calls for the monkey to assist the perturbation by pushing (left), there is short-latency increase in dentate activity at about 10ms. When instructed to resist the disturbance (right, pull), this activity is not present and there is a reduction in dentate firing. Corresponding position records from the handle are shown below (from Strick, 1978; Figure 1, with permission)

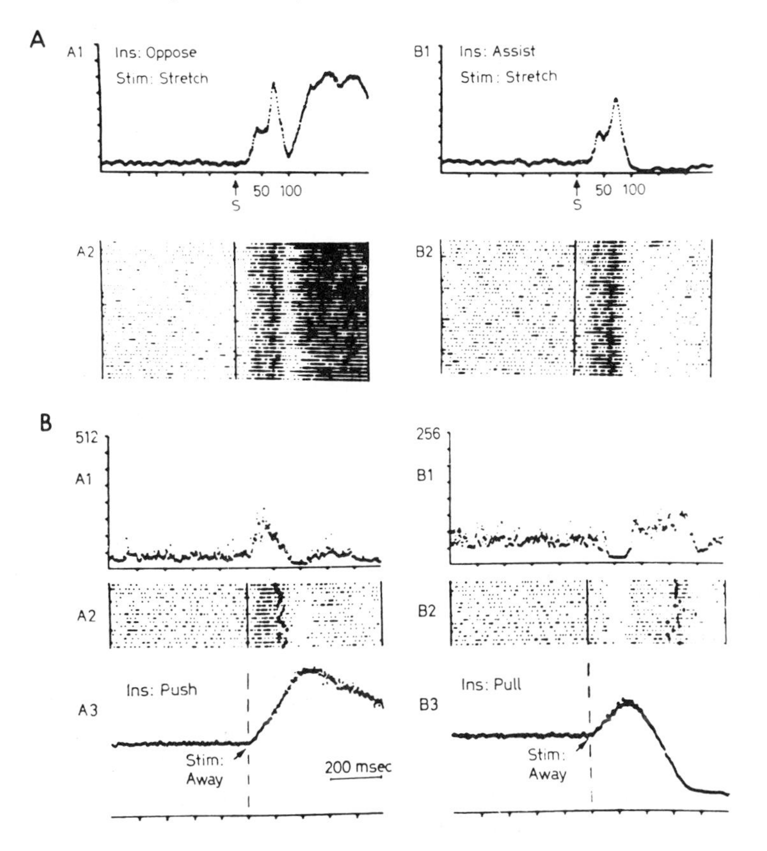

suggested that such a combination would allow comparison of the central command to move with the present limb position and provide a possible basis on which to make corrective adjustments to the movement in progress (Brooks and Thach, 1981; Spidalieri, Busby and Lamarre, 1983).

Relation of Discharge to Movement Parameters. Timing considerations alone provide us with no information on which aspect of movement is related to the cerebellar discharge. Since the cerebellum has no direct connexions with spinal or bulbar motoneurones, it is possible that analysis of its neuronal discharge could reveal aspects of 'higher level' control of muscle activity. For example, is the discharge related to muscle force, direction of movement, or overall pattern of muscle activity?

Experiments have been devised to dissociate several of these parameters. Most recordings have been made from neurones in the cerebellar nuclei rather than Purkinje cells. In one experiment, monkeys were trained to make wrist flexion and extension movements through three different joint positions, starting in extension, going into flexion, and back again. The monkey maintained the wrist at each position until signalled to move to the next by a light signal. Figure 9.14 shows the details. The force opposing the movements could be changed so that movement from one position to the next could be made either with activity in flexor muscles or activity in extensor muscles. For example, the wrist flexor movement from A to B would be made with activity in flexor muscles if there was a load opposing flexion. In contrast, if the load assisted flexion, the movement would be made with activity in the antagonist extensor muscles which would act as a brake during the movement. The experiment could therefore distinguish three different variables: the pattern of muscle activity, the joint position at the wrist and the direction of the next intended movement.

Interpositus neurones discharged well during tonic holding of the wrist at each intermediate position. The activity of any one cell was often related to the amount of either flexor or extensor muscle activity. In effect, the interpositus neurones behaved very like spinal alphamotoneurones, being related to the pattern of muscle activity, and (because of the length-tension relationship of muscle) to the angle of the joint (Figure 9.15). This relationship was disrupted transiently during the phasic transition from one wrist position to the next. In these circumstances, a neurone which showed, for example, a strong *extensor* coupled activity in the holding task would fire most strongly when the monkey made rapid intended *flexor* movements. The reason for this dissociation is not known. However, it is a warning not to take a correlation between neuronal activity and muscle activity as evidence for a causal connexion between the two. It is quite possible that interpositus activity is directly related not to muscle activity but to activity in spinal interneuronal systems or even gamma motoneurones.

The activity of dentate neurones was more difficult to classify. There were approximately equal numbers of dentate neurones related to each of the tested

Figure 9.14: Paradigm for Dissociating the Parameters of Joint Position (JPOS), Pattern of Muscle Activity (MPAT) and Direction of the Next Intended Movement (DSET). The monkey was instructed to move his wrist through extension and flexion to three different positions, A, B, C, moving first from A to B, then B to C, and then in the reverse direction C to B, and so on (top right). At the middle position (B), the next movement could be either in the extension or flexion direction (see DSET). To dissociate the pattern of muscle activity, the whole sequence was carried out with a load which either opposed flexion or opposed extension. Thus, with a load opposing flexion, the cycle of flexion movements would require flexor muscle activity, whereas if there was an extensor load, the same flexion movements would require braking activity in the extensor muscles (MPAT) (from Thach, 1978; Figure 1, with permission)

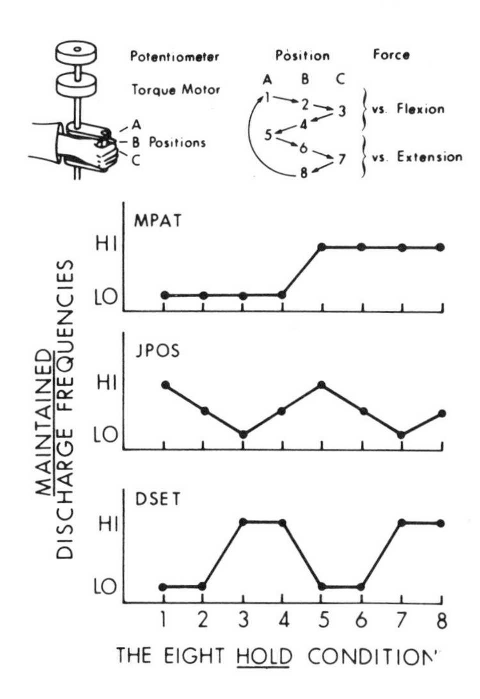

variables: pattern of muscle activity, joint position and direction of the next movement. Almost all dentate neurones discharged prior to the voluntary movement from one position to the next, and could possibly have formed part of a movement command signal. The nature of that command signal might then be slightly different for each class of dentate neurones.

The main conclusion is that dentate neurones are not linked specifically to any one feature of the task, and are better envisaged as performing some role in 'high level' control of movement. In contrast, interpositus neurones discharge more closely in relation to the amount of peripheral muscle activity, and may therefore be seen as being closer to the final motor output than dentate neurones (Brooks and Thach, 1981; Thach, 1978).

Figure 9.15: Relationship of Neural Discharge and Muscle Activity to Direction of Next Intended Movement, Joint Position and Muscle Activity in the Paradigm Illustrated in Figure 9.14. The graphs show the degree of fit to each of these three variables with a perfect fit being given a score of 1. On the top row, the pattern of muscle activity obviously is best related to the pattern expected (all the points fall along the horizontal plane in the first two graphs), and not related at all to DSET (all values less than 0.2) and only slightly related to JPOS (all values less than 0.3). (The fact that the muscle activity does not always score 1.0 on its relation to MPAT is because the actual muscle activity needed to maintain a wrist position depends upon the length of the muscle and hence on the joint angle). Interpositus neurones behave very similarly to muscle. They score reasonably high on their relation to MPAT and JPOS, but are poorly related to DSET (all values less than 0.3). Dentate neurones behave differently. On each variable there are some neurones out of the total which score reasonably well. Dentate neurones may therefore be related to either MPAT, JPOS or DSET (from Thach, 1978; Figures 4, 13, and 15, with permission)

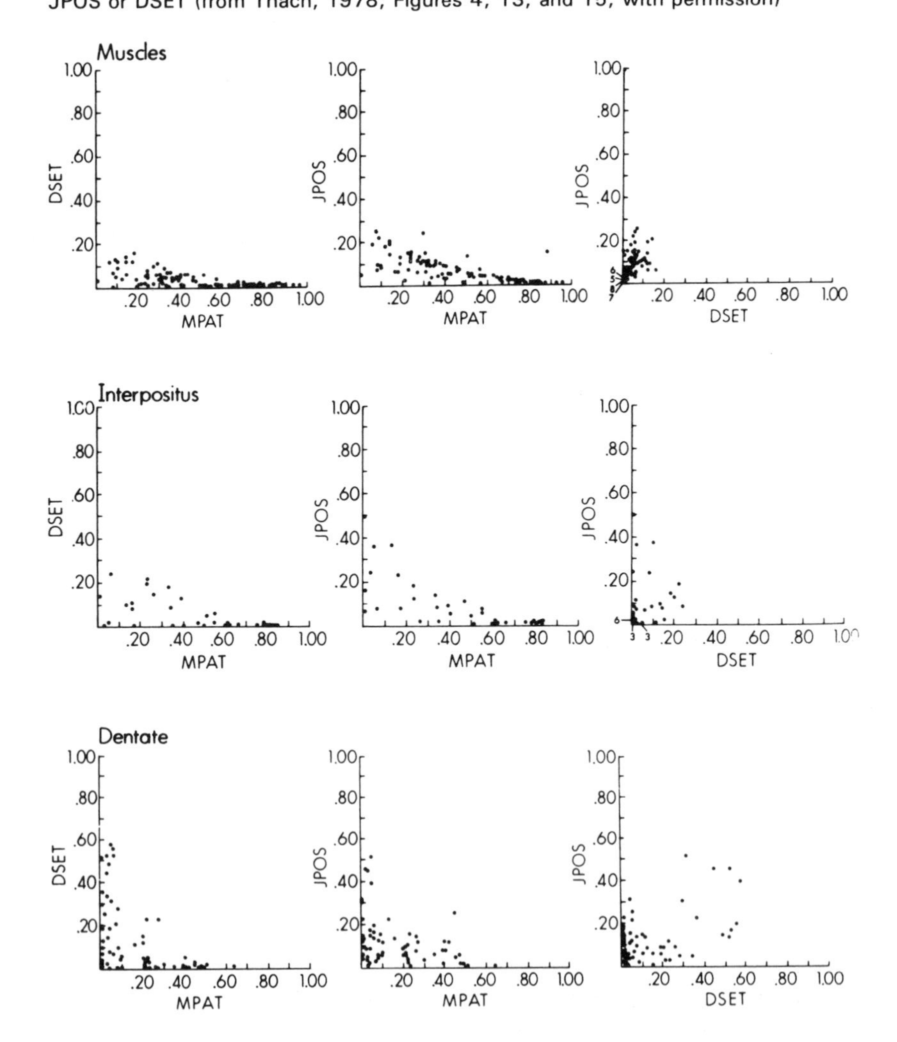

Relationship of Cerebellar Firing to Activity in Antagonist Muscles. So far, it has been considered that cerebellar activity might be related to the control of the prime mover, or agonist, muscle. However, there is no reason why the cerebellum might not be involved in control of antagonist muscles. Evidence supporting this possibility comes from studies in which monkeys were trained to co-contract antagonist muscles, rather than to use them reciprocally as in the usual type of experiment. Recordings from identified Purkinje cells in the posterior part of the anterior lobe revealed that about 70 per cent of cells decreased their firing rate during an isometric grip of thumb and forefinger which involved co-contraction of antagonist wrist flexor and extensor muscles in the forearm. In contrast, almost all of these cells showed a reciprocal activation to isometric flexion or extension of the wrist which did not involve co-contraction of antagonists. When recordings were made from dentate or interpositus nuclei, many cells were found to behave in a similar fashion except that these neurones increased their firing during co-contraction, rather than decreasing it as did the Purkinje cells.

The hypothesis is that one role of Purkinje cell discharge is to control activity of antagonist muscles. In situations requiring reciprocal activation of antagonist muscles, increased Purkinje cell discharge might cause inhibition of antagonists. Conversely, in situations in which co-contraction is required, a decrease in Purkinje cell firing might cause an increase in antagonist activity.

The interest in these studies is the finding that there is a reciprocal relationship between the discharge of Purkinje and nuclear cells. Although it is not proven, it may be that the decrease in Purkinje output during co-contraction reduces inhibition of nuclear cells and causes an increase in nuclear firing. Additionally, it has also been found that other inhibitory interneurones of the cerebellar cortex actually increase their discharge during co-contraction. Such discharge may then be directly responsible for inhibition of Purkinje cells (Brooks and Thach, 1981; Wetts, Kalaska and Smith, 1985).

Activity of Cerebellar Neurones during Slow Movements. Slow movements do not have a precisely defined time of onset, so that it is not possible to investigate time relations between cerebellar activity and movement. However, it is possible to investigate the activity of cells during the movement. Unlike the reciprocal activity of many cells which is seen during rapid flexion and extension tasks, almost all task-related cells in dentate and interpositus discharge equally well in slow movements in either direction. They are said to be *bidirectional* neurones. Many of the cells are the same as those which have unidirectional discharge during rapid movements, or in response to torque perturbations. In monkeys that have been examined in this way, there was no antagonist co-contraction or excessive postural muscle activity which might explain such relatively nonspecific responses, so that other explanations have been sought.

One hypothesis which has been put forward is that the activity of these cerebellar neurones is more related to the activity of gamma motoneurones than alpha-motoneurones. It is further suggested that in slow tracking tasks, gamma discharge to both agonist and antagonist muscles is increased, making the spindles more sensitive in both muscles. In accordance with this idea, spindle recordings in the monkeys did show an increase in spindle activity from both agonist and antagonist muscles at the onset of movement. The increased antagonist activity persisted even during slight unloading of the muscle. At least in this instance, gamma discharge to the antagonist increased even though there was no sign of activity in the extrafusal muscle.

This is an example of alpha–gamma dissociation which was discussed in more detail in Chapter 6. However, it should be remembered that alpha–gamma dissociation has not yet been documented in man, particularly during similar slow movements of the wrist. The idea that the cerebellum may be involved in control of gamma motoneurones is supported by evidence from lesion studies (see below) (Brooks and Thach, 1981).

Role of Cerebellum in Learning New Movements

In addition to studying the discharge of cerebellar neurones during active movements, several investigators have examined the activity of the cerebellum during the process of learning a new task. The essential feature of these experiments is the emphasis that they place on the role of the climbing fibres.

During many types of active movement, there is little modulation in the firing rate of climbing fibres. Complex spikes appear in the Purkinje cells at a steady slow rate of between 0.5 to 2Hz. Despite the low level of activity in the climbing fibre system, the climbing fibres are essential for proper operation of the cerebellum. Removal of the inferior olive, with subsequent degeneration of the climbing fibres produces symptoms in experimental animals that are indistinguishable from removal of the cerebellum itself. Climbing fibre input, although infrequent, and with little immediate effect on Purkinje cell discharge rates, must have some vital long-term influence on cerebellar function.

Limb Movements. The most likely possibility is that occurrence of a complex spike can increase or decrease the effectiveness of other synaptic inputs onto the Purkinje cell. This does not occur immediately, but takes place gradually over many trials as part of a 'learning process'. There is one experiment involving limb movements that supports this idea (Figure 9.16). Recordings were made from Purkinje cells, so that the frequency of both complex (climbing fibre) and simple (parallel fibre) spikes could be monitored while a monkey maintained a constant wrist position. At unpredictable intervals, a known flexor or extensor load was applied alternately and the monkey had to restore the original wrist position as rapidly as possible.

When the task had been learned, the appearance of complex spikes was irregular and infrequent as usual. At this stage, the extensor load was suddenly changed to a new value while the flexor load remained the same. The wrist movements after the introduction of the new extensor load were initially

Figure 9.16: Possible Role of Climbing Fibre Discharges in Learning New Limb Movements. A monkey was trained to hold his wrist in a constant stationary position in the face of alternating flexor and extensor loads (lasting 300ms) which moved the handle away from target. With practice, when the same flexor and extensor loads were given each time, the monkey was able to restore the target position rapidly and repeatedly. The left part of the figure shows the performance of the monkey in terms of records of wrist position. The traces should be read from left to right and top to bottom. The first trace shows the response to the known flexor load. There is a pause and then a known extensor load is given, and so on: both sets of responses are very repeatable. At the arrow, the extensor load was changed from 300g to 450g. The monkey performed badly at first in response to this novel load, although his responses to the flexor load (which had not been changed) were relatively unaffected. Over several trials, performance and repeatability increased with the new extensor load. Recordings from a single Purkinje cell in the cerebellar cortex are shown on the right for a similar series of trials. The discharge is shown as raster displays of simple spike (small dots) and complex spike (large dots) activity. This Purkinje cell was load-related, with a higher simple spike frequency during the flexor response. At the arrow, the known extensor load was unexpectedly changed from 300 to 450g, while the flexor load remained constant. After the load change, there was a greatly increased frequency of complex spike activity in the extensor task, although the corresponding activity in the flexor task was unchanged. This increased activity persisted for about 70 trials and was accompanied by changes in simple spike activity (from Gilbert and Thach, 1977; Figures 1 and 3, with permission)

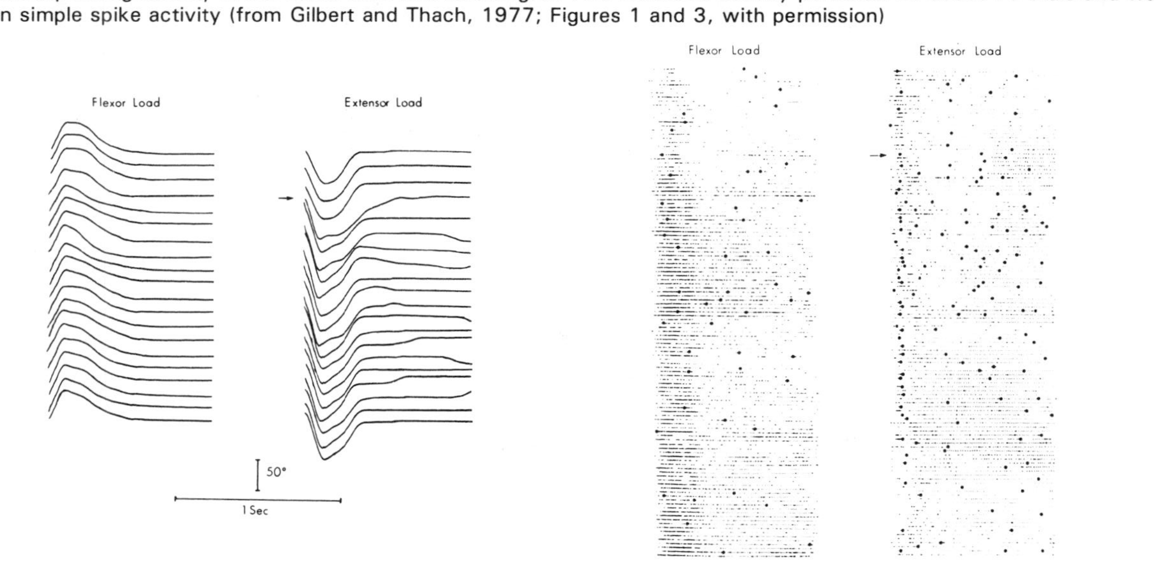

Figure 9.17: Changes in the Response of a Purkinje Cell to Mossy Fibre Input After a Period of Conjunctive Mossy and Climbing Fibre Stimulation. This page: Experimental arrangement. A Purkinje cell was recorded in the flocculus and three of its inputs were stimulated (S_1, S_2, S_3) in the ipsilateral and contralateral vestibular nuclei (iVN and cVN) and in the dorsal cap of the inferior olive (DC). Mossy fibres (MF), climbing fibres (CF) and cerebellar cortical interneurones are shown.
Opposite: Records of simple spike Purkinje cell discharge. A shows part of a continuous recording period. Over the period marked by dashed lines (b and c) stimuli were given at 2Hz to the ipsilateral vestibular nerve. Each stimulus produced, in b, a short duration increase in spike discharge which is shown more clearly in B. B is a time histogram of the number of simple spikes seen in the Purkinje cell following a single stimulus to the vestibular nerve (at arrow). On this short time scale, an early increase in excitability of the Pukinje cell is quite obvious. During the period marked by the solid horizontal line in A, there was conjunctive stimulation of the ipsilateral vestibular nerve at 20Hz and the olive at 4Hz. This gave a greatly increased rate of simple spike firing. After this period, the spike rate decreased again but became much more variable. The response to the ipsilateral vestibular nerve stimulation at 2Hz over time period c was now considerably diminished (see histogram C) (from Ito, Sakurai and Tongroach, 1982; Figures 1 and 5 with permission)

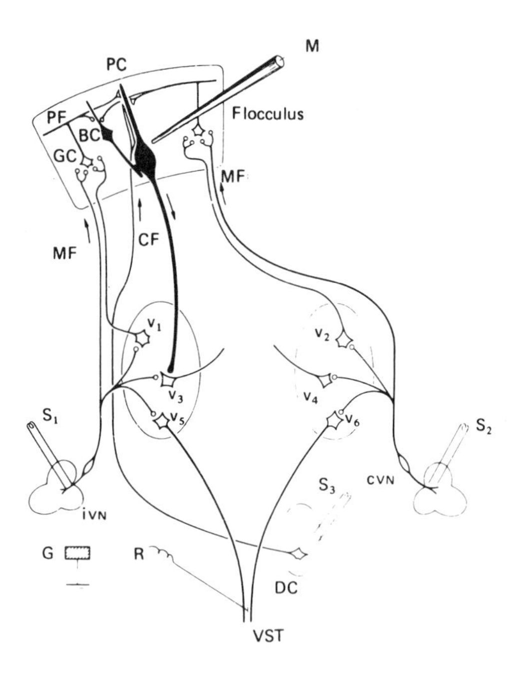

inaccurate, but improved over the succeeding 40 to 60 trials as the monkey learned the new task. After the appearance of the new load, the number of complex spikes following the extensor, but not the flexor, load increased, and the number of simple spikes decreased. As performance improved, the number of complex spikes gradually reverted back to its original value, but the simple spikes remained at the new lower frequency. The suggestion is that first, the cerebellum played a role in adapting the movement to the novel task. Second,

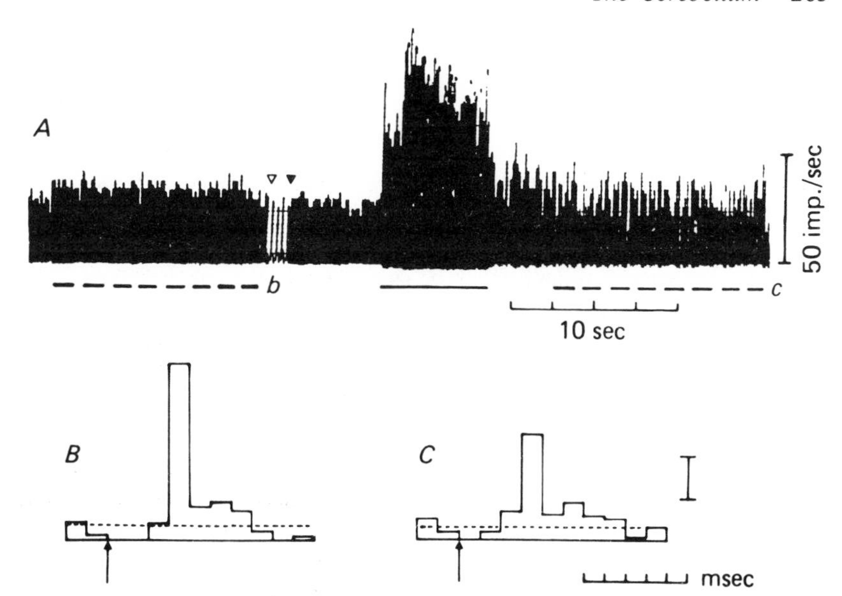

that the climbing fibre activity accompanying the novel stimulus modified the response of the Purkinje cells to mossy fibre input, seen in this experiment as a decrease in the number of simple spikes that the cell discharged (Gilbert and Thach, 1977).

Over short time periods, climbing fibre discharge can change the responsiveness of Purkinje cells to other inputs. Natural stimulation of cat forelimb normally evokes a simple spike discharge in some Purkinje cells. If the stimulus is given 20 to 70ms after a complex spike in the cell, the simple spike response is augmented.

Vestibular-ocular Reflex. Evidence for more long-term changes in cerebellar circuitry comes from other sources. Experiments on the vestibular–ocular reflex have provided strong arguments in favour of a role of cerebellum in learning processes, and of the ability of climbing fibre inputs to modulate the responsiveness of Purkinje cells over long periods.

The vestibular-ocular reflex in the external ocular muscles maintains the axis of the eyeball in a constant position with respect to the visual scene during movements of the head. An afferent signal from the vestibular apparatus is analysed to provide information about head position and velocity. This is then used to stabilise the eyes on an object in space by producing movements of the eyeball which oppose and compensate for the change in head position. The reflex is extremely important for stabilised vision, and can be modulated in various ways. For example, if reversing prisms are worn as spectacles so that the directions of left and right are reversed in the horizontal plane, the reflex gradually *reverses* in sign over four to five days in order to compensate for the spectacles. In animal experiments, this compensation is abolished by ablation of the floccular–nodular lobe of the cerebellum.

Evidence for the theory of climbing fibre function comes from a remarkable experiment performed on the vestibular input to cat cerebellum (Figure 9.17). A Purkinje cell was recorded while its input from three separate sources was stimulated: (1) mossy fibre input from the left vestibular nerve; (2) mossy fibre input from the right vestibular nerve; (3) climbing fibre input from the olive. If the climbing fibre input was stimulated at 4Hz at the same time as stimuli were given to the mossy fibre input from the right (but not the left) vestibular nerve, then later tests showed a specific influence on the effectiveness of the right mossy fibre input. Stimulation of the right vestibular nerve alone (without any concomitant climbing fibre input) evoked a *reduced* simple spike discharge in the Purkinje cell. The response to stimulation of the left vestibular nerve was unaffected. This is excellent evidence that climbing fibre inputs can produce changes in the effectiveness of specific categories of inputs to Purkinje cells. The mechanism is unknown. One possibility is that a climbing fibre discharge may modify the effectiveness of parallel fibre–Purkinje cell synapses. Alternatively, the Purkinje cell collateral to the glomerulus may influence transmission at the mossy fibre–granule cell junction (Ito, 1984; Ito, Sakurai and Tongroach, 1982).

Visually Triggered Movements. The role of the cerebellum can also be seen during visually triggered movements. It takes many days before animals first learn to associate the movement of the arm with the onset of a visual signal, and many more days before they learn to react as rapidly as possible to the appearance of the signal. Recordings in motor cortex show that the gradual decrease of reaction time which is seen in such experiments is associated with the appearance of surface positive–depth negative field potentials. These potentials represent EPSP activity in superficial dendrites of pyramidal cells, generated by thalamo-cortical input to these layers, and they increase in size as reaction time becomes shorter. This input is produced by activity in cerebellar efferent fibres, and is abolished by removal of the cerebellum. This procedure also increases the latency of reaction time. It is thought that this is another reflection of the role of the cerebellum in the initiation of movements via motor cortex (Sasaki *et al.*, 1982).

Effects of Cerebellar Lesions

Cerebellectomy in Cats and Dogs

Complete cerebellectomy has quite different results when performed in cats and dogs than it has in monkeys and man. Most studies have been performed on decerebrate cats in which the heightened level of background extensor activity produces a useful basis for the evaluation of changes in extensor tone. Cerebellectomy in these animals increases the extensor tone still further in all

Figure 9.18: Extreme Extensor Rigidity in All Four Limbs and Neck of a Decerebrate Cat in Which the Anterior Lobe of the Cerebellum Also Had Been Inactivated by Ligation of the Internal Carotoid Arteries (from Pollock and Davis, 1927; Figure 3, with permission).

four limbs and neck (opisthotonus). The effect can be produced by ablation of the anterior lobe alone, indicating that this is the principal structure responsible for increasing the hypertonus. (See Figure 9.18, in which ligation of the internal carotoid arteries has produced both decerebration and inactivation of the anterior lobe of the cerebellum.)

The mechanisms responsible for this effect are as follows. The anterior lobe of the cat and dog exerts a strong and direct inhibitory influence over the lateral vestibular nucleus of Deiters, which is the origin of the vestibulospinal tract. Removal of this influence results in excessive activity in the vestibulospinal projection, which is directed principally towards extensor muscles. Consistent with this, unilateral section of the vestibular nerve reduces rigidity on the same side. Any remaining increased tone in extensor muscles is thought to be mediated by reticulospinal projections from the cerebellar nuclei.

Rigidity in decerebellectomised cats is also abolished by ablation of the fastigial nucleus. The reason for this is that besides the direct projection from anterior lobe to Deiter's nucleus, there is an indirect projection via the fastigial nucleus. This pathway exerts an influence on activity in Deiter's nucleus which is opposite to that produced by the direct projection from Purkinje cells of the cerebellar cortex. Hence, removal of this excitatory influence can reduce activity in the lateral vestibulospinal tract and decrease extensor tonus.

The rigidity does not depend upon stretch reflex activity since it is not abolished (as is decerebrate rigidity) by sectioning the dorsal roots of the spinal cord. It is said to be an *alpha* rigidity, produced by direct activation of alpha-motoneurones in spinal cord from the vestibulospinal tract. Indeed, recordings from gamma-motoneurones show that in contrast to the excitation of alpha-motoneurones, there is a decrease in fusimotor activity and a decrease in sensitivity of the monosynaptic stretch reflex similar to that seen in primates (see below). This decrease in fusimotor activity is due to the removal of tonic input from all three cerebellar nuclei. Some days after cerebellar ablation (the time depends on the species of animal involved) there is a gradual decrease in extensor tonus, accompanied by a gradual increase in spindle sensitivity to normal values.

Cerebellectomy in Primates

The situation is quite different in primates. Cerebellectomy produces a *hypotonia* of all muscles which recovers only very slowly, if at all. More pronounced deficits are in the rate and force of voluntary muscle contractions, together with tremor (especially if the cerebellar nuclei are damaged) (see sections below on pathophysiology). Limbs tend to be held in flexion and there is nystagmus. All deficits are seen especially well if active movements are made which require postural support. As in cats and dogs, there is a general decrease in gamma drive to muscle spindles, which is probably responsible for hypotonia.

The reason for the difference between primates and non-primates probably lies in the difference in output projections from the anterior lobe vermis and in the predominance of the cerebellar hemispheres in primates. In sub-primates, the direct projection from the anterior lobe to the lateral vestibular nucleus of Deiter's is strong, whereas it is not so well developed in primates. In contrast, the output of the hemispheres is very well developed in primates, and it is removal of this part of the cerebellum which is responsible for major symptoms of dysmetria seen in primates (Gilman *et al.*, 1981; Kornhauser, Bromberg and Gilman, 1982).

Reversible Cooling of Cerebellar Nuclei

Studies of the effects of lesioning the output of the lateral cerebellum have been made using the technique of reversible cooling of the dentate nucleus. It has already been mentioned that cooling delays the onset of movement in a reaction time paradigm, suggesting a role of the dentate nucleus in initiation of movement. In addition, deficits have been noted in three other types of task during combined cooling of dentate nucleus and interpositus: (1) rapid arm movements from one position to another; (2) responses to perturbations introduced while the monkey tries to hold a constant position of limb; and (3) slow movements.

If normal monkeys are required to move their forearm repetitively from flexion to extension between two mechanical end stops, each movement is very rapid and little time is spent at each turnaround point. However, when the dentate nucleus is cooled, the amount of time spent at each end point is greatly increased, although the speed of movement is relatively unaffected. The critical deficit is a delay in onset of antagonist muscle activity needed to provide rapid reversal of movement. This is compounded by prolongation of the burst of activity in the agonist muscle.

In a slightly different task, monkeys were trained to perform single, self-terminated movements of the arm as rapidly as possible. This type of movement requires the animal to halt at the end point, without the aid of a mechanical stop.

Figure 9.19: Tracings of Rapid Pronation-Supination in a Patient with Unilateral Cerebellar Symptoms. The movements of the affected arm (below) were for a time regular though slower and of smaller amplitude than normal, but grew irregular and the arm became more or less fixed in supination. Muscular contractions appeared to spread aimlessly over the whole of the affected limb including unwanted shoulder movements and finger movements. One patient commented to Holmes, 'The movements of my left arm are done subconsciously, but I have to think out each movement of the right (affected) arm. I come to a dead stop on turning and I have to think before I start again' (from Holmes, 1939; Figure 16, with permission)

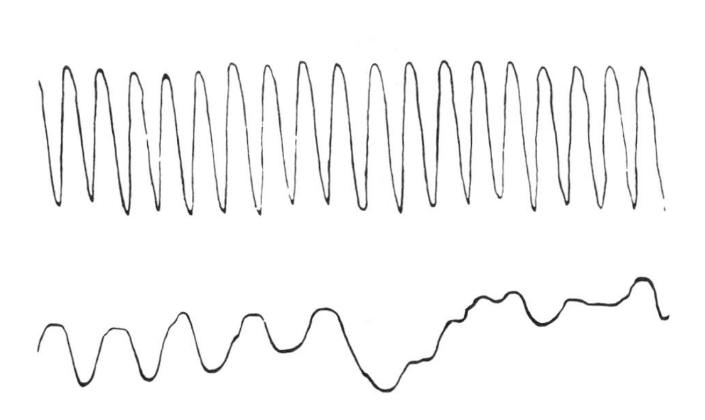

Under these conditions, the movements also tended to be a little slower than normal during nuclear cooling (see records of rapid alternating movements of cerebellar patient of Figure 9.19).

In normal, well practised monkeys, short-lasting perturbations evoke a stretch reflex in the agonist muscle (which helps to restore the wrist towards its start position), followed by a burst of EMG activity in the antagonist muscle. This latter helps to damp out the mechanical oscillations of the wrist which follow a pulsatile torque disturbance. Such antagonist activity is not a reflex response to muscle stretch since it occurs before the wrist begins to return towards its original start position (when stretch of the antagonist muscle will occur) (see Figure 9.20). It is thought to be a 'set-dependent' response triggered by the initial stretch of the agonist muscle. That is, a response which occurs only when the monkey has been trained to expect a pulsatile wrist displacement. If the 'set' of the monkey is changed, by giving long-lasting wrist displacements, this antagonist muscle response is no longer present. Instead, there is an EMG burst in the agonist muscle at the same latency, which follows the usual stretch reflex response in that muscle. Under these conditions, antagonist activity is not needed to damp out oscillations. It is more important that extra agonist activity be called up to restore the original wrist position against the maintained load.

The responses to pulsatile disturbances have been investigated in some detail

Figure 9.20: Cerebellar Control of 'Set-dependent' EMG Activity in Antagonist Muscles, Possibly Via a Precentral Motor Cortical Pathway. A monkey was trained to hold its elbow in a fixed position and at random times it received brief, pulsatile disturbances which stretched either biceps or triceps and took the elbow away from the target position. The monkey was well trained to restore the position as rapidly as possible. The upper part on this page shows the normal EMG responses in triceps (middle traces) and biceps (bottom traces) to this type of disturbance (thin lines), together with superimposed responses recorded during cooling of interposed and dentate nuclei (thick lines). When the biceps is stretched, there is a stretch reflex response in biceps. This helps to restore the original elbow position. It is followed by a burst of activity in triceps which in the normal condition (thin lines) begins before the elbow has started to return to the control position. This predictive activity in triceps helps damp out the oscillation in the elbow position produced by the disturbance. The same happens when triceps is stretched, but now it is biceps which shows the 'predictive' activity. The lower part on the opposite page shows another experiment in the same monkey, in which a biceps-linked precentral cell also was recorded at the same time as the EMG activity. In the control state, stretch of biceps generates a reflex EMG plus short-latency excitation of the precentral unit which occurs at about the same time. Stretch of triceps (C) generates the usual 'predictive' responses in biceps which are preceded by a burst of activity in the precentral cell (see dotted line at t = 50ms). Cooling the cerebellar nuclei does not affect either the initial stretch reflex EMG or precentral response to biceps stretch (B). However, when triceps is stretched, onset of biceps EMG activity is delayed as is the precentral unit activity D. The precentral activity no longer precedes this burst of biceps activity, as it did in the intact animal (from Vilis and Hore, 1980; Figures 4 and 5, with permission)

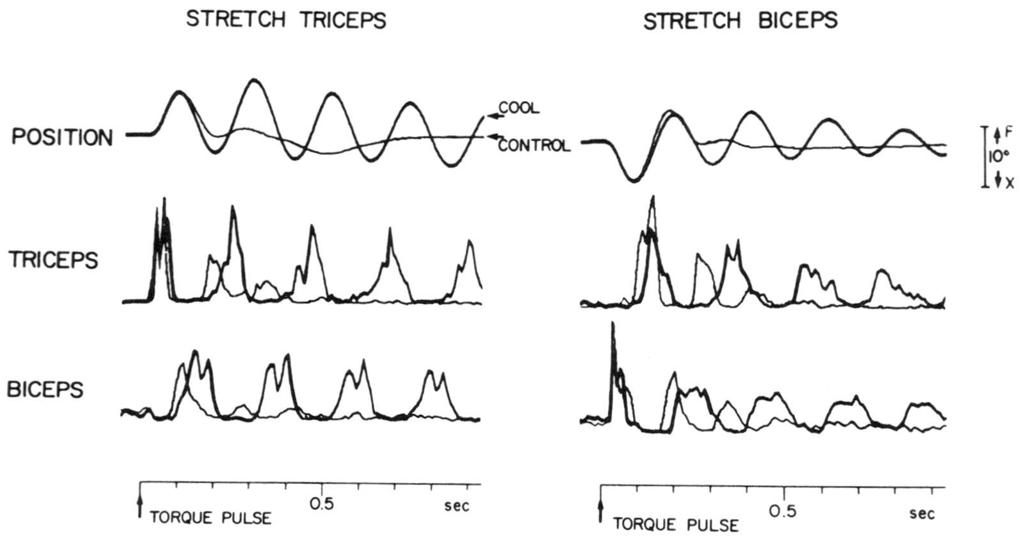

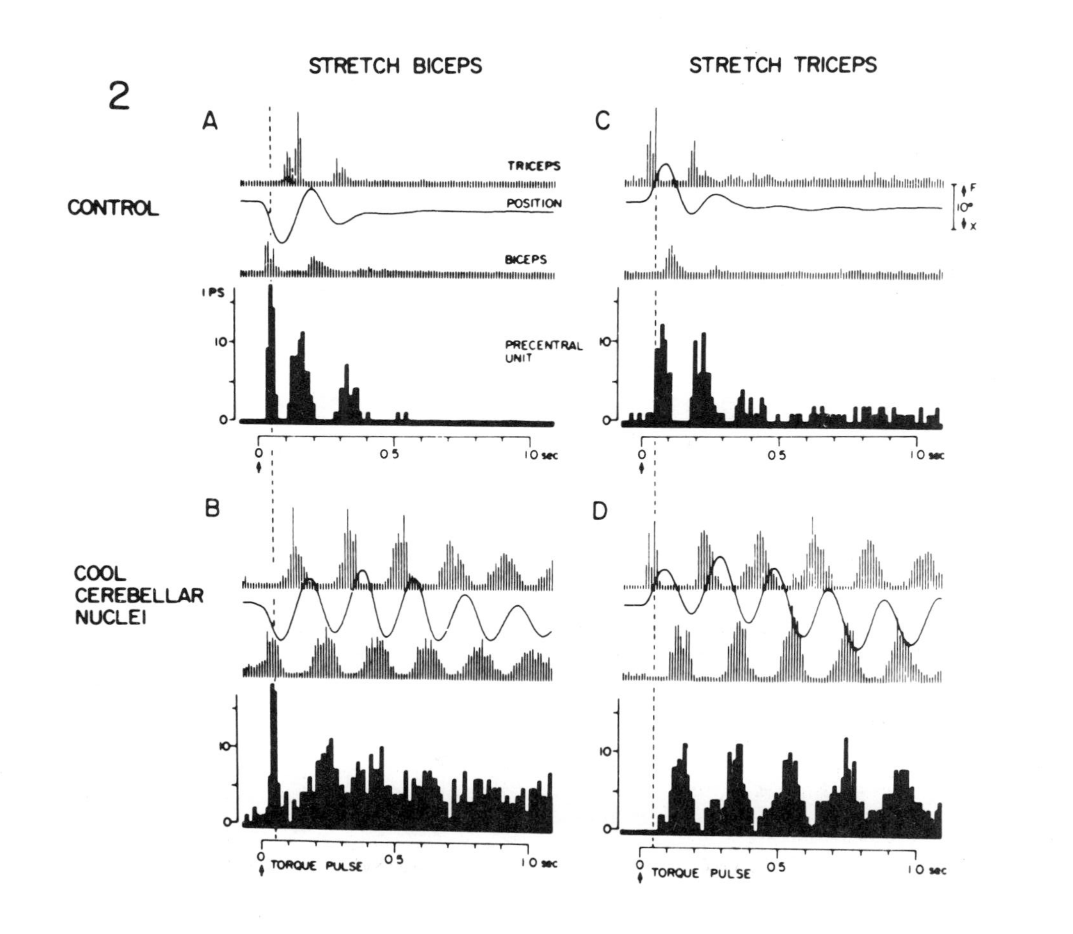
2
STRETCH BICEPS
STRETCH TRICEPS
A
C
B
D
CONTROL
COOL
CEREBELLAR
NUCLEI
TRICEPS
POSITION
BICEPS
PRECENTRAL
UNIT
IPS
10
0
0.5
1.0 sec
F
10°
X
TORQUE PULSE

(see Figure 9.19). Cooling of the cerebellar dentate and interposed nuclei has no effect on normal stretch reflex responses of the agonist muscle. It is the 'set-dependent' EMG activity of the antagonist which disappears, to be replaced by later activity driven purely by stretch reflex pathways. This stretch reflex antagonist activity now occurs later than usual, and no longer helps to dampen out oscillations in limb movement. It may even make them worse.

In the normal animal, 'set-dependent' activity in the antagonist muscle is preceded by changes in motor cortical activity in neurones linked to the contraction of that muscle. During cerebellar cooling, this linked precentral activity also becomes delayed and no longer occurs before the EMG onset. It appears that the cerebellum may drive these 'set-dependent' EMG responses via its connections with precentral motor cortex. Such observations after cooling suggest that the tremor seen after cerebellar lesions may be due to the operation of stretch reflex mechanisms which are no longer supplemented by the predictive 'set-dependent' activity of the cerebellum.

During nuclear cooling, smooth movements can no longer be performed at constant speeds. Instead, they break up into jerky, intermittent segments (see Figure 9.22, for a human example), which in a monkey have a predominant frequency of about 3–5Hz. Records from motor cortical neurones also reveal a predominant oscillation at a frequency of about 3Hz. Possible reasons for this are discussed below in connection with pathophysiological studies in man. In monkey, the deficit does not occur with cooling of interpositus alone. The dentate nucleus must also be involved (Brooks and Thach, 1981; Flament, Vilis and Hore, 1984; Hore and Vilis, 1984).

Electrical Stimulation of the Cerebellum

Consistent with the effect of anterior lobe ablation, stimulation of the anterior lobe can decrease extensor rigidity in decerebrate cats. Stimulation probably produces an increase in activity in the inhibitory output of Purkinje cells to the vestibular nuclei. This results in a decrease of activity of vestibulospinal pathways to the extensor muscles.

Few studies have been performed on primates. One general finding emerges. Stimulation of the cerebellar nuclei, especially the interposed nuclei, produces flexion, particularly at proximal joints. In the monkey, this effect is not mediated via the nucleo-cortical projection or via the red nucleus. It has been suggested that projections to the reticular formation are involved. In the cat, the situation is different. The similar effects of stimulation are mediated via the red nucleus (Schultz, Montgomery and Marini, 1979).

Studies of Cerebellar Dysfunction in Man

Clinical Symptoms

There are two main categories of symptoms. One involving muscle tone, the other involving co-ordination of muscle activity. The classic descriptions of these deficits were summarised by Holmes in 1939.

Hypotonus. This is a reduction in the resistance of the limbs to passive movements. However, it may be difficult to detect in most patients, especially if accompanied by inco-ordination. It is probably due to a reduction of background spindle discharge, which reduces the stretch reflex contribution to muscle tone.

Inco-ordination. This is known by clinicians as *ataxia*. Patients have difficulties in specifying the extent, velocity and rhythm of muscle contractions, and in co-ordinating the activity of several muscle groups involved in a task (*asynergia*).

If the legs and trunk are affected this leads to ataxia of gait. Patients are unsteady and stabilise themselves by walking with a broad-based gait. At first sight they often appear as if they were drunk. Walking heel-to-toe is very difficult, if not impossible. Running the heel of one leg down the shin of the other while seated or lying (heel-shin test) also is very poorly performed.

Ataxia of the arms has various manifestations. Normally, smooth movements such as reaching out to touch an object become decomposed into several steps. If the patient is asked to point from one stationary target to another (for instance, finger–nose test), the movement overshoots the target (hypermetria) and has to be corrected by additional movements. Towards the end of such movements there is a coarse tremor of the arm that is not present at rest or when the subject ordinarily sustains a fixed posture of the arms. It is known as *intention* tremor, to indicate its dependence on the intent of the subject to perform a specific pointing task. Ataxia in the arms also is seen as an inability of patients to execute movements of constant force and rhythm. This is best appreciated by asking subjects to tap the thigh alternately with the dorsum and palm of one hand. The rhythm is poor, and the force of each tap varies through the sequence (Figure 9.19) (dysdiadochokinesia).

Cerebellar deficits can be seen in bulbar muscles. This leads to slurring of speech and to a coarse nystagmus of the eyes, especially prominent at the extremes of gaze.

Not all patients with cerebellar lesions have all these symptoms. However, most of them have some form of gait ataxia. As a general rule, if symptoms predominate in the trunk and legs, then the lesion is likely to be near the midline. Damage to the hemisphere produces more obvious symptoms in the arms. If only one side is affected, the symptoms are unilateral. Speech disturbances occur only with bilateral damage.

Pure lesions of the cerebellum are very rare and often are due to the presence of a tumour. Three types of syndrome are recognised, which affect function in the three major subdivisions of the cerebellum:

1. Floccular–nodular syndrome produced in children by growth of a special type of glioma, the medulloblastoma, usually begins in the nodulus. The symptoms include nystagmus and a difficulty in maintaining equilibrium which gives rise

to a broad-based gait. If the body is supported, there is no obvious ataxia of the legs in the heel-shin test;
2. Anterior lobe damage sometimes is seen in alcoholic patients. This is limited to the anterior and vermal regions, and results in a broad-based gait and ataxia of leg movements. The arms are relatively spared;
3. Neocerebellar damage results in ataxia of the arms as well as difficulties in gait, plus hypotonia.

A cerebellar 'picture' can be seen in other conditions which affect the afferent or efferent connections of the cerebellum. The spinocerebellar ataxias are a group of diseases with a number of common clinical features. The most common, although still exceedingly rare, is Friedreich's ataxia. The disease produces degeneration of the spinocerebellar tracts and dorsal columns, with some involvement of the pyramidal tract and dorsal roots. Multiple sclerosis also may produce a cerebellar picture by interfering with cerebellar connections.

Physiological Investigations of Cerebellar Patients

Stretch Reflexes. Detection of hypotonus in cerebellar disease is difficult because passive manipulation of the joints in totally relaxed normal individuals meets with relatively little resistance to movement. Indeed, such normal subjects would be described as hypotonic on clinical criteria (see Chapter 6). Another sign of cerebellar hypotonia is the pendulous knee jerk (Chapter 6), which is seen following the initial extension of the leg after tapping the patellar tendon. Undamped pendular decay of the movement is due to lack of resistance by the thigh muscles to passive displacement of the knee joint.

As with manual assessment of muscle tone, pendular knee jerks are in practice rather difficult to spot. Other objective methods for detecting hypotonia are said to be more accurate. The tonic vibration reflex (TVR, Chapter 6) is decreased in cerebellar disease as can be seen clearly in patients with unilateral symptoms. The reflex builds up steadily on the non-affected side but is much smaller or absent on the other. Lack of a TVR cannot be due to the lack of gamma drive to the muscle spindles which has been demonstrated in cerebellectomised animals. This is because vibration is such a powerful stimulus that it can 'drive' the spindle output in the absence of any tonic fusimotor input. During vibration the spindle discharge is thus believed to be maximal. Cerebellar hypotonia detected in this way must therefore be due to central mechanisms. One possibility is that cerebellar disease produces increased presynaptic inhibition of spindle Ia afferent fibres, although there are as yet no comparable animal data that indicate that this would be possible.

Long-latency stretch reflexes have not been examined extensively in cerebellar disease, but in the cases so far examined the reflex is of normal latency but decreased in size and duration, consistent with the clinical observations of hypotonia.

Ballistic Limb Movements. These are voluntary movements of a limb made as rapidly as possible from one target position to another. The movement is made quite freely without end stops in the apparatus, so that the subject has to reach and maintain the final end position himself. The movements are completed in 200ms or less, which is so short that the subject is unable to make any voluntary corrections to the ongoing movement. Corrections can only be made on completion of the initial movement. The movements are made with a relatively stereotyped pattern of muscle activity. This begins with a burst of activity in the agonist muscle which lasts for 100ms or so. This is followed by a burst of activity in the antagonist and then by a small second burst in the agonist. The agonist activity provides the initial impulsive force for the movement, while the antagonist activity brakes the motion. The function of the second agonist burst of activity is unknown.

This pattern of EMG activity has been studied extensively since it represents a package of nervous commands which is relatively immune to modification during execution. Because of this, the EMG activity in a ballistic movement is said to be 'preprogrammed', meaning that the size and timing of the agonist and antagonist bursts of activity are decided upon somewhere in the CNS prior to the initiation of the movement. That such a pattern of activity can be seen in patients with no afferent input from their limbs (due to peripheral neuropathies of various causes), emphasises their central origin.

There are a number of abnormalities in the ballistic movements of patients with cerebellar disease, all of which involve errors in the timing of muscle contraction.

First, if the movements are made in response to a visual reaction stimulus, the reaction time is some 70ms or so longer than normal. This is the same result as seen in monkeys with cooling probes implanted in their cerebellar nuclei, and illustrates the importance of the cerebellum in initiating movements.

Second, the movements tend to overshoot their target, as expected from the clinical observations of hypermetria. However, unlike the performance of monkeys in such experiments (see section on cooling cerebellar nuclei), the movements were not slower than normal. The difference is probably due to the way that monkeys were trained in the task. They only achieved reward if the movements were accurate, whereas the patients were allowed to overshoot the target. Since large movements are usually executed faster than smaller ones, it is possible that the hypermetric movements of the patients would have been slow if compared with movements of the same size made by normal individuals.

Third, consistent with the overshoot in the movement, the duration of the first burst of agonist EMG activity is longer in patients with cerebellar disease than normal. This presumably results in increased muscle force and contributes to hypermetria. Interestingly, the patients seem unable to compensate for this disturbance by aiming, for example, at an imaginary target slightly nearer than that indicated.

Fourth, cerebellar patients show a disorder of reciprocal inhibition of antagonist muscles. In normal subjects, if there is any prior activity in the

Figure 9.21: Effect of Ingestion of Alcohol Sufficient to Cause Cerebellar Ataxy on the EMG Patterns Underlying Fast Ballistic Thumb Flexion Movements. Records in (a) were taken in the morning prior to alcohol. Those in (b) were obtained 1.5h after ingestion of 200ml of 50 per cent ethanol by mouth. The movements were made in response to a visual signal to move, given 50ms before the sweep in (a), and 100ms before in (b). Note: (1) the delay in onset of movement; (2) the overshoot of target and inability to hold the end position; (3) prolongation of EMG bursts in the agonist (flexor pollicis longus, F.P.L.) and antagonist (extensor pollicis longus, E.P.L.) muscle. Vertical calibration is 20°, 250°/s and 100μV (from Marsden, Merton, Morton, *et al.*, 1977; Figure 11.6, with permission)

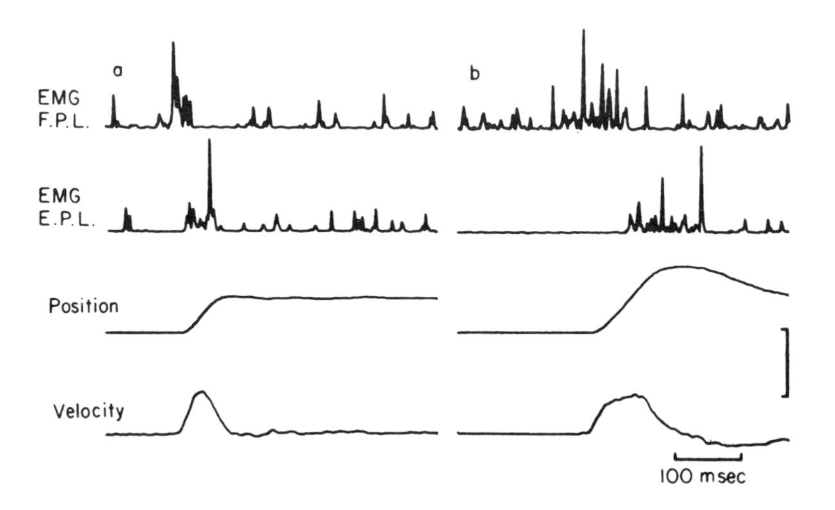

antagonist muscle before a ballistic movement, then this is silenced before activation of the agonist muscle. This prevents the antagonist from opposing the action of the agonist during initiation of the movement. In cerebellar disease, antagonist activity sometimes persists until after EMG onset in the agonist, a phenomenon never seen in normal subjects.

All four of these typical cerebellar deficits can also be seen in normal subjects rendered ataxic by ingestion of large quantities of alcohol (see Figure 9.21) (see also review by Brooks and Thach, 1981).

Slow Movements. A slow movement can be defined as a movement made at a speed which allows the subject time to make feedback corrections to the trajectory using visual or somaesthetic information. The movement is said to be under feedback control. Such slow movements are more variable than rapid ones and have for that reason been investigated in less detail.

The usual experiment is to ask subjects to track a visual target on a screen before them by moving the elbow or wrist (Figure 9.22). In the simplest task, subjects start at rest and track a target which begins to move at a steady speed, but at an unpredictable time, across the display screen. In normal

Figure 9.22: Slow Tracking Movements in Cerebellar Ataxia. Subjects had to track a slowly moving target by flexing their elbow at the appropriate rate. Both target and elbow position were displayed to the subjects on an oscilloscope screen. Targets were moved at 7.5, 15 or 30deg/s, and began to move at random times. The performance of a normal subject is shown in the top panel. Shortly after the target (T) moves, the subject makes a rapid 'catch-up' movement (trace D) and then continues to track the target perfectly well until the end of the movement. The error between target and actual movement (T-D) and the velocity of elbow movement (Ḋ) are shown, together with EMG records from biceps (Bi) and triceps (Tri). Performance of a cerebellar patient is shown in the bottom panel. The initial 'catch-up' movement is delayed and is bigger than necessary (overshoot). The smooth phase of tracking is broken up into a series of steps (from Beppu, Suda and Tanaka, 1984; Figures 2 and 3, with permission).

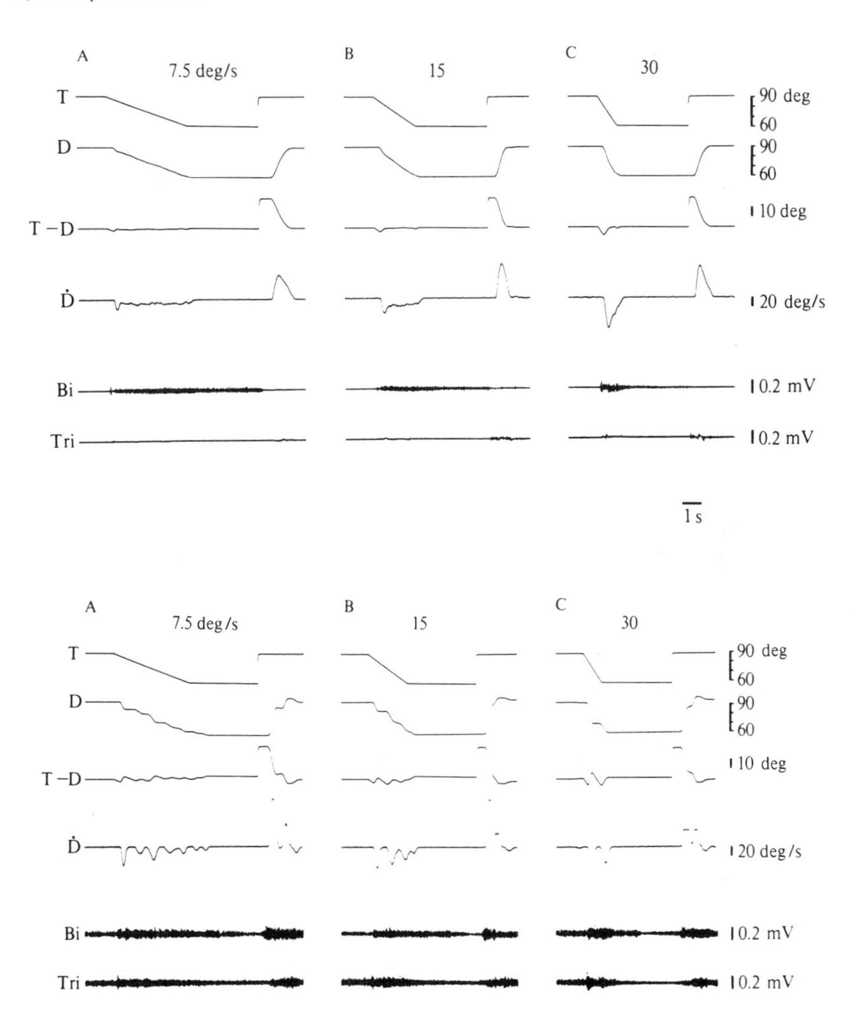

individuals, there is an initial pause before the subject moves due to the finite reaction time. The movement begins with an initial 'catch-up' phase which corrects for the lag introduced by the pause. After this, it continues steadily at a speed equal to that of the target.

In cerebellar disease, there is, as usual, an increase in the reaction time to the onset of movement. The 'catch-up' phase which follows is not scaled appropriately to the lag which has evolved. Thus, patients overshoot or undershoot the target during this phase. Finally, and most dramatically, patients cannot then switch into the smooth tracking mode seen in normal subjects. The rest of the tracking movement is made up in a series of 'saccadic' steps: rapid movements interrupted by pauses, as if the patient was making a series of corrections throughout the movement, rather than tracking the target smoothly. There are delays also in the termination of movement, as expected from problems in controlling antagonist muscle activation.

The discontinuities in smooth movement are not due to a superimposed tremor. The movement is quite clearly 'stepwise' (that is, including definite pauses) and not oscillating, as expected from tremor. Neither was the breakdown of attempted smooth movements into 'saccadic' tracking due to a fundamental inability of the movement control system in cerebellar patients to produce such movements. Patients could perform smooth, continuous movements if they moved their arms as they wanted in their own time, rather than following a moving visual target. The defect seems to be a problem in moving from the 'catch-up' (or truly saccadic) phase to the smooth pursuit phase of movement. Such discontinuous movements are similar to those seen in monkeys with cooling probes implanted into the interposed and dentate nuclei. Both have a similar frequency of from 1–3Hz (Beppu *et al.*, 1984).

Complex Movements about More Than One Joint. Such movements have yet to be studied using modern physiological techniques. The only observations in man of relevance are those made originally by Holmes (1939). He argued that complex movements involving more than one joint require close control over the timing and extent of postural muscle activity in order that the proximal parts of a limb remain fixed while the distal parts perform an action. If delays in execution and errors of force occur in both fixator and prime moving muscles, then they will become compounded such that performance of complex movements is more affected in cerebellar patients than simple movements. Holmes (1939) demonstrated that a simple movement of flexion/extension at the elbow was better performed when the elbow was supported than when free to move. He was of the opinion that no additional deficit other than those observed in simple movements needed to be invoked to explain the difficulties encountered in complex movements.

Posture. Unsteadiness of balance is one of the most common features of cerebellar disease. Accordingly, there have been a number of investigations into the possible abnormalities which might be responsible for this. Unfortunately, there are little data from animal experiments (there are no other *bipedal* animals) to compare with the description of human cerebellar patients. Because of this, the mechanisms of the reflexes described below and the precise role of the cerebellum in each of them is unknown.

One way of investigating unsteadiness of balance is to examine postural reflexes in standing subjects. In order to maintain balance, normal subjects employ input from three sources: vestibular, visual and somatosensory. The most rapidly acting of these are the somatosensory mechanisms, and it is these which have been analysed in most detail. Although visual and vestibular input can also have strong influences on posture, the latency of their action (over 200ms in the absence of any other inputs) is roughly twice that of the fastest somatosenory routes. However, they *modulate* somatosensory evoked activity at rather shorter latencies.

Postural disturbances can be given to the body in two ways; by moving the surface on which the subject is standing or by applying forces to the trunk or arms to move subjects standing on a stable base. If a subject stands on a platform which can be moved suddenly in the horizontal direction, then backwards displacement will result in a forward sway of the body. This stretches the calf muscles and evokes a stretch reflex which tends to oppose the direction of sway and compensate for the disturbance. In addition, other muscles of the leg are activated in a synergistic fashion. The hamstring muscles contract, again helping restore the equilibrium of the trunk. The timing of the contractions is very precise. The triceps surae contracts with a latency of 100–120ms, while the hamstrings always contracts some 10–12ms *later.* This order of contraction, distal to proximal, ensures that the stabilisation of the body begins at the base and works its way up the legs.

There are two mechanisms which could explain this timing: (1) the responses in each muscle might be driven only by input from muscle spindle afferents in the calf muscle. Horizontal motion of the platform might stretch the calf muscles before those in the thigh; (2) the responses in all muscles might be driven by the *same* afferents, in which case a central delay must be invoked to explain the later onset of activity in the most proximal muscles. It is likely that the latter is true since the delays in the same subject from trial to trial, and from subject to subject are relatively constant.

In cerebellar patients, these responses are disrupted. The order of muscle activation varies considerably from the stereotyped pattern seen in normal. In addition, the relative amount of activity in the calf and thigh muscles is also variable and different from normals. Thus, these reflex responses are affected in the same way as simple voluntary movements: there are errors in timing and force of muscular contraction.

Besides moving the platform horizontally, rotation about the axis of the ankle also has been used to investigate postural reflex function (Figure 9.23). Rotation of the supporting surface poses an interesting problem for the body. Consider

Figure 9.23: Postural Responses Evoked by Rotation of the Supporting Surface About the Ankles. The left panel shows the responses in a normal subject to repeated dorsiflexion of the foot. Upper traces on the right are the EMG records from triceps surae for the first four trials. On the first trial, there is a large reflex EMG response in triceps surae, which can be seen clearly in the integrated EMG records. This is an inappropriate destabilising response, which increases the backwards sway of the body (see records of body sway). By the fourth trial, the EMG response has adapted and body sway is much reduced. The right panel shows the triceps surae EMG responses from two ataxic cerebellar patients (A and B) in the same type of experiment. In these cases, there is no adaptation of the inappropriate reflex response, even over five to ten trials (from Nashner, 1976; Figures 3 and 6, with permission)

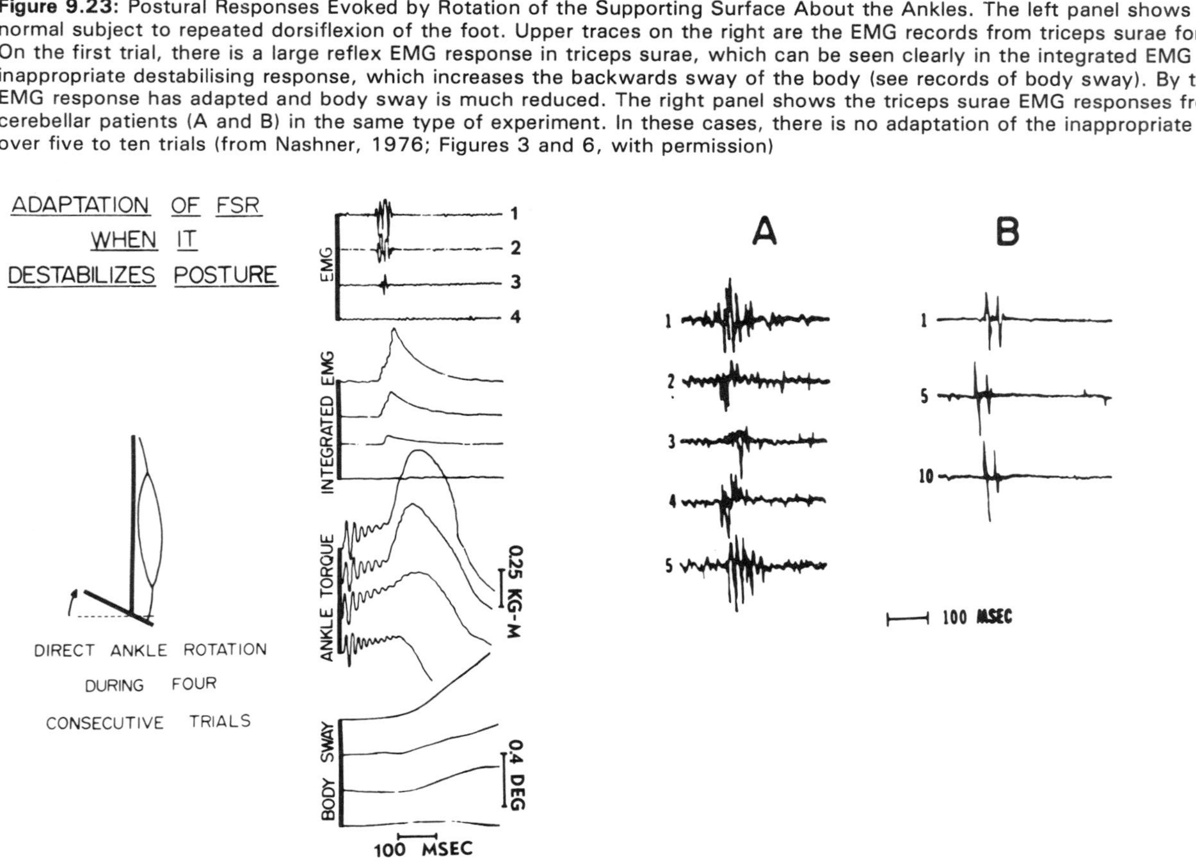

dorsiflexion of the foot in the standing subject. This will induce backwards sway of the trunk. However, the motion at the ankle will initially produce stretch of the triceps surae muscles. If this stretch evokes a stretch reflex, contraction of the triceps surae will tend to pull the subject backwards and hence further off balance, rather than correcting for the disturbance. In normal subjects, a strong triceps surae stretch reflex can be recorded on the first occasion that platform dorsiflexion is given. It is a destabilising reflex, which gradually adapts if the same stimulus is given again and again. The CNS appears to be able to control this inappropriate response. However, in cerebellar patients, the response never adapts. It seems that there is a failure to learn the new situation.

Platform rotation at the ankle also produces very strong responses in muscles shortened by the disturbance. Following platform dorsiflexion, the pretibial muscles show a very strong reflex contraction with a latency of about 130ms. This is the response which actually *compensates* for the body sway. Although the response is large, it is not clear which sensory receptors are involved. Some authors believe that proprioceptors in the legs detect the disturbance and provide the most important input for the reflex. Others suggest that the major input comes from the semicircular canals since the tibialis anterior response is absent in patients with bilateral labyrinthectomy.

Patients with atrophy of the cerebellar anterior lobe have enlarged responses in the tibialis anterior muscles following dorsiflexion of the ankles. These overcorrect for the displacement by bringing the body too far forward of the centre of foot pressure, and set up stretch reflexes in the triceps surae. Repetitive cycles of anterior tibial and calf muscle activity produce a postural tremor at about 3Hz. Other categories of cerebellar patient showed little abnormality in these tests, except for a group with Friedreich's ataxia. In this group, the responses were delayed. This suggests that the responses may have been mediated by the spinocerebellar or pyramidal tracts which are involved in this disease.

A different method of producing postural disturbances involves applying forces directly to the trunk or arms, and then measuring the reflex responses in the muscles of the legs. In normal subjects, a small pull forwards to the arm at the wrist can evoke EMG responses in the triceps surae muscles at a latency of 80–100ms. These responses are in such a direction to compensate for any induced sway of the body. However, the responses are actually initiated *before* the body sway reaches the leg. They appear to be leg muscle responses which are 'driven' by input from the moving arm. These reflexes are absent or decreased in cerebellar patients with postural instability.

Thus, unsteadiness of cerebellar patients is reflected in a number of abnormal postural reflexes. There may be a lack of organisation or timing of the muscle responses, lack of adaptation of responses to new situations or even absence of responses altogether. The mechanism of the responses produced by the different types of disturbance is unknown. It is likely that they involve some supraspinal mechanisms. If so, the cerebellar influence may act directly on the reflex pathway or indirectly via its influence on other brain structures (Allum and

Pfaltz, 1985; Diener, Dichgans, Bacher *et al.*, 1984; Nashner, 1983; Traub, Rothwell and Marsden, 1980).

Theories of Cerebellar Function

Despite the large mass of data on the anatomy and physiology of the cerebellum, there is little agreement on the precise function of the cerebellum in control of movement. Three main functions have been attributed to the cerebellum in different theories: (1) the cerebellum has been supposed to work as a timing device; (2) to be involved in learning new movements; and (3) to co-ordinate movements by transforming the idea of where to move into the appropriate changes in joint angle necessary to make the movement. It is not known whether any of these theories is correct. Perhaps all of them are.

Cerebellum as a Timing Device

This was one of the earliest theories of cerebellar function. It proposes that the cerebellum is used to time the duration of agonist muscle activity and the latency of antagonist activity, so that any movement is halted at the correct point. Hypermetria on this theory is simply an inability to stop a movement at the correct point in space due to a defect in timing of agonist and antagonist muscle activity. As already seen, there is some evidence from the study of rapid limb movements that suggests that defects in timing of EMG activity do exist in cerebellar patients.

In detail, the theory suggests the following mechanism. The cerebral cortex would initiate activity in the agonist muscle to start the limb moving, and it would be the function of the cerebellum to stop the movement at the correct point (by agonist inhibition and antagonist excitation). Mossy fibre input would be activated by a cortical control signal and produce a pattern of activity on the surface of the cerebellum corresponding to the desired end position of the limb in space. In contrast, the pattern of climbing fibre input would reflect its actual position. Thus at the beginning of a movement, the two patterns would represent the start and stop positions of the limb.

It was suggested that the time taken for impulses set up by mossy fibre activity (that is, in the 'stop' position) to travel down the parallel fibre system to the area of climbing fibre input would be related to the timing of EMG activity needed in the movement. The larger the distance between the mossy and climbing fibre patterns of activity, the longer the EMG burst would last. Conjunction of parallel and climbing fibre activity in a Purkinje cell would signal the command to halt.

Cerebellum as a Learning Device

In this theory, the pattern of mossy fibre input on the cerebellar cortex is supposed to represent the position of the limbs. During a novel voluntary task, the climbing fibres are supposed to activate Purkinje cells, which are then suggested to have some role in producing movement. In a proportion of Purkinje cells, there will be a conjunction of parallel fibre activity (from the mossy fibre input) reflecting the present limb position, with climbing fibre input responsible for the active movement. It is suggested that if the same movement is made in the same context over and over again, then this conjunction of activity will recur. In so doing, it is supposed that the strength of the synaptic connection between the parallel fibres and Purkinje cells will be increased such that, in time, the occurrence of the same pattern of mossy fibre activity will be able to activate these Purkinje cells with very little aid from climbing fibre input. At this stage, the cerebellum would have learned to make a particular movement in a particular context.

As already seen, there is some evidence that such synaptic changes might occur during learning. There is also evidence that climbing fibre activity changes during learning.

Cerebellum as a Co-ordinator

The last of the theories is the most recent. It differs from the others in that less emphasis is placed on the specificity of patterns in neuronal activity in the cerebellar cortex. This is in keeping with the current opinion that there is a great deal more divergence of inputs from any one source and convergence of inputs from many sources in cerebellar afferent systems than previously was supposed.

The idea is that several brain regions contribute (in parallel) separate components of an intended movement. For example, in a pointing task involving the shoulder, elbow and wrist, there will be instructions to move the shoulder, extend the elbow and rotate the wrist. The role of the cerebellum is suggested to be that of scaling the amplitude of these components so that the movement is accomplished by the correct (and complementary) amounts of movement at each joint. It can be viewed as tranforming what we want to do into how we do it. In this scheme, the mossy fibre system describes both the status quo and also carries the details of the intended movement. The climbing fibres are supposed to adjust the operation of the cerebellum according to the subject's intention or to fit in with the particular position of the body at the start of the movement (for further discussion of cerebellar theories see Llinas, 1981).

References and Further Reading

Review Articles and Books

Bloedel, J.R., and Courville, J. (1981) 'Cerebellar Afferent Systems' in V.B. Brooks (ed.), *Handbook of Physiology,* sect.1, vol.II, part 2, Williams and Wilkins, Baltimore, pp. 735–830

Brodal, A. (1981) *Neurological Anatomy in Relation to Clinical Medicine*, Oxford University Press, Oxford

Brooks, V.B. and Thach, W.T. (1981) 'Cerebellar Control of Posture and Movement' in V.B.

Brooks (ed.), *Handbook of Physiology,* sect.1, vol. II, part 2, Williams and Wilkins, Baltimore, pp. 877–945
Gilman, S., Bloedel, J.R. and Lechtenberg (1981) *Disorders of the Cerebellum,* F.A. Davis, Philadelphia
Holmes, G. (1939) 'The Cerebellum of Man', *Brain, 62,* pp. 1–30
Ito, M. (1984) *The Cerebellum and Neural Control,* Raven Press, New York
Llinas, R. (1981) 'Electrophysiology of the Cerebellar Networks' in V.B. Brooks (ed.), *Handbook of Physiology,* sect. 1, vol. II, part 2, Williams and Wilkins, Baltimore, pp. 831–75
Marsden, C.D., Merton, P.A., Morton, H.B. *et al.* (1977) 'Disorders of Movement in Cerebellar Disease in Man' in F. Clifford Rose (ed.), *Physiological Aspects of Clinical Neurology,* Blackwell, Oxford

Original Papers

Allum, J.H.J. and Pfaltz, C.R. (1985) 'Visual and Vestibular Contributions to Pitch Sway Stabilisation in the Ankle Muscles of Normals and Patients with Bilateral Vestibular Deficits', *Exp. Brain Res., 58,* pp. 82–294
Beppu, H., Suda, M. and Tanaka, R. (1984) 'Analysis of Cerebellar Motor Disorder by Visually Guided Elbow Tracking Movement', *Brain, 107,* pp. 787–809
Combs, C.M. (1954) 'Electroanatomical Study of Cerebellar Localisation. Stimulation of Various Afferents', *J. Neurophysiol., 17,* pp. 123–43
Diener, H.C., Dichgans, J., Bacher, M. *et al.* (1984) 'Characteristic Alterations of Long-loop "Reflexes" in Patients with Friedreich's Disease and Late Atrophy of the Cerebellar Anterior Lobe', *J. Neurol. Neurosurg. Psychiat., 47,* pp. 679–85
Flament, D., Vilis, T., and Hore, J. (1984) 'Dependence of Cerebellar Tremor on Proprioceptive But Not Visual Feedback', *Exp. Neurol., 84,* pp. 314–25
Ghez, C. and Fahn, S. (1985) 'The Cerebellum' in E.R. Kandel and J.H. Schwartz (eds.), *Principles of Neural Science,* 2nd edn., Elsevier, New York
Gilbert, P.F.C. and Thach, W.T. (1977) 'Purkinje Cell Activity During Motor Learning', *Brain Res., 128,* pp. 309–28
Groenewegen, H.J., Voogd, J. and Freeman, S.L. (1979) 'The Parasagittal Zonal Organisation within the Olivocerebellar Projection. II. Climbing Fibre Distribution in the Intermediate and Hemispheric Parts of Cat Cerebellum', *J. Comp. Neurol., 183,* pp. 551–602
Harvey, R.J., Porter, R. and Rawson, J.A. (1979) 'Discharges of Intracerebellar Nuclear Cells in Monkeys', *J. Physiol., 297,* pp. 559–80 (see also pp. 271, 515–36)
Hore, J. and Vilis, T. (1984) 'Loss of Set in Muscle Responses to Limb Perturbations During Cerebellar Dysfunction', *J. Neurophysiol., 51,* pp. 1137–48
Ito, M., Sakurai, M., and Tongroach, P. (1982) 'Climbing Fibre Induced Depression of Both Mossy Fibre Responsiveness and Glutamate Sensitivity of Cerebellar Purkinje Cells', *J. Physiol., 324,* pp. 113–34
Jansen, J. and Brodal, A. (1958) 'Das Kleinhirn' in Mollendorff (ed.), *Handburch der Mikroskopischen Anatomie des Menschen IV/8,* Springer–Verlag, Berlin
Kornhauser, D., Bromberg, M.B. and Gilman, S. (1982) 'Effects of Lesions of Fastigial Nucleus on Static and Dynamic Responses of Muscle Spindle Primary Afferents in the Cat', *J. Neurophysiol., 47,* pp. 977–86
Martinez, F.E., Crill, W.E. and Kennedy, T. (1971) 'Electrogenesis of Cerebellar Purkinje Cell Responses in Cats', *J. Neurophysiol., 34,* pp. 348–56
Nashner, L.M. (1976) 'Adapting Reflexes Controlling the Human Posture', *Exp. Brain. Res., 26,* pp. 59–72
Nashner, L.M. (1983) 'Analysis of Movement Control in Man Using the Movable Platform' in J.E. Desmedt (ed.), *Advances in Neurology,* vol. 39, pp. 607–19
Pollock, L.J. and Davis, L. (1927) 'The Influence of the Cerebellum Upon the Reflex Activities of the Decerebrate Animal', *Brain, 50,* pp. 277–312
Sasaki, K., Gemba, H. and Mizuno, N. (1982) 'Cortical Field Potentials Preceding Visually Initiated Hand Movements and Cerebellar Actions in the Monkey', *Exp. Brain. Res., 46,* pp. 29–36
Schultz, W. Montgomery, E.B., and Marini, R. (1979) 'Proximal Limb Movements in Response to Microstimulation of Primate Dentate and Interpositus Nuclei Mediated by Brain-stem

Structures', *Brain, 102,* pp. 127–46

Spidalieri, G., Busby, L. and Lamarre, Y. (1983) 'Fast Ballistic Arm Movements Triggered by Visual, Auditory and Somesthetic Stimuli in the Monkey. II. Effects of Unilateral Dentate Lesion on Discharge of Precentral Cortical Neurons and Reaction Time', *J. Neurophysiol., 50,* pp. 1359–79

Strick, P.L. (1978) 'Cerebellar Involvement in "Volitional" Muscle Responses to Load Changes' in J.E. Desmedt (ed.), *Prog. Clin. Neurophysiol.,* vol. 8, Karger, Basel, pp. 85–93

Thach, W.T. (1978) 'Correlation of Neural Discharge with Pattern and Force of Muscular Activity, Joint Position and Direction of Intended Movement in Motor Cortex and Cerebellum', *J. Neurophysiol., 41,* pp. 654–76

Traub, M.M., Rothwell, J.C. and Marsden, C.D. (1980) 'Anticipatory Postural Reflexes in Parkinson's Disease and Other Akinetic-Rigid Syndromes and in Cerebellar Ataxia', *Brain, 103,* pp. 393–412

Vilis, T. and Hore, J. (1980) 'Central Neural Mechanisms Contributing to Cerebellar Tremor Produced by Limb Perturbations', *J. Neurophysiol., 43,* pp. 279–91

Wetts, R., Kalaska, J.F. and Smith, A.M. (1985) 'Cerebellar Nuclear Cell Activity During Antagonist Cocontraction and Reciprocal Inhibition of Forearm Muscles', *J. Neurophysiol., 54,* pp. 231–44.

Williams, P.L. and Warwick, R. (1975) *Functional Neuroanatomy of Man,* Churchill Livingstone, Edinburgh

10 THE BASAL GANGLIA

At one time, the term 'basal ganglia' was used to describe all the large nuclear masses in the interior of the brain, including the thalamus. Gradually, its use has become restricted to five closely related nuclei: caudate, putamen, globus pallidus, subthalamic nucleus and sustantia nigra (Figure 10.1). The basal ganglia receive no direct sensory inputs and, like the cerebellum, send no direct motor output to spinal cord. However, there is no doubt that these structures are involved in the control of movement. All diseases of the basal ganglia in man have some disorder of movement as their primary symptom. These range from an excess of involuntary movements (for instance, chorea) to a poverty and slowness of voluntary movement (for instance, Parkinson's disease). Important as their role may be, there is to date no agreement on the precise function or the mechanism of action of the basal ganglia (see articles in Massion, Paillard, Schultz *et al.*, 1983).

Figure 10.1: Anterior Aspect of a Coronal Section Through the Right Cerebral Hemisphere Showing the Main Structures of the Basal Ganglia (from Williams and Warwick, 1975; Figure 7.156, with permission)

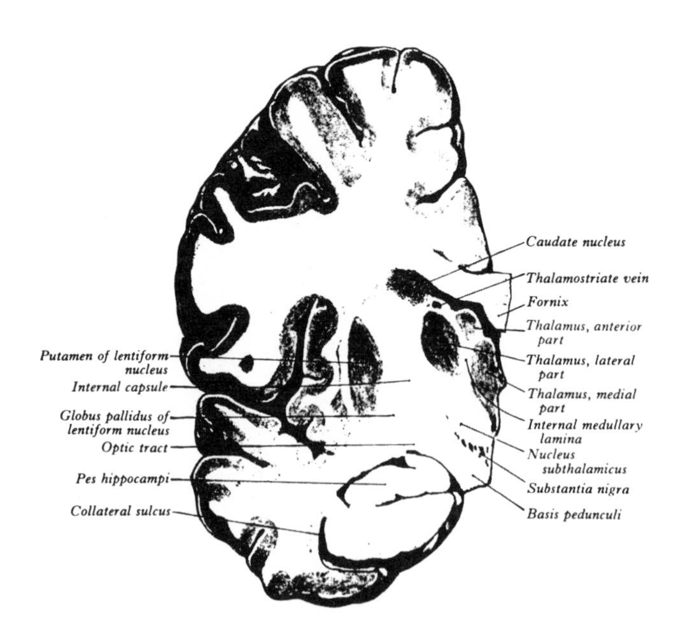

Anatomy

The major part of the basal ganglia is made up of three nuclei: the caudate, putamen and globus pallidus. These three nuclei lie in the medulla of the cerebral cortex, lateral to the thalamus and separated from it by the internal capsule. The globus pallidus is phylogenetically the oldest of the group and is known as the paleostriatum. Its name (pallidum) derives from the fact that in unstained sections it is paler than caudate and putamen. The phylogenetically newer parts of the striatum (caudate and putamen) are known as the neostriatum. Most books refer to these two nuclei simply as the striatum. In man they are separated from each other by the internal capsule, whereas in the rat they form one single structure. The term *striatum* is descriptive of their appearance in myelin stained sections. A number of fibre bundles known as Wilson's pencils traverse the nuclei, giving them a striped appearance.

The globus pallidus is divided into two parts, lateral (or external, the GP_e) and medial (internal, or GP_i), by the medial medullary lamina. The lateral or external part is larger than the medial or internal part, and has connections with quite different parts of the brain.

The other two nuclei of the basal ganglia, the substantia nigra and subthalamic nucleus, lie in the midbrain. The subthalamic nucleus, is a small lens-shaped nucleus situated ventral to the thalamus and on the dorsomedial surface of the internal capsule at the point where the fibres group together to form the cerebral peduncle. Ventral and caudal to the subthalamic nucleus, and continuous with it at this point, is the substantia nigra. It is the largest nucleus of the midbrain, and is particularly well developed in man. The cells are densely packed with granules of melanin and, in coronal sections of brain, the substantia nigra can be seen with the naked eye as a darkly coloured arched stripe above the cerebral peduncle.

Like the globus pallidus, the substantia nigra is also divided anatomically into two parts which have quite separate connections with other areas of brain. There is no clear demarcation line, but the cells are more densely packed in the dorsal part of the nucleus than in the ventral part, and the area appears darker when examined by eye. The dorsal region is known as the pars compacta, and the ventral region as the pars reticulata.

These nuclei of the putamen, caudate, pallidum, nigra and subthalamus are the most usually accepted members of the basal ganglia. However, in recent years, additional nuclei have been included and are said to constitute the ventral striopallidal system. The concept is based on similarities between the connections and histochemical features of the neostriatum and both the nucleus accumbens septi and the medium-celled portion of the olfactory tubercule. These two areas are the receiving nuclei (or 'ventral striatum') of the ventral striopallidal system, and receive input principally from limbic areas of cerebral cortex. They have connections with a rostral and ventral extension of the globus pallidus and rostral substantia nigra pars reticulata. The nucleus accumbens is large in rats, but small in man (see comprehensive reviews of basal ganglia anatomy in Ciba Foundation Symposium, 1984; Carpenter, 1981).

Input–Output Connections of the Basal Ganglia

The basal ganglia receive a major input from the cerebral cortex, and send most of their output back, via the thalamus, onto specific areas of cortex (Figure 10.2). In addition to this cortico-basal ganglia-cortex loop, there are two other circuits within the basal ganglia themselves. One is the projection from caudate and putamen to substantia nigra and back again. The other is the loop from the GP_e to subthalamic nucleus and back to the GP_i. Other areas of the brain which provide a smaller number of inputs to the basal ganglia are the midline thalamic nuclei and the dorso-medial Raphé. Other areas which receive basal ganglia output are the superior colliculus, reticular formation and the pedunclopontine nucleus.

Cortex–Basal Ganglia–Cortex Loop (Figure 10.3)

All areas of the cerebral cortex project onto the basal ganglia, sending terminals to the receiving nuclei of the caudate and putamen. From here the major flow of information is to both segments of the globus pallidus and the substantia nigra pars reticulata. The GP_i and SN_{pr} send outputs via the thalamus back to the cortex. The output of the GP_e enters the subthalamic loop, as discussed below.

One of the major points of interest at the present time is the extent to which inputs from different areas of cortex intermingle, and how much they remain separate in their pathways through the basal ganglia. At the level of the caudate and putamen, there is a clear topographical demarcation between inputs from different cortical areas. The infero-temporal regions project to the tail of the caudate, the sensori-motor and premotor areas to the putamen, and the prefrontal areas to the head of the caudate. However, the outputs of striatum are compressed into the successively smaller regions of pallidum and substantia nigra. Even though anatomical studies have shown that the efferent fibres of the caudate and putamen are topographically organised, the dendritic tree of single neurones in the globus pallidus is large and may spread to cover up to 30 per cent of the cross sectional area of the nucleus. Thus it would be in a position to receive inputs from a fairly wide area of the caudate and putamen and provide an opportunity for convergence of inputs from different cortical areas. Despite this opportunity for anatomical convergence, single cell recording studies by DeLong and colleagues (1985) show that there is relatively little physiologial convergence between inputs from various cortical areas. This suggests that despite its apparently homogenous structure, there are many parallel pathways through the basal ganglia which process their inputs quite separately.

The Motor Loop. The arrangement of the various cortical loops through the basal ganglia is best illustrated by the 'motor loop'. This loop involves the largest

Figure 10.2: Principal Intrinsic and Extrinsic Connections of the Basal Ganglia. The main loop pathway is from neocortex through the basal ganglia and thalamus, back to cortex. Note the other intrinsic loops from GP_e to STN, and from striatum to SN_{pc} and back. Dopamine pathways are dashed. Acc, accumbens; GP_e and GP_i, external and internal segments of globus pallidus; STN, subthalamic nucleus; SN_{pc} and SN_{pr}, pars compacta and pars reticulata of substantia nigra; VTA, ventral tegmental area; SC, superior colliculus; RF, reticular formation; TPC, pedunculopontine nucleus pars compacta; VA and VL, ventroanterior and ventrolateral thalamic nuclei; CM, centromedian nucleus of thalamus. Inputs from the raphe nuclei to striatum and substantia nigra have been omitted (from DeLong and Georgopoulos, 1981; Figure 1, with permission)

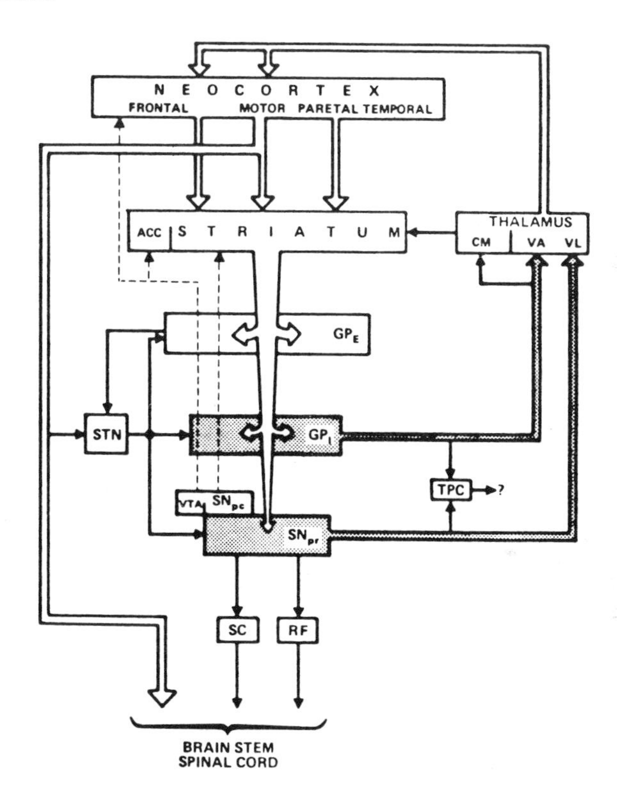

flow of information through the basal ganglia, and is the circuit most directly involved in the control of movement.

A large proportion of the total input to the basal ganglia comes from the sensorimotor areas of cortex. These include the precentral motor, premotor, supplementary motor, and postcentral sensory areas (including area 5) of cortex. Their fibres terminate almost exclusively (in primates) on the putamen, which is the receiving nucleus of the 'motor loop'. The inputs form a topographic body map, following the arrangement of the cortical mantle above. Dorsal parts of the putamen receive from medial (or leg) areas of cortex, while ventral parts receive from lateral (or face) areas of cortex. Within the gross area of arm

Figure 10.3: Postulated Segregation of Cortico-Basal Ganglia Connections into a 'Complex' and Motor Loop. The 'complex' loop involves input from frontal, association areas of cortex, whereas the motor loop input is from sensorimotor areas. VA_{pc}, pars compacta of ventroanterior thalamic nucleus; VA_{mc} magnocellular part of ventroanterior nucleus; VL_o and VL_m, pars oralis and pars medialis of ventrolateral thalamic nucleus. It should be noted that the final thalamic output projects only to a limited region of the cortical areas which contribute input to these loops (see text) (from Delong, Georgopoulos and Crutcher, 1983; Figure 1, with permission)

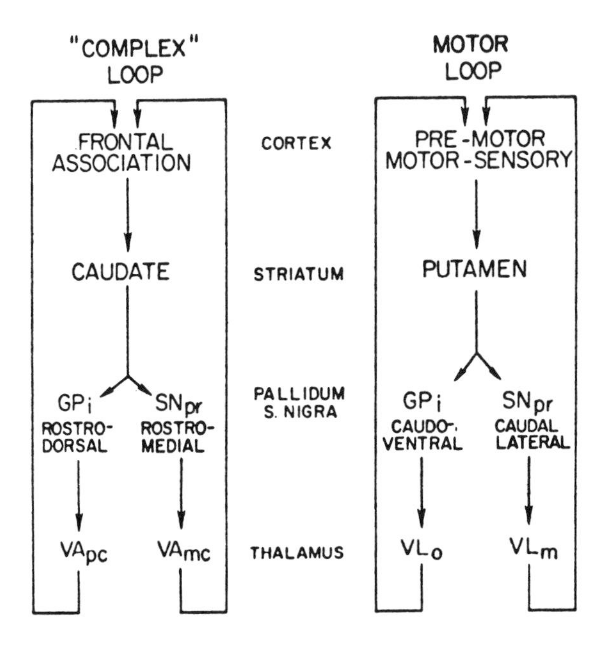

representation, cortical projections from somatosensory and motor areas may overlap.

The cortical terminals form discrete patches in the putamen which probably correspond with patchy physiological representation found in microrecording and stimulation studies (see below). This type of patchy termination has also been revealed when sections have been stained for various neurotransmitters such as dopamine, enkephalin, and substance P, or for enzymes such as acetylcholinesterase. All have a heterogenous distribution within the striatum, which may reflect an underlying cytoarchitectural structure which is not visible using conventional microscopy. These histochemically defined patches have been termed 'striosomes' or 'islands'.

Outputs from the striatum group together in the bundles known as Wilson's pencils, and project to the caudo-ventral parts of both segments of the globus pallidus, and caudo-lateral substantia nigra pars reticulata. The projections are topographically organised, preserving the dorso-ventral orientation within the putamen. The final stage is from the internal segment of the globus pallidus and substantia nigra pars reticulata to the ventro-lateral thalamus. Thalamic efferents of the GP_i travel in two separate fibre bundles through the descending axons of the internal capsule. The bundles are known as the ansa lenticularis, which arises from the outer part of the GP_i, and the lenticular fasciculus, from the inner part of the GP_i. After crossing the internal capsule, they join and meet the ascending thalamic afferents from substantia nigra pars reticulata to form a single fibre bundle called the thalamic fasciculus. Pallidal fibres in the 'motor loop' end on distal dendrites of cells in the VL_o. Fibres from the 'motor' part of the substantia nigra pars reticulata are believed to end in VL_m. The major output of these thalamic nuclei is to the supplementary motor area, which is thus the final destination of the basal ganglia ''motor loop'' (see Chapter 8).

Within the 'motor loop' there is a convergence of inputs from a fairly wide area of sensorimotor cortex onto a relatively specific motor area, the SMA. This type of organisation is seen in the other cortical circuits through the basal ganglia.

'Complex Loop'. The 'complex loop' of Figure 10.3 is probably not a single entity. In a recent review, Alexander, DeLong and Strick (1986) have considered that it may be composed of four or more separate pathways. The caudate is the receiving nucleus for these pathways, and projects onto different areas of the pallidum and substantia nigra pars reticulata than the fibres in the 'motor loop' from the putamen. From here, efferents project to several thalamic nuclei and thence to cerebral cortex.

One of the components of this 'complex loop' which is directly relevant to the physiological data discussed below is the 'oculomotor loop'. The input to this loop comes from three cortical areas known to be involved in the control of eye movements, the frontal eye fields (area 8), dorsolateral prefrontal cortex (areas 9 and 10), and posterior parietal cortex (area 7). These regions send axons to the body of the caudate nucleus, and from there to the 'oculomotor region' of the pallidum and the ventrolateral substantia nigra pars reticulata. The final output is through the VA nucleus of the thalamus back onto the frontal eye fields. Thus, like the 'motor loop', this pathway focuses inputs from several areas back onto a more limited region of cerebral cortex. The pathway also has a large output from the nigra to the superior colliculus, another area of the brain known to be involved in the control of eye (and perhaps head) movements.

A final loop running parallel to these 'motor' and 'complex' loops is the 'limbic loop' of the ventral striopallidal pathway. The limbic input to the nucleus accumbens and olfactory tubercle (the ventral striatum) travels to the ventral extension of the pallidum and substantia nigra pars reticulata, and back via the thalamus to the limbic system. In the rat, where the nucleus accumbens is large, there is an additional input to the loop from a group of dopaminergic cells adjacent and slightly caudal to the substantia nigra in the ventral tegmental area (known in rat terminology as the A10 group). This ascending projection of the dopamine system is called the mesolimbic dopamine system, and may be an important influence on motivational behaviour in the rat. Its function in man is unknown.

Other Outputs of the Basal Ganglia

Although the cerebral cortex appears to be the main output target for the basal ganglia, its fibres project to several other regions. The output to the superior colliculus has been mentioned above in relation to the 'oculomotor loop'. Outputs also end in the centromedian nucleus of the thalamus, which together with all other midline thalamic nuclei sends some fibres back to the striatum. These are collaterals of axons which also project to diffuse areas of cortex. There are also projections from GP_i to the habenular nucleus and the pedunculopontine nucleus. The latter is the largest and in the cat can account for 8 per cent of basal ganglia output. The axons travel in the pallido-tegmental tract and terminate in a small mass of cells partly embedded in the superior cerebellar peduncle. Unfortunately, little is known about the connections of the pedunculopontine nucleus. It receives some direct afferents from cerebral cortex, and has been implicated in the control of walking in the cat.

Strio-nigro-striatal and Strio-nigro-thalamic Loops

This is one of the best known internal loops of the basal ganglia. Reciprocating the striatal projections to the substantia nigra pars reticulata is the nigro-striatal pathway. This arises from the pars compacta of the substantia nigra and is a dopaminergic pathway. Cells of SN_{pc} have dendrites which descend into the pars reticulata. Terminals of striatal efferent fibres probably synapse only with cells of the pars reticulata, so that dendrites of cells in the pars compacta only receive indirect striatal input. Efferent fibres of the pars compacta project topographically onto the same areas of caudate and putamen as gave rise to the nigral afferents. This forms the strio-nigro-striatal loop.

GP_i-Subthalamic Nucleus-GP_e Loop

The connections of the internal segment of the globus pallidus are very different from those of the external segment. Efferent fibres terminate topographically on the subthalamic nucleus and provide the main source of its input. A small number of fibres to subthalamic nucleus also arise in precentral, premotor, and prefrontal areas of cerebral cortex. Output from the subthalamic nucleus is mainly to both segments of the globus pallidus, with a smaller projection to the substantia nigra.

Cytology and Pharmacology of Basal Ganglia Connections

Ninety per cent of cells in caudate and putamen are medium-sized, densely spiny neurones. They are both receiving and output neurones; 1–2 per cent of the cells are large multipolar neurones, which were at one time thought to be the output cells of the striatum. Nowadays, they are believed to be interneurones. Cortical efferents to striatum are glutaminergic, and terminate (together with the smaller number of thalamic inputs) on the dendritic spines of striatal cells. They are excitatory connections, and have the usual asymmetric membrane thickenings when viewed under the electron microscope.

Nigral inputs to the striatum synapse on the base of the dendritic spines and are dopaminergic. They are inhibitory, and the synapses have symmetrical thickenings when seen under the electron microscope. It is believed that input to the base of a single dendritic spine has little effect on input from other spines. So, it may be that in areas where their terminals overlap, specific nigral efferents can inhibit the action of specific cortical efferents at the level of dendritic spines. Finally, output neurones of the striatum send axon collaterals back into caudate and putamen. These form synapses onto smooth proximal dendrites of the striatal cells and are in a position to modulate any potentials produced by input to the spiny parts of the dendrites. Many of the interneurones of the caudate and putamen are cholinergic. Staining for acetycholine esterase reveals that there are patches of less-dense staining throughout the striatum produced by higher levels of activity. The less dense patches are called *striosomes.*

Striatal output to globus pallidus and SN_{pr} is GABAergic, and inhibitory. In the pallidum, other chemicals are released together with GABA at striatal synapses. Terminals in GP_e are rich in enkephalins, and terminals in GP_i are rich in substance P. The output from GP_e to thalamus is GABAergic and inhibitory. There is an ill-understood serotonergic (5HT) input to all parts of the putamen from the dorsal and median raphe nuclei.

A summary of excitatory and inhibitory connections within the basal ganglia is shown in Figure 10.4.

Electrophysiological Recordings from Behaving Animals

Resting Discharge

Neurones in different parts of the basal ganglia have different levels of resting discharge. This ranges from slow or absent in the cells of the caudate, putamen and substantia nigra pars compacta, to high frequencies of 60–100Hz in cells of the globus pallidus and substantia nigra pars reticulata. Neurones of the subthalamic nucleus have a maintained but slow rate of spontaneous firing. Since the output of the caudate and putamen is inhibitory onto the cells of the globus pallidus (Figure 10.4), it may be that the high firing frequencies of globus pallidus neurones are caused by lack of inhibition from slowly firing cells of the striatum. Further subdivision of cell firing patterns has been seen in the globus pallidus. Cells in the GP_i have a sustained high frequency firing; GP_e cells fire in high frequency bursts and are of two subtypes, one with relatively long bursts and the other with relatively short bursts of firing. (See review by DeLong and Georgopoulos, 1981.)

Relationship of Discharge to Movement

As electrophysiologists study neurones which are more and more distant from the peripheral motor apparatus, it becomes more difficult to define the precise relationship of their firing patterns to movements being performed. Some

Figure 10.4: Main Excitatory (Open Arrows) and Inhibitory (Filled Arrows) Connections in the Basal Ganglia. DR, dorsal raphe; CMP, centromedian and parafascicular complex; TE, midbrain tegmentum, ST, subthalamic nucleus (from Kitai, 1981; Figure 15, with permission)

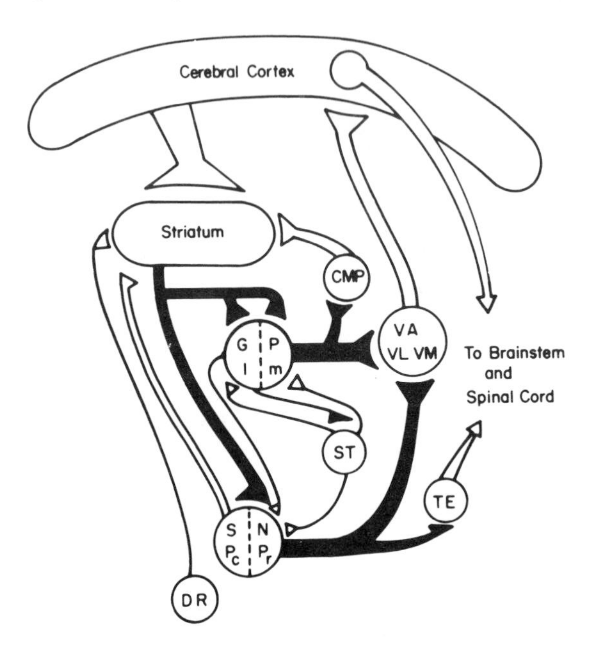

neurones of the basal ganglia show phasic changes in their discharge frequency in relation with movements of the body. Such cells are easily studied to reveal the relationship of discharge to various parameters of movement such as speed, amplitude or force. Other neurones are apparently silent and unrelated to movement performance. Their discharge is not related to movement *per se* but is dependent on both movement and the context in which the movement is made. They are said to be 'context-dependent' cells.

Movement-related Cells: Somatotopy. Many cells in putamen, globus pallidus, substantia nigra pars reticulata and subthalamic nucleus show phasic changes in their firing frequencies in association with movements of the contralateral side of the body. As expected on the basis of the anatomical projections, there is a somatotopic organisation of these cells: in the putamen, globus pallidus and subthalamic nucleus, cells related to the movement of the legs lie dorsal to those related to movements of the face. Arm-related neurones are found between these two populations. Neurones of the SN_{pr} are related only to orofacial and saccadic eye movements (Figure 10.5). Within the putamen, this dorso-ventral somatotopic arrangement parallels the medio-lateral organisation of the somatomotor cortex lying above.

The rostrocaudal extent of the putaminal movement related neurones is much

Figure 10.5: Composite Figure Made From the Results of Several Experiments Showing Somatotopic Grouping of Movement-related Neurones in the Internal and External Segments of Globus Pallidus (GP_i, GP_e), the Substantia Nigra Pars Reticulata (SN_{pr}) and Subthalamic Nucleus (STN). Representative coronal sections of each structure are shown, although each is taken at slightly different anteroposterior levels of the brain. Filled triangles, leg; filled circles, arm; open circles, face. The representation of each body part extends over most of the anterior–posterior extent of each structure. Note the clustering of orofacial neurones in SN_{pr} and absence of any neurones related to limb movements. Cells of SN_{pc} did not show any phasic discharges during movement of any part of the body (from DeLong and Georgopoulos, 1979; Figure 1 with permission)

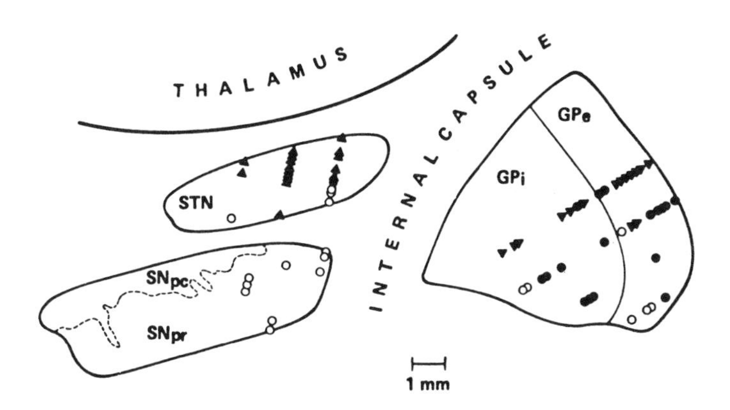

larger than that of the motor cortex. This is because the putamen receives input from premotor, motor and sensory areas of cortex, all three of which may contribute to the discharge of putaminal neurones. Movement related neurones of the putamen are found in clusters, like the clusters of cortico- or nigro-striatal terminals in anatomical studies. Cells lying between these clusters and the cells of substantia nigra pars compacta and the caudate nucleus are not directly related to movement (Alexander and DeLong, 1985; DeLong, Crutcher and Georgopoulos, 1985; Liles and Updyke, 1985).

Movement-related Cells: Timing Relationships. As with motor cortex and cerebellum, there have been several studies recording the activity of basal ganglia neurones before and after onset of a voluntary movement made in response to a visual or auditory reaction signal. Because it is easy to define the onset of the movement, such experiments give an indication of the sequence of activity in different brain areas. Cells in putamen tend to discharge after the onset of movement (see Figure 10.6): only 20 per cent or so fire before the onset of EMG. This compares with a figure of 50 per cent for motor cortex cells in similar tasks.

This finding has two implications: first, the late discharge of putaminal neurones probably reflects an input from motor cortex neurones involved in the movement. Second, in this type of reaction task, basal ganglia probably are

Figure 10.6: Example of Single Unit Activity in the Putamen of a Monkey to Active and Passive Flexion and Extension Movements of the Wrist. The top four records in each column are sample original records of unit activity, forearm extensor EMG, forearm flexor EMG and wrist position. These are followed by 10 raster displays of unit activity (each upstroke represents one action potential), and the response averages of unit activity, forearm extensor EMG (EX), forearm flexor EMG (FX) and wrist position (P) (± 1.S.D., dotted lines). These traces are aligned to the time of movement onset (vertical dotted line). The firing of this cell was related to active extension movements of the wrist. Discharge changed after EMG onset in the muscle, at about the same time as the wrist began to move. Passive flexion of the wrist also evoked activity in the cell about 100ms after onset of movement (time calibrations are 100ms apart) (from Liles, 1985; Figure 4A and 4B, with permission)

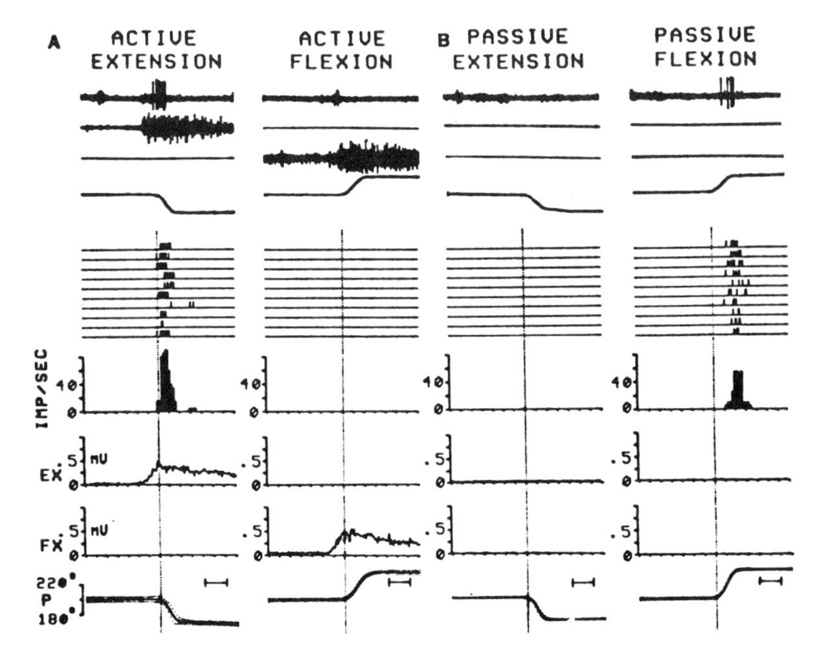

involved little in the initiation of movement, although they may be important in its *execution.* This conclusion is reflected in the clinical observation that patients with diseases of the basal ganglia have relatively little impairment in their simple reaction times. However, rather more difficult to interpret is the akinesia which is a prominent symptom of Parkinson's disease. Although some patients may perform relatively well in reaction time movements to an external stimulus, they may have a more pronounced poverty of spontaneous movements (see below). This is a warning that the visual and auditory triggered movements studied so far in monkeys may not give the whole story of basal ganglia involvement in initiation of movement (Crutcher and DeLong, 1984; Liles and Updyke, 1985).

Movement-related Cells: Relationship to Movement Parameters. Unlike the corticomotoneurone discharge of precentral neurones, discharge of many basal ganglia units is not related to the pattern of EMG activity used to perform a

Figure 10.7: Relationship Between Firing Frequency and Direction of Elbow Movement in Three Different Neurones Encountered During the Course of a Single Microelectrode Penetration (Left) Through the Monkey Putamen. The task was to flex (right) or extend (left) the elbow rapidly in response to movement of a visual target. Movements were made either with no load (NL), with a constant load (150g) opposing flexion (HF) or with a load opposing extension (HX). Histograms of average unit activity for each cell in the six different tasks are shown. Vertical dashed line indicates time of onset of movement. The first neurone (1) was active during extension movements, whereas the other two neurones (2 and 3) were active during flexion movements. Thus, activity was related to direction in these cells, but not the force of the movement (from Crutcher and DeLong, 1984; Figure 3 with permission)

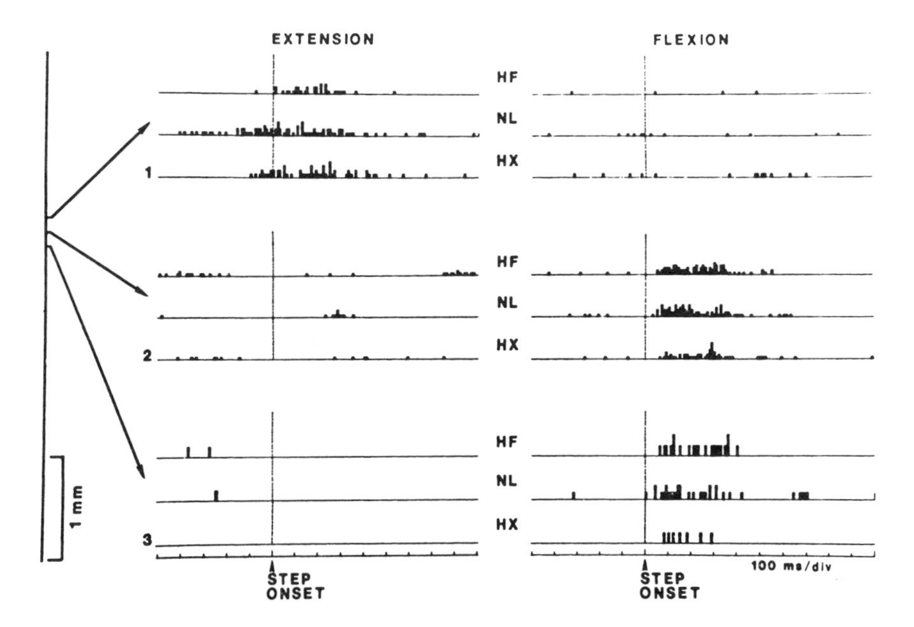

movement. In a task involving, say, elbow flexion movements against either an opposing or assisting load, the pattern of EMG activity can be dissociated from the direction of movement. With the opposing load, the movement is made using agonist activity, whereas with an assisting load the movement is made using the braking action of the antagonist. In the putamen, only 13 per cent of neurones had a discharge which reflected the pattern of EMG activity, while half of the neurones had a discharge related to the direction of movement (Figure 10.7). This was true also for globus pallidus and subthalamic nucleus, although in the globus pallidus (either segment), many neurones were affected by the amplitude of movement as well as its direction. The figure of 13 per cent EMG-related cells in this study is a little lower than in some other studies. Liles (1985) found 59 per cent neurones in putamen had a firing frequency which was related to the pattern of EMG activity, whereas only 41 per cent were not related.

At one time, it was believed that the basal ganglia were involved primarily in the performance of slow, rather than fast movements. Studies like those above have shown that this is not true. A large proportion of units do fire during slow movements, but this may not be because they are selectively activated by movements below a certain speed. It is possible that these units are related to greater levels of postural muscle activity during slow, precise tasks.

Somatosensory Inputs to Basal Ganglia

Many of the movement-related cells in putamen, globus pallidus and subthalamic nucleus also respond to somatosensory stimuli. The most effective inputs come from muscle or joint particularly from the region related to discharge during active movement. There is little effective input from superficial cutaneous receptors. In general, somatosensory responses recorded when the animal is at rest are much weaker than movement-related discharges. Many of the inputs to putamen, globus pallidus and subthalamic nucleus are directional: that is, they give opposite responses to opposite directions of joint rotation. As in the motor cortex, there are a number of directionally selective movement-related cells which also receive directionally specific sensory input (see Figure 10.6).

The fact that most cells with sensory input also respond to movement implies a convergence of sensory and motor input from different cortical areas. However, there are a small number of cells which respond either to somatosensory input or to movement. There is then *incomplete* convergence within the basal ganglia. Within the whole class of somatosensory-related cells there is a certain somatotopy of face, arm and leg representations, although this does not compare with the discrete representation seen in the primary sensory areas of cortex. In the basal ganglia, there is intermixing between, say, the representation of the shoulder, elbow and wrist within the general area for the arm, although areas of arm and leg input remain separate.

Timing of discharge to sensory inputs suggests that the main source of afferent information comes from the cerebral cortex. Putaminal neurones discharge 30 to 50ms after onset of a torque pulse disturbance to the wrist. Globus pallidus neurones discharge slightly later with a latency of 40 to 60ms. This compares with about 30ms in motor cortex.

Context-dependent Cells

The discharge pattern in the remainder of the neurones in globus pallidus, putamen, substantia nigra pars reticulata and subthalamic nucleus is not influenced by movement alone. In this respect they resemble the majority of neurones in caudate and substantia nigra pars compacta. It is now believed that some of these neurones respond to behaviourally significant external stimuli, rather than to movement *per se*. There are neurones within the caudate nucleus with complex properties which respond to more than one modality of sensory stimulation. In cat experiments, phasic increases in activity can be produced by a tone or a visual stimulus which warns of an impending electrical shock. The animal can avoid the shock by jumping to another part of the cage, but the phasic changes in activity occur if the stimulus does not require any relevant movement of the animal.

Figure 10.8: Context-dependent Cell Discharge in Putaminal (A) and Globus Pallidus (B) Neurones. Responses of tonically active neurones were recorded in three conditions: (1) self-paced movements (SPM) in which a series of elbow movements resulted in an auditory click and a fruit juice reward; (2) free-reward (FR) in which click and juice occurred at regular (6s) intervals while arm position was fixed; (3) no-reward (NR), which was similar to FR except that the juice reward was omitted. Animals performed long sequences of each type of trial before the next type was studied. The putaminal neurone in A did not respond to arm movement (see SPM condition) or to licking alone (not shown). However, it did respond to the click (arrow) when it was followed by a juice reward (licks in SPM and FR). There was no response to a click which no longer signalled a reward. In B, the pallidal neurone likewise responded to the click when it signalled a reward, but not when reward was no longer given. Note how the putaminal neurone increases its discharge rate in response to the stimulus, whilst the pallidal neurone decreases its rate. The vertical calibration for part A should read 0 and 30 (not 0 and 100 as for B) (from Evarts and Wise, 1984, in *Ciba Foundation Symposium, 107*; Figures 2A and 3B, with permission)

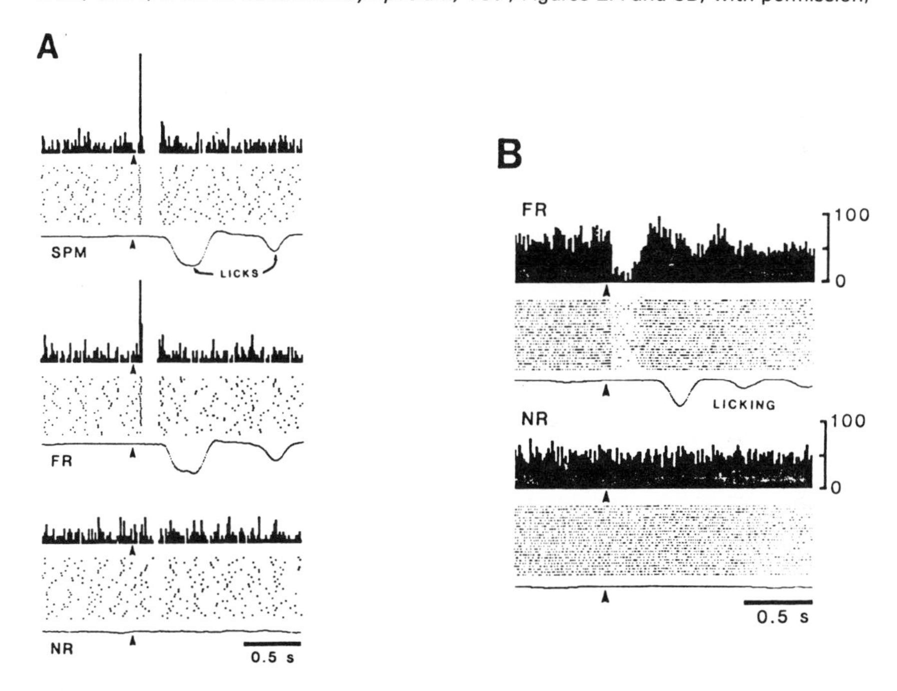

Similar cells are found among the non-movement related cells in the putamen. They may change firing rate in response to an auditory click which signals the appearance of a fruit juice reward, but will not fire to the movement alone (that is licking the lips), or to the same click if it is no longer associated with the reward (Figure 10.8). As with the electric shock, such cell activity signals the presence of a behaviourally significant stimulus in the environment. It is possible that these responses are related to the input to basal ganglia from association areas of cortex (see section on premotor neurones in Chapter 8).

Neurones of the substantia nigra pars compacta discharge slowly, less than 8Hz, compared to the 30 to 100Hz of SN_{pr} cells. They do not change their

Figure 10.9: Memory-contingent Saccade-related Neurone in the Substantia Nigra Pars Reticulata of a Monkey. The monkey was trained to look at an illuminated fixation point (F). No eye movements were allowed while the fixation light remained on. In A, a target light (T) was presented in the peripheral visual field at the same time as the fixation light was extinguished. The monkey then made a horizontal saccade (H) to fixate the new target. The raster display shows the firing pattern of the neurone during successive trials. The histogram shows the average response pattern. Trials on the left are aligned in time to the onset of target movement; those on the right to onset of eye movement. Both ways of presentation show a small decrease in firing frequency at around the time of the saccade. This is seen best in the integral of the histogram response, shown as the diagonal line between histogram and raster display. In B, the monkey again fixated an illuminated spot, but the peripheral target light in this case was presented only briefly. Since the fixation light remained on, the monkey did not make a saccade. However, on extinguishing the fixation spot, the monkey made a saccade to the position of the previously remembered target. In this instance, the neuronal firing changed far more than it did when the monkey had made the same saccade whilst the target light was on (A). As in A, the right panel shows the same data aligned to the onset of eye movement, rather than to onset of target movement (from Hikosaka and Wurtz, 1983; Figure 3, with permission)

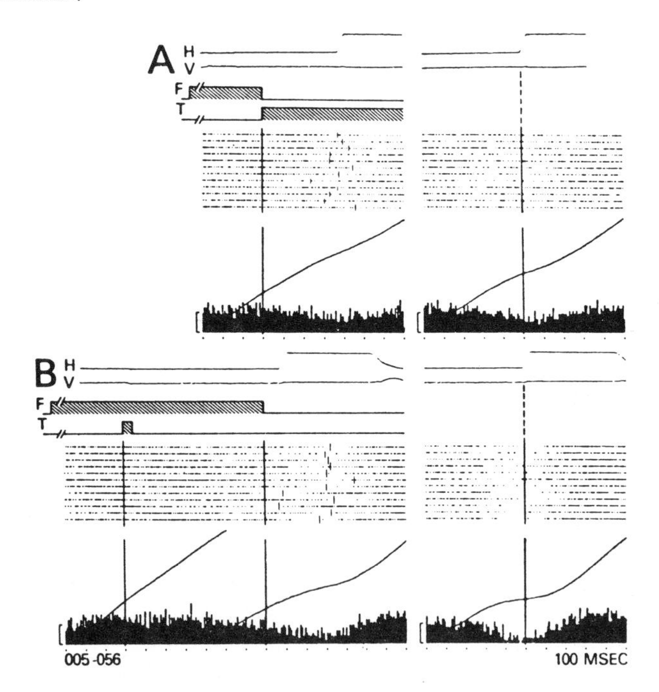

discharge rate during the small elbow movements such as those studied in the experiments above or in response to peripheral stimuli. However, they do begin to fire more rapidly (up to 20Hz) during large reaching movements of the contralateral arm. They are not related to the pattern of EMG activity in any one muscle and, therefore, may code some more general aspect of movement. A possibility is that they set a tonic level of activation in other systems involved in the movement (see articles in Ciba Foundation Symposium for further information).

Eye Movements and the SN_{pr}

The output of the SN_{pr} travels to two separate areas of brain concerned with movements of the head and eyes (1) via thalamus to the frontal eye fields; and (2) directly to the superior colliculus. It is therefore of no surprise that many neurones in the SN_{pr} have a discharge which is related to visual stimuli or to movements of the eyes (or both). However, the relationship is not simple. The discharge of the cells is context-dependent and, in almost all cases, is seen as a decrease in the rate of ongoing activity. For example, cells may not change their firing rate in response to saccades made in darkness, but may have a vigorous response to the same saccades made to visual targets. Similarly, although some cells respond to the appearance of a visual stimulus with a latency of about 120ms, the response may be enhanced if the stimulus provokes a saccadic eye movement to the target.

There are other cells with even more complex relationships between stimulus and response. One type has a memory-contingent discharge. The experiment which reveals this type of cell is as follows. A monkey is required to fixate a visual stimulus while a target light is presented briefly in the peripheral visual field. The monkey is trained not to make a saccade to the target light while it is presented, but has to remember its location within the field. A later saccade is rewarded if it is made to the position of the no-longer-present stimulus. Some SN_{pr} cells showed vigorous responses (usually a decrease of firing frequency 70 to 240ms before the saccade) under such conditions even though they were unresponsive when saccades were made with the stimulus present (Figure 10.9) (Hikosaka and Wurtz, 1983).

In summary, many neurones in the basal ganglia are phasically active in relation to movements of specific body parts. Because of their specificity, these neurones cannot be providing a tonic or modulating influence over some general aspect of movement control. However, since their firing rate is only rarely directly related to the pattern of EMG activity in specific muscles, they are unlikely to provide any direct input to control muscle activity. Their action must be at some intermediate level of command (for instance, in specifying direction, rather than the pattern of muscle activity in a movement). This would be consistent with their receiving input from pyramidal neurones of the precentral cortex which do not contribute to the corticospinal tract, and with their output via VA/VL thalamus to the supplementary motor area. The role of the context dependent neuronal discharges is of some interest since they seem to form a distinct subset of basal ganglia neurones which are not involved in movement *per se*.

Electrical Stimulation of the Basal Ganglia

Almost without exception, electrical stimulation studies have given little insight into the function of the basal ganglia. There are the usual problems of spread of stimulating current to nearby structures and fibre tracts, as well as the lack of direct connections with the spinal motor apparatus. Despite these difficulties, attempts have been made to stimulate basal ganglia. The most common effects from stimulation of caudate or putamen are the interruption of ongoing movements and contraversive turning of the head and body away from the stimulated side of the brain. Stimulation of the substantia nigra produces similar effects, as does the globus pallidus, although since both the strio-nigral and nigro-striatal tracts pass through the pallidum, the effects of pallidal stimulation may conceivably have been due to stimulation of these pathways rather than to specific activation of pallidal cells. Flexion of the contralateral arm also has been observed, but this may have been due to spread of stimulating current to the adjacent internal capsule.

Effects of Lesions of the Basal Ganglia

Because the basal ganglia lie deep within the cerebral hemispheres, it is difficult to produce localised lesions without producing damage to other surrounding structures such as the hypothalamus, internal capsule and overlying white matter. One particular fibre connection that may contribute to the symptoms seen after some lesions is the ascending dopaminergic connection from the ventral tegmental area of the mid brain (slightly caudal to the substantia nigra pars compacta) to frontal cortex and nucleus accumbens. It is known as the mesolimbic dopamine system. Through its connections with limbic system structures, it plays an important role in motivational behaviour, rather than having any direct part in movement control. This connection has been extensively studied in the rat, an animal in which the nucleus accumbens is far larger than in man. Whether the mesolimbic dopamine system is important in man is unknown (DeLong and Georgopoulos, 1981, provide a comprehensive reveiw of lesion studies).

Globus Pallidus

The inner segment of the globus pallidus, together with the substantia nigra pars reticulata, forms the major output for the basal ganglia. Lesions of this structure might therefore have been expected to produce the most effective interruption of basal ganglia function. In fact, unilateral pallidal lesions (even those involving both segments) have not been observed to produce motor deficits except when special testing procedures were used (see below). Bilateral lesions, if limited to the globus pallidus, do produce a mild but transient hypokinesia. However, it is a smaller effect than might have been expected. There are two possible reasons for this: first, there may have been incomplete destruction of the globus pallidus; second, such lesions leave intact the nigrothalamic output of the basal ganglia.

Attempts to destroy the output more thoroughly, by making larger lesions or by destroying the pallidal efferent fibres where they are gathered together in a small bundle near the lateral hypothalamus, have produced larger effects. Animals became extremely hypokinetic, lose their righting reactions and hold their body in a permanently flexed posture. Unfortunately, these lesions encroach upon other structures, particularly the ascending dopaminergic fibres from the midbrain to frontal cortex and limbic system structures, which may contribute to the profound behavioural deficits seen in the animals. Such large lesions do not contribute any conclusive evidence concerning the role of the globus pallidus alone.

Recent experiments have examined the effect of unilateral pallidal lesions in specially trained monkeys and have revealed deficits very similar to those described in patients with Parkinson's disease. The animals were taught to make rapid flexion and extension movements at the wrist or elbow, either in their own time or in response to a visual trigger signal. The globus pallidus was then inactivated (both segments) by one of two methods. In one method, cooling probes were implanted unilaterally for reversible cooling of the pallidal tissue. Neurones were inactivated at temperatures of about 28°C. Although this produced little obvious effect on the overall behaviour of the monkey, there was a dramatic change in performance of the arm movement task. The most impressive effects were a decrease in the speed and extent of movement, sometimes associated with a smaller increase in reaction time to the visual stimulus (Figure 10.10). During experiments in which the monkey was allowed no visual feedback of arm position, the elbow joint tended to assume a flexed position.

Similar effects were seen with the second method of pallidal inactivation, injection of kainic acid. This is a neurotoxic substance which first inactivates and then destroys cell bodies with relative sparing of fibres of passage. In these experiments it was observed that despite the slowing of movement, there was no change in the *order* of muscle activation in complex tasks involving postural fixation of joints as well as prime movement about one single joint.

Lesions of the globus pallidus therefore lead to rather mild effects on the control of movement. Whether this is due to incomplete lesions or the remaining function of the nigrothalamic output is not known. No changes in muscle tone and no involuntary movements have ever been observed. The effects seen in trained animals (slowness of movement, flexed posture, and increased reaction time) are all characteristics of patients with Parkinson's disease (see below) (Anderson and Horak, 1985; Hore and Vilis, 1980; Iansek, 1980).

Substantia Nigra

Unilateral lesions of the substantia nigra have few long-lasting effects on gross motor behaviour. Deficits can be revealed only in animals trained to perform specific tasks. In baboons trained to point to an illuminated target, unilateral lesions produced an initial period of contralateral hypokinesia and flexed posture during which they could no longer make the appropriate pointing movements. As they recovered, pointing became possible, and the animals were found to

Figure 10.10: Effect of Basal Ganglia Cooling (Primarily the External Segment of the Globus Pallidus) on Extension Movements at the Elbow Made in Response to a Visual 'Go!' Signal at t = 0ms. Elbow position and rectified EMG activity from triceps and biceps are shown for consecutive single trials. In the control state, the movements are fast and repeatable, being performed mainly with a phasic burst of activity in triceps, followed by a lower level of sustained activity. After cooling, the responses are far more variable; some are slow (4th, 5th and 7th movements), some are smaller than required (1st), some do not start properly (2nd, 3rd and 6th), and one rebounded back towards the original position shortly after starting (8th). All the movements were accompanied by co-contracting EMG activity in biceps and triceps (from Hore and Vilis, 1980; Figure 5, with permission)

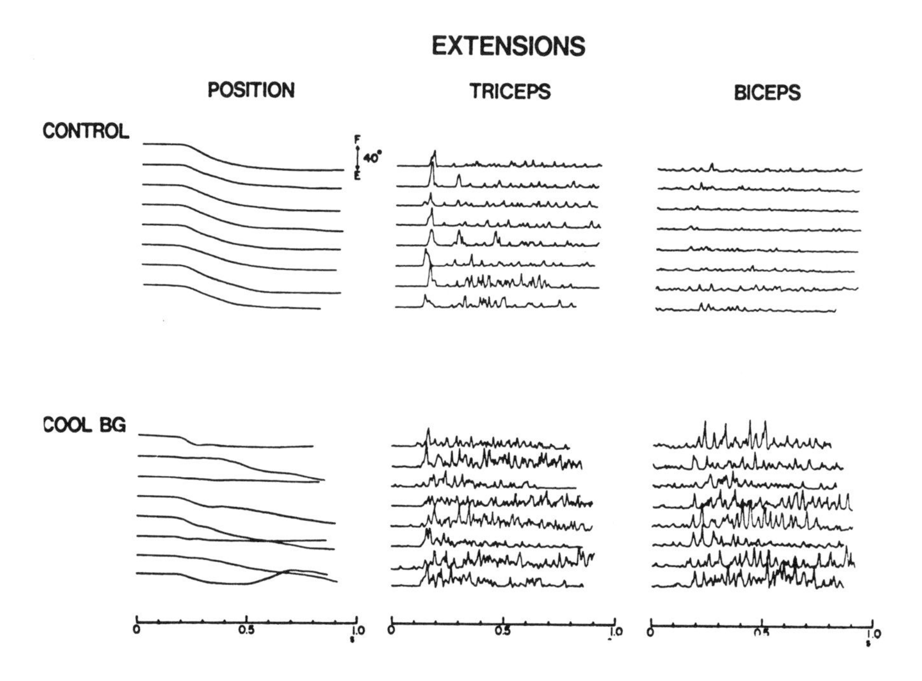

move much more slowly and to have an increased reaction time in the task (Figure 10.11). These deficits recovered very slowly. Similar deficits can be observed in patients with Parkinson's disease.

Bilateral lesions of the substantia nigra produce profound changes in movement: animals become bradykinetic and akinetic, and adopt a flexed posture. Electrolytic lesions invariably destroy both the pars compacta and the pars reticulata, so that it is unclear which effects arise from lesions of the nigro-thalamic or the nigro-striatal tracts. Injection of 6-hydroxydopamine (6-OHDA) into the ascending dopamine systems can be used to make a more selective lesion. This is a neurotoxic substance which is taken up selectively by dopaminergic neurones. Injection produces lesions in the nigrostriatal and the mesolimbic dopamine systems.

Monkeys with unilateral 6-OHDA lesions adopt an asymmetrical posture

Figure 10.11: Effect of a Unilateral Electrolytic Lesion of the Substantia Nigra on a Pointing Task in a Baboon. The animal was trained to hold a lever pressed down until a stimulus light appeared (upper part of figure). This was the signal to release the lever and point as rapidly as possible to the target. Traces below compare the average performance of the contralateral arm before (thick lines) and some days after (thin lines) the lesion. Arm position, velocity and acceleration are shown (from Viallet, Trouche, Beaubaton *et al.*, Figures 1 and 6, with permission)

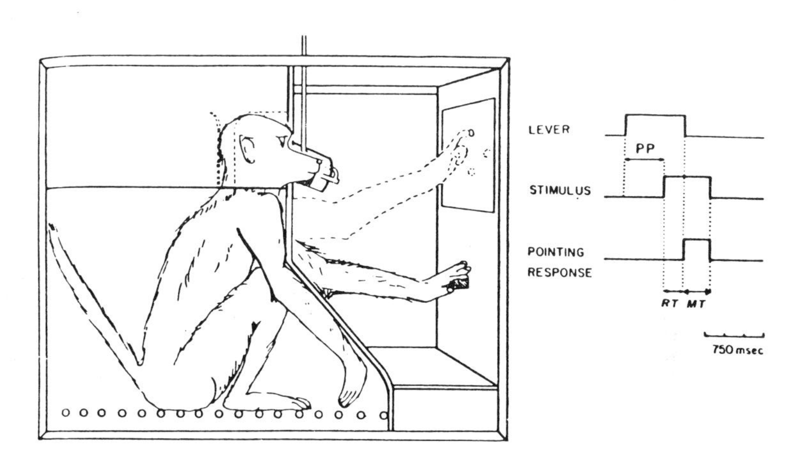

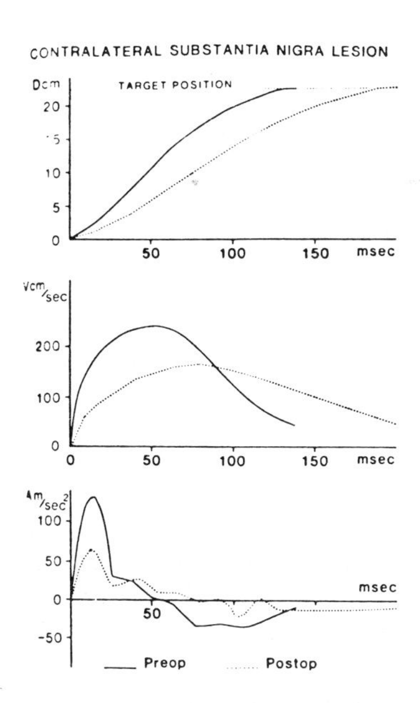

with their head and neck turned to the side of the injection (ipsilateral torticollis). They tend to move spontaneously in a circular fashion towards the same side (ipsiversive circling). There is a degree of bradykinesia. From experiments in rats, these symptoms are believed to be caused by an imbalance in the activity of the basal ganglia on the two sides of the brain. In rats, the effect of 6-OHDA lesions is similar to that in the baboon but reversed. After 24 hours, unilateral lesions produce a contraversive posture and circling, accompanied by hypokinesia and neglect of stimuli given to the contralateral side of the body. The turning behaviour is due to release of dopamine from terminals of degenerating neurones on the ipsilateral side, onto supersensitive (denervated) striatal dopamine receptors. It is enhanced by giving directly acting dopamine agonist drugs, such as apomorphine. In contrast, adminstration of amphetamine, which induces release of dopamine from intact terminals (contralateral to the lesion) causes ipsiversive turning because of its action on the undamaged neurones of the intact side.

Bilateral 6-OHDA lesions result in very severe deficits, including akinesia, adipsia and aphagia. This state is known as the lateral hypothalamic syndrome, and usually proves fatal for the animal. It can be reversed temporarily by adminstration of apomorphine. Its severity correlates well with the degree of dopamine deficiency in the neostriatum.

At first sight it seems paradoxical that three of the symptoms of unilateral 6-OHDA lesions (torticollis, circling and unilateral neglect) are not seen after destructive lesions of the substantia nigra. There are two possible explanations for this. First, these symptoms may be due to a secondary interruption of the ascending dopamine pathways of the mesolimbic system. Alternatively, the nigrothalamic output from the pars reticulata may be necessary for the appearance of such behaviours. Damage to this part of the nigra when electrolytic lesions are made may prevent their occurrence.

As with lesions of the globus pallidus, nigral lesions (chemical or electrolytic) do not produce tremor or changes in muscle tone (Viallet *et al.*, 1983).

Subthalamic Nucleus

In contrast with other parts of the basal ganglia, unilateral lesions of the subthalamic nucleus produce gross disorders of movement. Electrolytic lesions in monkeys, and the vascular lesions sometimes seen in man, both result in involuntary movements of the contralateral limbs. These usually affect proximal more than distal muscles. They may consist of wild, flinging movements of the arms and legs (hemiballismus) or less violent, irregular and random writhing movements (chorea). In man, the arm is more regularly affected than the leg. The movements are abolished by lesioning the inner segment of the globus pallidus (to which the subthalamic nucleus projects) or the ventrolateral nucleus of the thalamus or the corticospinal tract. Because of this, it is thought that the subthalamic nucleus may normally provide a tonic excitatory input to the GP_i. Because the output of the pallidum is inhibitory, removal of this excitatory influence would reduce the amount of inhibition, effectively providing an excitatory drive to motor cortex, via the thalamus and supplementary motor area which could produce the involuntary movements. Hypotonia is associated with lesions of the subthalamic nucleus.

Caudate and Putamen

The putamen receives input from sensorimotor areas of the cortex, while the caudate receives input from frontal and association areas of cortex. Since these represent the major input nuclei of the basal ganglia, it is very surprising to find that unilateral lesions of either or both structures produce very little effect on gross movements. Very large bilateral lesions produce hyperactivity and compulsive following of visual stimuli. Some investigators have described hypotonus.

Recently, lesions of the caudate nucleus have been analysed more carefully. Bilateral lesions result in behavioural deficits similar to those seen after removal of frontal areas of cortex. These include difficulties with delayed response, delayed attention and object-reversal tasks. More directly related motor deficits occur after cooling the putamen. This produces an increase in reaction times to a visual stimulus, with very little change in movement velocity.

Chemical lesions of the striatum have been attempted. Injection of bicuculline, a GABA antagonist, into the ventral part of the putamen in monkeys can produce a contralateral chorea. This is of particular interest, since studies of the postmortem brains of patients with Huntington's chorea show a reduction in the number of GABA receptors and a decrease in the GABA-synthesising enzymes GAD in the striatum. GABA (like ACh) may be a transmitter at some interneuronal synapses within the striatum (Crossman, Sambrook and Jackson, 1984).

Summary

Despite difficulties, some lesion studies have been successful and have provided clues as to the function of the basal ganglia in movement. The main findings are that damage to substantia nigra or globus pallidus reduces the speed of voluntary movement (bradykinesia) and produces a lack of spontaneous movement (akinesia). In contrast, lesions of the putamen or subthalamic nucleus produce an excess of spontaneous movement (hyperkinesia). Lesions of the caudate produce few direct deficits in movement; they affect complex behaviours and cognitive functions, probably because they interrupt the processing of afferents from frontal association cortex. A finding common to all lesion studies, except those involving the subthalamic nucleus, is that unilateral lesions rarely produce any substantial deficits, at least in freely moving animals. Subtle effects on movement control can however be seen in animals performing specially trained movements. Clear effects are seen only after bilateral lesions.

Pathophysiology of Diseases of the Basal Ganglia in Man

Diseases of the basal ganglia encompass a wide variety of different movement disorders. Common to all these conditions are either an excess of (abnormal) involuntary movements or a lack of spontaneous movement. In addition, there may be changes in muscle tone and defects of postural reflexes. (General reviews are provided by DeLong and Georgopoulos, 1981; Marsden, 1982.)

Some brief definitions may be useful. *Hyperkinesia* refers to an excess of movement; *akinesia* or *hypokinesia* to a lack of spontaneous movement. Akinesia may also be used to refer to a lack of normal associated movements such as swinging of the arms when walking. In contrast to these terms which describe the amount of movement, *bradykinesia* defines the speed of movement: bradykinetic movements are slow (whether or not there is an excess, or a lack of them).

Parkinson's Disease (Figure 10.12)

Parkinson's disease is the most common disorder of the basal ganglia. Histological examination of the postmortem brains shows, in most cases, a degeneration of neurones in the substantia nigra and locus coeruleus. There is a concomitant dramatic decrease in the dopamine content of the striatum due to degeneration of the nigro-striatal tract. This decrease is more pronounced in the putamen than in the caudate. Interestingly, for symptoms to appear, it has been estimated that there must be at least an 80 per cent decrease in the dopamine content of the striatum. If more than 20 per cent of the nigro-striatal terminals are left intact, then the nervous system seems capable of compensating for the deficiency. The dopaminergic projection from the ventral tegmental area to the accumbens (the mesolimbic dopamine system) is not greatly affected in Parkinson's disease. Degeneration is seen in other areas of the brain (for instance, dorsal motor nucleus of the vagus, substantia innominata, hypothalamus, sympathetic ganglia and many others), but it is not so obvious as in the nigra. Because of this it is believed that the abnormalities of movement, which are the main symptom of the disease, are due mainly to striatal dopamine deficiency. The success of L-DOPA replacement therapy in the treatment of Parkinson's disease supports this view.

The classic symptoms of Parkinson's disease are akinesia, tremor and rigidity. To this might be added bradykinesia and postural instability. Patients have a poverty of associated and spontaneous movements (akinesia): their face has a mask-like expression, their arms do not swing while walking and they sit or stand motionless, with their body flexed at the waist, elbows and knees. All movements are slow (bradykinesia), giving the impression of a very deliberate and careful action. Despite the difficulties with voluntary movement, there may be a conspicuous tremor, especially of the hands which only disappears during sleep. This is known as a *rest* tremor and has a frequency of 4–6Hz. Usually, it appears early in the disease, beginning in the hands and later affecting the lips

Figure 10.12: Typical Posture of a Patient with Parkinson's Disease: Flexed Trunk, Neck and Arms, With a Shuffling Gait.

and tongue and then the feet. In the hands, this tremor has a very characteristic motion, consisting of opposition and adduction of the thumb towards the index finger, usually described as a 'pill-rolling' tremor. In the early stages of the disease, this tremor may disappear when the patient moves his hands. However, as time passes the tremor persists during voluntary movement. Such *action* tremor has a higher frequency (6–8Hz) then the rest tremor.

In relaxed patients, passive manipulation of the joints meets with more resistance than normal (rigidity). The increased tone is evenly distributed between extensor and flexor muscles (unlike the increased tone in spasticity), and may be accompanied by a phenomenon known as 'cog-wheeling'. This is due to superimposition of tremor onto the increased muscle tone, and gives the examiner the feeling of moving the joint through a ratchet-like device. Finally, patients have postural difficulties, especially in the later stages of the disease. A firm, unexpected push to the shoulders may easily overbalance such individuals. Stepping reactions are lost, and there is a failure to throw the limbs out to protect the body during falling.

The question which arises from these clinical observations is, what do they tell us about the function of the basal ganglia in normal man? To answer this question it is important to go back to an important concept of CNS function first put forward by the British neurologist John Hughlings Jackson over 125 years ago. His ideas was that CNS lesions give rise to two types of symptom: positive and negative. Positive symptoms are those which involve more nervous activity than in the normal state; negative symptoms involve less activity. The positive symptoms of Parkinson's disease are tremor and rigidity; the negative symptom is akinesia (and bradykinesia). The lesioned area of brain cannot play any direct role in producing positive symptoms. They are produced secondarily by the action of remaining areas of brain which have been released from the influence of the lesioned part. They are known as *release phenomena.* In contrast, the negative symptoms are likely to be due directly to loss of function of the damaged part.

It can be argued, therefore, that the principal basal ganglia deficit revealed in Parkinson's disease is akinesia. Tremor and rigidity are secondary symptoms due to loss of basal ganglia influence on other structures. The interpretation of the postural abnormalities is not entirely clear. It is possible that they are a secondary result of rigidity and akinesia. Alternatively, since postural deficits are not usually seen early in the disease, they may be the result of the disease process spreading to areas of the brain other than the basal ganglia. Certainly, it can be shown that the movements of distal parts of the body (such as the thumb) are slow, or bradykinetic, even if well supported and requiring no active postural maintenance. This suggests that postural difficulties are not the primary symptom of the disease.

Further evidence that tremor and rigidity are *release phenomena* comes from the surgical treatment of Parkinson's disease by stereotactic surgery. This was a popular operation before the advent of L-DOPA therapy, and involved lesioning the output of the basal ganglia in the globus pallidus or ventrolateral thalamus. Lesions of either area were effective in abolishing tremor and rigidity but had little effect on akinesia. Such operations were extremely dramatic to watch. Appropriate lesions *immediately* abolished tremor on the contralateral side of the body without affecting the patient's voluntary control of the limbs. The interpretation is that loss of dopamine input into parts of the basal ganglia releases some cells from an inhibitory control. Their excessive output which is responsible for the tremor and rigidity can be interrupted by surgical lesions. A topographic organisation of globus pallidus and thalamus was observed in these operations. In the pallidus, this was similar to that seen in electrophysiological recording from monkeys, with a rostro-caudal representation of the arm and leg. A different type of organisation was seen in the thalamus. Lesions of rostral VL were more effective in controlling rigidity, while caudal lesions improved tremor (see review by Marsden, 1982).

Physiological Studies in Parkinson's Disease

Rigidity. The possible mechanisms of Parkinsonian rigidity have been discussed in Chapter 6. Briefly, the increased tone felt on passive manipulation of the joints is due to a combination of the patients' difficulty in relaxing their muscles, and also to abnormal stretch reflex activation of the muscles. There is no evidence for increased excitability in the spinal monosynaptic pathway which might explain the increased stretch reflexes. However, long-latency stretch reflexes often are enhanced and excessive activity of these reflexes may be partly responsible for the rigidity of Parkinson's disease (Figure 10.13). There is no universal agreement on which anatomical pathways mediate these long-latency reflexes in man. However, there is one attractive theory which explains the influence of the basal ganglia on such responses: (1) the long-latency reflexes may use a transcortical reflex pathway (via thalamus, sensory cortex, motor cortex and spinal cord); (2) electrical stimulation of the supplementary motor area (Chapter 8) is known to suppress activity in monkey motor cortex evoked by muscle stretch; (3) much of the basal ganglia output is directed towards the SMA. Thus it is possible that defects of basal ganglia function produce rigidity via their action on the SMA and long-latency reflexes.

However, this hypothesis is not proven. Other workers favour an entirely different explanation. They suggest that slowly conducting group II afferents from secondary muscle spindle afferent terminals contribute strongly to the long-latency stretch reflex. The long latency of the response would then be due to slowly conducting afferent pathways, rather than to the long distance of the supraspinal pathway. They suggest that the basal ganglia may have some influence on the excitability of this pathway (Berardelli, Sabra and Hallett, 1983; Rothwell *et al.*, 1983; Tatton, Eastman, Bedlingham, *et al.*, 1984).

Tremor. Parkinsonian rest tremor has a frequency of 4–6Hz, and is produced by alternating activity in antagonist muscles. In those patients in whom the tremor persists during voluntary movement (action tremor), the frequency increases to 6–8Hz, and the EMG activity sometimes becomes synchronous in antagonists. Because the rest and action tremor are believed by some workers to have different mechanisms, they will be discussed separately.

Tremor cannot be produced in experimental animals by lesions limited to structures of the basal ganglia. Neither electrolytic lesions of the substantia nigra nor 6-OHDA lesions of the ascending dopamine pathways from substantia nigra and the neighbouring ventral tegmental area produce tremor. It is only when lesions of the ascending cerebellar efferent fibres are combined with nigro-striatal lesions that tremor appears. The usual place at which lesions are placed is in the ventromedial tegmental area of the midbrain. At this point, the lesions interrupt four pathways involving the substantia nigra and red nucleus: nigro-striatal, cerebello-rubral, cerebello-thalamic (which travel through the red nucleus) and rubro-olivary tracts. Lesion of the superior cerebellar peduncle alone (or dentate nucleus, from which many of the fibres arise), which severs the ascending cerebellar efferents, does not produce rest tremor. Instead, ataxia and intention tremor (Chapter 9) are seen.

The conclusion is that Parkinsonian rest tremor is not due to a pure striatal dopamine deficiency, but that damage to other structures (in particular, the cerebello-thalamic and cerebello-rubral tracts) must also be involved.

In both lesioned monkeys and in patients with Parkinson's disease,

Figure 10.13: Average Stretch Reflex EMG Responses in the Flexor Pollicis Longus Following Ramp Stretches of Different Rates Applied to the Thumb (Top Traces, 1–6). On the left are responses from a severely rigid patient with Parkinson's disease; on the right are the responses from a normal subject. In the patient, only the long-latency component of the reflex (latency of 50ms) can be seen; in the normal, at high-stretch velocities, a small, short-latency component can be distinguished. Note how the sensitivity to stretch is greatly increased in the patient, as is the maximum size of the reflex response (from Rothwell, Obeso, Traub *et al.,* 1983; Figure 3, with permission)

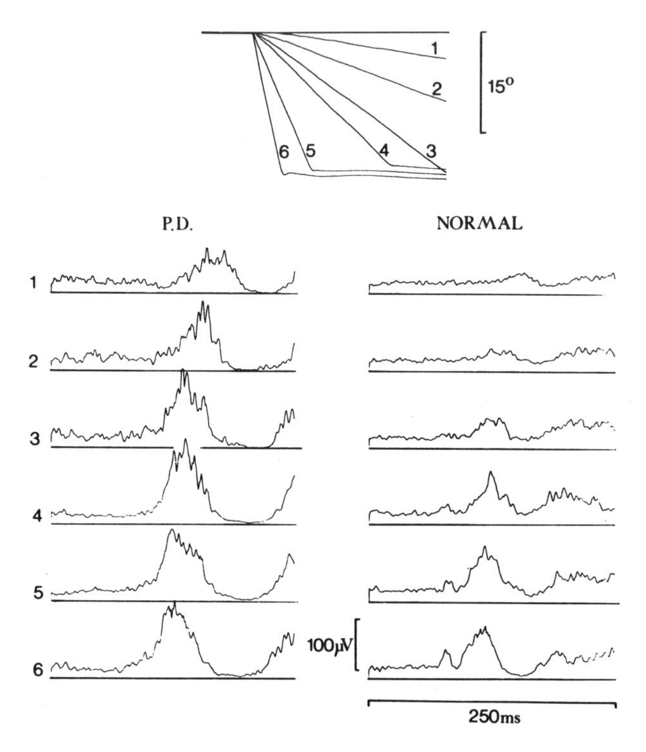

microelectrode recordings from globus pallidus, ventrolateral thalamus and motor cortex reveal neurones which discharge in phase with the ongoing tremor. Such observations do not reveal whether the neuronal firing produces the tremor, or whether the firing is due simply to afferent feedback generated in the periphery by the tremor. Further work in experimental animals has shown that peripheral deafferentation by sectioning the dorsal roots of the spinal cord does not abolish tremor. The bursts of neuronal activity in the thalamus and cortex are also unaffected. This suggests that tremor is produced by rhythmic activity in these areas of brain, rather than being due to oscillation in a feedback loop. Lesions in both ventrolateral thalamus and cortex can abolish rest tremor. Interestingly, pyramidal tract lesions are ineffective in experimental animals, suggesting that cortico-reticulo-spinal or cortico-rubro-spinal pathways are involved.

Work on animal models of rest tremor has emphasised the central nature of the tremor-producing mechanism. Studies of human patients have shown that the true picture may be more complicated. The argument runs as follows. If the tremor is produced solely by an internal oscillator within the CNS, then whatever happens to the limb, the tremor should carry on regardless since it is not reliant on peripheral feedback. If, on the other hand, the tremor were produced by oscillations in an unstable feedback loop, then it would be critically dependent upon all afferent feedback from the moving limb.

These theories can be put to the test in patients with Parkinson's disease. The EMG bursts of rest tremor can be recorded from extensor and flexor muscles at the wrist joint. At random intervals, an electric motor can forcibly disturb the tremor by applying a force to extend the wrist. If the perturbation has no affect on the tremor, then the result fits in with the idea of a central oscillator. If there is an effect, then a peripheral feedback mechanism seems more likely.

In fact, what happens depends very much on how large a perturbation is applied to the wrist. If the tremor is relatively small and the disturbance relatively big, then it is possible to 'reset' the phase of tremor. That is, forcible extension of the wrist interrupts the ongoing tremor cycles and causes the tremor to 'restart' with its timing time-locked to the timing of the disturbance. If tremor is large, then 'resetting' cannot be demonstrated.

These experiments show that the rest tremor of Parkinson's disease can be influenced by peripheral feedback, even though animal work has shown that feedback is not *necessary* for tremor to occur. It is now generally accepted that parkinsonian rest tremor is produced by a combination of activity in independent central neuronal oscillators and in peripheral feedback loops. Interference with either mechanism can affect the resulting tremor (Lamarre and Joffroy, 1979; Stein and Lee, 1981).

Action tremor of Parkinson's disease has not been so extensively studied. Some investigators believe that it has a different mechanism to that involved in rest tremor. The reasons for this are: (1) its higher frequency (6–8Hz); (2) the frequent appearance of co-contracting EMG activity in antagonist muscles, rather than alternating activity as in rest tremor; and (3) it is sometimes possible to record tremor which consists of *both* an 8Hz and a 5Hz component simultaneously.

A phenomenon related to action tremor is 'cog-wheeling'. This is the feeling of a ratchet-like resistance to movement which can be detected during passive manipulation of a patient's joints. It is produced by bursts of EMG activity in the appropriate muscles which have the same frequency as action tremor. It is believed to be due to superimposition of action tremor onto the increased muscle tone.

Akinesia. Akinesia is a term which is often used in a broad (but strictly incorrect) sense to refer to several types of movement deficit in Parkinson's disease. These are: (1) a lack of spontaneous movement, especially noticeable as an immobile and mask-like expression of the face; (2) a lack of associated movements (such as swinging the arms when walking); (3) an increase in reaction time to external stimuli; (4) slowness of movement; (5) a tendency to make smaller movements than required (for instance, small steps when walking, or small movements of a pen while writing — micrographia); (6) fatiguability of repetitive movements, seen, for example, in tapping the hands for any length of time; and (7) difficulty in performing two different movements at the same time (for instance, rising from a chair to shake hands is performed by normal individuals in one continuous action, whereas the patient with Parkinson's disease will first rise from his chair and only then will he extend his hand in greeting).

Such changes are very similar to those seen in monkeys after lesions of the substantia nigra or globus pallidus. Different books may refer to all or one of these phenomena as manifestations of akinesia. However, since the symptoms may be dissociated from one another in certain patients, it is likely that, in fact, they are not all different manifestations of the same deficit. Akinesia is more strictly defined as a lack of spontaneous or associated movement. Bradykinesia refers to the slowness of voluntary movements.

Poverty of movement is a deficit which is not only seen in patients with Parkinson's disease. Lack of spontaneous movement in conscious patients is seen in a variety of other conditions, involving damage to other parts of the brain, such as the frontal lobes. Pure akinesia is extremely difficult to measure since it represents the absence of movement. Counting the number of movements made under strictly controlled conditions (for example, through a two-way mirror) is the most direct way of measuring akinesia. Another method is to count the rate of spontaneous eye-blinking, which is reduced in patients with Parkinson's disease. Such measures of akinesia do not always correlate with the other deficits mentioned above. For example, in measures of reaction time in which patients have for example to press a key as rapidly as possible in response to an auditory or visual cue, patients are only slightly slower than normal. In any one individual, the increase in reaction time may or may not be related to measures of spontaneous movement.

The conclusion is that the process of making a spontaneous movement must differ from that involved in releasing a pre-planned, reaction time action. One feature that does correlate with measures of spontaneous movement is fatiguability in the performance of a repetitive task. Akinetic individuals with Parkinson's disease cannot tap their hand rapidly on a table for any length of time without considerable slowing of the rhythm.

Bradykinesia, the slowing of movements, is more easily measured. A simple experiment is to ask patients to flex their wrist as rapidly as possible through several different distances. Normal subjects perform such tasks in a very repeatable and stereotyped manner. The EMG consists of a burst of activity in the agonist, flexor muscles and is followed by a burst of activity in the antagonist extensor muscles. The agonist burst provides the impulsive force for the movement, and the antagonist helps to brake the movement so that it does not overshoot the end point. In large movements, the agonist burst is large and because of this the peak velocity is greater than in small movements. The duration of the EMG bursts and their relative latencies are remarkably constant from one individual to another. Because agonist and antagonist bursts are often complete before the limb has reached its final end point, and because the EMG pattern can be seen even in totally deafferented individuals, the 'ballistic burst pattern' is thought to represent a single central programme of muscle activation.

When patients with Parkinson's disease attempt these tasks, two striking deficits are seen (Figure 10.14). Their movements are much slower than normal and the size of the movement is smaller than normal. Because of this latter effect, patients often make large movements in a series of smaller steps. In the EMG this is reflected as a decrease in size of the first burst of agonist EMG activity which results in a smaller impulsive force for the movement than normal. If the movement is made in a number of small steps, then there is an accompanying number of small EMG bursts in the agonist. The important feature of the EMG pattern is that although it is reduced in amplitude, the duration of the EMG bursts and the latency of the antagonist burst (when present) are quite normal. Because of this it is thought that the parts of the basal ganglia affected in Parkinson's disease do not play a direct role in producing the basic 'ballistic burst pattern'. The deficit is in scaling the amplitude of this pattern to match that required in the movement.

A striking feature of the akinesia of patients with Parkinson's disease is the way in which visual input can affect movement. Patients may find it easier to walk along a pattern of stripes painted on the ground than on a plain surface. In contrast, many patients 'freeze' when trying to walk through a doorway. Even if the door is held open, they will halt on the threshold and find it impossible to pass through. Their feet appear to be stuck on the ground. Frequently, the only way to get the patient moving again will be to use another visual input. Placing one's foot before the patient so that he has to step over it to carry on usually proves to be an effective stimulus (Evarts, Teravainen and Calne, 1981; Marsden, 1982; Schwab, Chafetz and Walker, 1954).

Postural Difficulties. The disorders of posture seen in Parkinson's disease are beautifully described by Martin in his classic book on the basal ganglia and posture (1967). Patients walk with a flexed posture, bent at the knees, elbows and trunk. They lose their righting reactions: seated on a tilt table, they fail to keep their trunk vertical when the table is tilted from side to side. They also lose stepping reactions, usually seen following an unexpected push to the shoulders, and protective reactions of the arms when falling. The result is that not only do patients with Parkinson's disease have frequent falls, but they also damage themselves more than normal when they do fall. These postural reactions are triggered by proprioceptive and/or labyrinthine input.

Another type of postural reaction is abnormal in Parkinson's disease. These are the anticipatory postural reactions described in Chapter 9. They are evoked

Figure 10.14: Deficits in Rapid Self-paced Thumb (A) and Wrist (B) Movements in Bradykinetic Patients with Parkinson's Disease. The upper panel shows a representative rapid thumb flexion movement of 20° in a normal subject and a patient with Parkinson's disease. The thumb was firmly clamped at the interphalangeal joint so that no postural activity was needed to steady it. The normal subject produces the movement in a single step, using a large burst of EMG activity in the flexor pollicis longus (FPL) muscle. The patient (P.D.) makes the movement in a series of three small steps, with correspondingly small bursts of EMG activity. The bottom panel shows an analysis of similar rapid flexion movements made at the wrist in a group of eight normal and eight patients with Parkinson's disease. Only the initial step of the patients' movements was analysed. The graph on the left shows that movements of a given extent are much slower in patients than normals. However, the duration of the first burst of agonist EMG activity (AG1) was the same (left). This indicates that the basic instructions for the movement were intact (unpublished observations of Berardelli, Rothwell, Dick *et al.*)

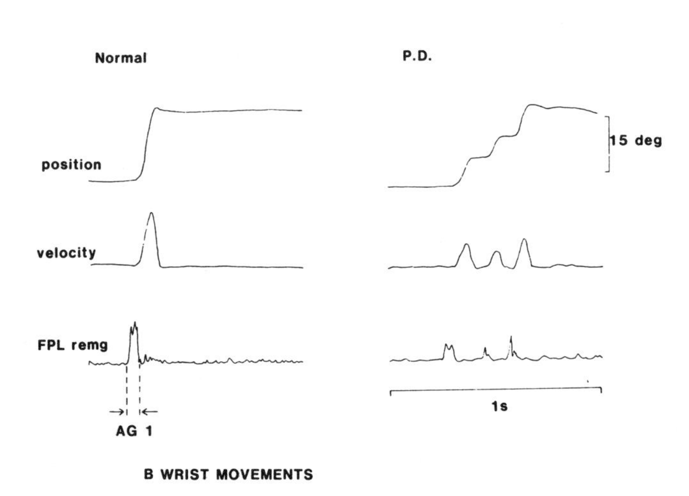

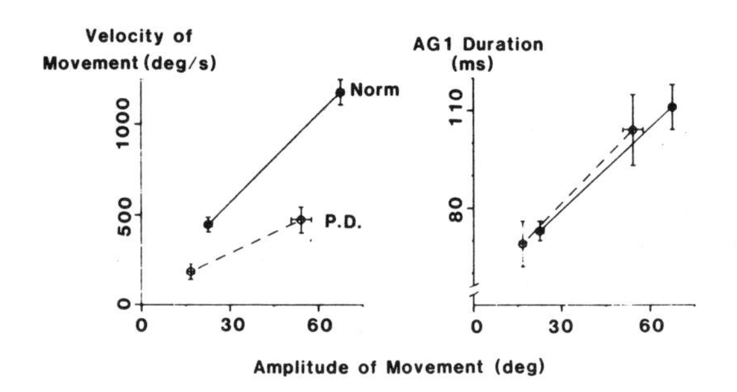

at short latency in postural muscles by proprioceptive input from a distant part of the body. For example, short forward pulls to the arm can evoke reflex activation of the soleus muscle with a latency of 80–90ms. These reflexes are used to prevent the anticipated body sway that the pull to the arm is likely to generate. They are not as powerful as the stepping and protective reactions referred to above and might be envisaged as providing a 'fine tuning' mechanism for postural stability. Anticipatory postural reflexes are small or absent in patients with Parkinson's disease who have a history of postural instability (Traub, Rothwell and Marsden, 1980).

The flexed posture typical of Parkinson's disease is also seen in animals with lesions of the basal ganglia. However, in animals the lesions must be large and bilateral for this to happen. In Parkinson's disease, too, the postural instabilities are prominent only in the later stages of the illness, when the disease may spread beyond the substantia nigra. Martin (1967) believed that postural difficulties were the principal symptom of the disease. They would be evident, he thought, not only in examining whole body posture but also in failure of postural fixation at proximal joints when making movements with a distal extremity. As pointed out above, this idea is now believed to be wrong. Distal movements may be slow even in circumstances when postural fixation is not needed (Figure 10.14). Akinesia and bradykinesia are not produced by defects in postural control. It is more likely that they contribute to postural instability by affecting operation of postural reflex systems.

Chorea and Athetosis

Chorea and athetosis are two common symptoms of basal ganglia disease. They describe involuntary, purposeless and irregular movements of the body which occur in several different conditions. Chorea usually is applied to movements at proximal or axial joints, while athetosis is used more frequently to describe movements of the distal parts of the limbs. Choreic movements are relatively slow (compared with hemiballismus, see below), and produce a writhing motion of the body. Almost all muscle groups may be affected, but the chorea will seem to move randomly from one part of the body to another. Facial and limb movements are equally affected. Grimacing, rolling of the eyes, twitching or writhing of the limbs and head all occur unpredictably.

Chorea may result from many different causes. Sydenham's chorea is seen mainly in children and is caused by an attack of rheumatic fever. Recovery occurs after several weeks. There is no fever and no rheumatism, and the rheumatic origin of the disease is only betrayed later by the typical cardiac lesions which ensue. Chorea is also seen in polycythaemia and following the use of oral contraceptives. However, the main evidence for its association with the basal ganglia comes in Huntington's disease. This is a dominantly inherited disease with symptoms involving not only chorea but also progressive dementia. The symptoms do not appear until about the age of 30 to 40 years, by which time the next generation has usually been born; 50 per cent of these children will have the disease. Involuntary movements and psychiatric changes often begin together, and are seen at first only as increased restlessness and eccentricity. However, chorea gradually becomes more florid until the movements are almost incessant. Patients become incontinent, unable to feed themselves and severely demented. Death follows within 10 years of the onset of symptoms. In the final stages of the disease, the involuntary movements become less frequent and the muscle tone greatly increased.

The pathology is widespread, but the principal affected areas of brain are the caudate, putamen, globus pallidus and cerebral cortex. There is a loss of small cells in the striatum and relative sparing of large cells. Because there are decreases in activity of enzymes related to GABA metabolism, selective loss of GABAergic interneurones has been suggested. In the cerebral cortex, frontal and precentral regions are affected with cell loss mainly in the deeper layers. It is believed that chorea is the result of basal ganglia deficit, whereas dementia is caused by cortical cell loss.

Very little is known about the physiology of Huntington's disease. Cortical somatosensory-evoked potentials (the response of the cortex to peripheral nerve stimuli) are absent or reduced, as are long-latency stretch reflexes in the small hand muscles. Short latency, spinal stretch reflexes are present. If long-latency stretch reflexes use a transcortical pathway, then it is likely that their absence, and that of the somatosensory evoked potential, is due to cortical cell loss and loss of thalamocortical projection neurones.

Athetosis describes a distal writhing movement of the fingers, usually accompanied by similar slow movements of the tongue. A characteristic feature is the posture of the outstretched arm: adducted at the shoulder, partially flexed at the elbow and wrist with hyperextension of the fingers. The movements are superimposed upon this posture. Athetosis is uncommon, but often seen in children with cerebral palsy. Birth trauma or malformation is believed to produce damage to the putamen which is responsible for the athetosis.

Torsion Dystonia

Torsion dystonia is a term used to describe twisted (torsion) sustained postures of the limbs, neck or trunk. In contrast to the forever changing, fleeting muscle contractions of chorea, torsion dystonia describes a relatively fixed posture, maintained by abnormal muscle activity (dystonia). Torsion dystonia is rare and, in most instances, its cause is unknown. If all parts of the body are affected the condition is known as generalised torsion dystonia or dystonia musculorum deformans. This usually begins in childhood, perhaps affecting one leg while walking, and spreads gradually to affect all four limbs, the trunk and neck. The sustained muscle contractions contort the body into grotesque postures, making it impossible for the patient to walk, clothe or feed himself. If symptoms begin in adult life, the progression is likely to be much less severe. The most frequently affected part is the neck, which becomes twisted to one or other side of the body (torticollis). Sometimes the spasms are limited to one hand or

Figure 10.15: Single, Rapid, Extension Movements of the Elbow Through 20° in Three Subjects with Dystonia of Different Severity and Extent. A: segmental dystonia affecting the arm and shoulder; B: segmental dystonia of the arm, shoulder and neck; C: generalised dystonia. In A, the movement and the EMG pattern is normal, with the typical initial burst of EMG in the agonist (triceps), followed by a burst in the antagonist (biceps). In B and C, the EMG pattern becomes more disorganised. The bursts increase in duration, and there is pronounced co-contraction between biceps and triceps. The movements are slower than normal (from Rothwell, Obeso, Day *et al.,* 1983; Figure 4, with permission)

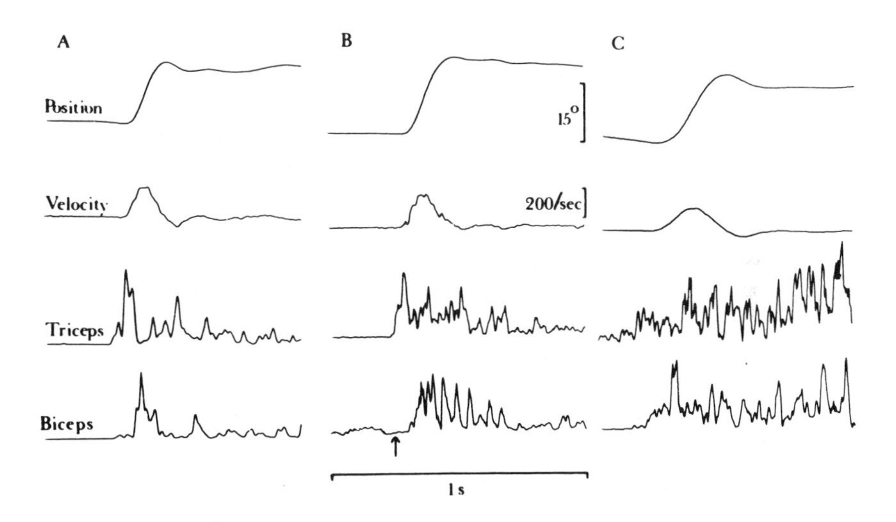

forearm, and only occur when the patient performs some delicate manual task such as writing or typing or playing a musical instrument (writers', typists' or musicians' cramp).

Even in severe cases of generalised torsion dystonia, investigation of the brain usually reveals no gross abnormality. However, there are examples where symptoms of torsion dystonia have been linked with basal ganglia damage. Wilson's disease, a condition resulting from an abnormality of copper metabolism, often presents with abnormal movements and dystonic postures and, in some cases, this is associated with damage to the putamen. In addition, since the advent of the CT brain scanner, several cases have been reported of individuals with localised lesions of the putamen and globus pallidus, and contralateral dystonia on the opposite side of the body. No specific neurochemical abnormalities have yet been revealed.

Little is known about the physiology of torsion dystonia. The muscle spasms are characterised by co-contraction of antagonist muscles, rather than the more usual reciprocal pattern seen in many normal voluntary movements. Because of this it has been suggested that there may be an associated disorder of reciprocal inhibition in these patients (see Chapter 6). Unlike the situation in Parkinson's disease, the 'ballistic burst pattern' of EMG activity in rapid limb movements may be severely disrupted in torsion dystonia. Burst duration, latency and the reciprocal relationship between antagonists all may be affected (Figure 10.15) (Rothwell, Obeso, Day *et al.*, 1983).

Hemiballismus

As in the monkey, hemisballismus in man results from lesions of the subthalamic nucleus, usually due to vascular infarction. Recovery occurs within a few weeks. The abnormal movements consist of gross, rapid, flinging movements of the contralateral arm (and sometimes leg), particularly at proximal joints.

References and Further Reading

Review Articles and Books

Alexander, G.E., De Long, M.R. and Strick, P.L. (1986) 'Parallel Organisation of Functionally Segregated Circuits Linking Basal Ganglia and Cortex', *Ann. Rev. Neurosci.*, *9*, pp. 357–81

Carpenter, M.B. (1981) 'Anatomy of the Corpus Striatum and Brain Stem Integrating Systems' in V.B. Brooks (ed.), *Handbook of Physiology*, sect.1, vol.II, part 2, Williams and Wilkins, Baltimore, pp. 947–95

Ciba Foundation (1984) *Functions of the Basal Ganglia*, Ciba Foundation Symposium 107, Pitman, London

De Long, M.R. and Georgopoulos, A.P. (1981) 'Motor Functions of the Basal Ganglia' in V.B. Brooks (ed.), *Handbook of Physiology*, sect.1, vol.II, part 2, Williams and Wilkins, Baltimore, pp. 1017–61

Kitai, S.T. (1981) 'Electrophysiology of the Corpus Striatum and Brain Stem Integrating Systems' in V.B. Brooks (ed.), *Handbook of Physiology*, sect. 1, vol.II, part 2, Williams and Wilkins, Baltimore, pp. 997–1015

Marsden, C.D. (1982) 'The Mysterious Motor Function of the Basal Ganglia: The Robert Wartenberg Lecture', *Neurology*, *32*, pp. 514–39

Massion, J., Paillard, J., Schultz, W., *et al.* (eds.) (1983) 'Neural Coding of Motor Performance', *Experimental Brain Research Supplement 7*, Springer-Verlag, Berlin

Stein, R.B. and Lee, R.G. (1981) 'Tremor and Clonus' in V.B. Brooks (ed.), *Handbook of Physiology*, sect.1, vol.II, part 2, Williams and Wilkins, Baltimore, pp. 325–43

Williams, P.L. and Warwick, R. (1975) *Functional Neuroanatomy of Man*, Churchill Livingstone, Edinburgh

Original Papers

Alexander, G.E. and De Long, M.R. (1985) 'Microstimulation of the Primate Neostriatum. Parts I and II', *J. Neurophysiol.*, *53*, pp. 1401–30

Anderson, M.E. and Horak, F.B. (1985) 'Influence of the Globus Pallidus on Arm Movement in Monkeys. Parts I, II and III', *J. Neurophysiol.*, *52*, pp. 290–304, pp. 305–22; *54*, pp. 433–48

Berardelli, A., Sabra, A.F. and Hallett, M. (1983) 'Physiological Mechanisms of rigidity in Parkinson's Disease', *J. Neurol. Neurosurg. Psychiat.*, *46*, pp. 45–53

Crossman, A.R., Sambrook, M.A. and Jackson, A. (1984) 'Experimental Hemichorea/Hemiballismus in the Monkey: Studies on the Intracerebral Site of Action in a Drug-induced Dyskinesia', *Brain*, *107*, pp. 579–96

Crutcher, M.D. and De Long, M.R. (1984) 'Single Cell Studies of the Primate Putamen. Parts I and II', *Exp. Brain. Res.*, *53*, pp. 233–58

De Long, M.R., Crutcher, M.D. and Georgopoulos, A.P. (1985) 'Primate Globus Pallidus and Subthalamic Nucleus: Functional Organisation', *J. Neurophysiol.*, *53*, pp. 530–43

——— and Georgopoulos, A.P. (1979) 'Motor Function of the Basal Ganglia as Revealed by Studies of Single Cell Activity in the Behaving Primate', *Adv. Neurol.*, *24*, pp. 131–40

——— Georgopoulos, A.P. and Crutcher, M.D. (1983) 'Corticobasal Ganglia Relations and Coding of Motor Performance' in J. Massion *et al.* (eds.), *Experimental Brain Research Supplement 7,* Springer-Verlag, Berlin, pp. 30–40

Evarts, E.V., Teravainen, H. and Calne, D.B. (1981) 'Reaction Time in Parkinson's Disease', *Brain, 104,* pp. 167–86

Hikosaka, O. and Wurtz, R.H. (1983) 'Visual and Oculomotor Functions of Monkey Substantia Nigra Pars Reticulata. Parts I to IV', *J. Neurophysiol., 49,* pp. 1230–301

Hore, J. and Vilis, T. (1980) 'Arm Movement Performance During Reversible Basal Ganglia Lesions in the Monkey', *Exp. Brain Res., 39,* pp. 217–28

Iansek, R. (1980) 'The Effects of Reserpine on Motor Activity and Pallidal Discharge in Monkeys: Implications for the Genesis of Akinesia', *J. Physiol., 301,* pp. 457–66

Lamarre, Y. and Joffroy, A.J. (1979) 'Experimental Tremor in Monkey: Activity of Thalamic and Precentral Cortical Neurones in the Absence of Peripheral Feedback' in L.J. Poirier, T.L. Sourkes and P.J. Bedard (eds.), *Advances in Neurology,* vol. 24, Raven Press, New York, pp. 109–22

Liles, S.L. (1985) 'Activity of Neurons in Putamen During Active and Passive Movements of Wrist', *J. Neurophysiol., 53,* pp. 212–36

——— and Updyke, B.V. (1985) 'Projection of the Digit and Wrist Area of Precentral Gyrus to the Putamen: Relation Between Topography and Physiological Properties of Neurones in the Putamen', *Brain, Res., 339,* pp. 245–55

Martin, J.P. (1967) *The Basal Ganglia and Posture,* Lippincott, Philadelphia

Rothwell, J.C., Obeso, J.A., Day, B.L., *et al.* (1983) 'Pathophysiology of Dystonias' in J.E. Desmedt (ed.), *Advances in Neurology,* vol. 39, Raven Press, New York, pp. 851–63

——— Obeso, J.A., Traub, M.M., *et al.* (1983) The Behaviour of the Long-latency Stretch Reflex in Patients with Parkinson's Disease', *J. Neurol. Neurosurg. Psychiat., 46,* pp. 35–44

Schwab, R.S., Chafetz, M.E., and Walker, S. (1954) 'Control of Two Simultaneous Voluntary Motor Acts in Normals and in Parkinsonism', *Arch. Neurol. Psychiat., 72,* pp. 591–8

Tatton, W.G., Eastman, M.J., Bedingham, W., *et al.* (1984) 'Defective Utilisation of Sensory Input as the Basis for Bradykinesia, Rigidity and Decreased Movement Repertoire in Parkinsons's Disease: A Hypothesis', *Can. J. Neurol. Sci., 11,* pp. 136–43

Traub, M.M., Rothwell, J.C., and Marsden, C.D. (1980) 'Anticipatory Postural Reflexes in Parkinson's Disease and Other Akinetic-rigid Syndromes and in Cerebellar Ataxia', *Brain, 103,* pp. 393–412

Viallet, F., Trouche, E., Beaubaton, D., *et al.* (1983) 'Motor Impairment After Unilateral Electrolytic Lesions of the Substantia Nigra in Baboons: Behavioural Data with Quantitative and Kinematic Analysis of a Pointing Movement', *Brain Res., 279,* pp. 193–206

INDEX